Recent Studies in Differential Geometry and Its Applications

Recent Studies in Differential Geometry and Its Applications

Guest Editor

Hristo Manev

MDPI

Basel • Beijing • Wuhan • Barcelona • Belgrade • Novi Sad • Cluj • Manchester

Guest Editor
Hristo Manev
Department of Medical
Physics and Biophysics
Medical University of Plovdiv
Plovdiv
Bulgaria

Editorial Office
MDPI AG
Grosspeteranlage 5
4052 Basel, Switzerland

This is a reprint of the Special Issue, published open access by the journal *Mathematics* (ISSN 2227-7390), freely accessible at: https://www.mdpi.com/journal/mathematics/special_issues/895VWV5AWL.

For citation purposes, cite each article independently as indicated on the article page online and as indicated below:

Lastname, A.A.; Lastname, B.B. Article Title. *Journal Name* **Year**, *Volume Number*, Page Range.

ISBN 978-3-7258-8327-1 (Hbk)
ISBN 978-3-7258-8328-8 (PDF)
https://doi.org/10.3390/books978-3-7258-8328-8

Contents

About the Editor

Hristo Manev

Hristo Manchev Manev is an Associate Professor at the Medical University of Plovdiv, with a PhD in Differential Geometry. His research interests are in the fields of Differential Geometry, Applications of Mathematics, Biostatistics, and Electronic Education.

mathematics

MDPI

icle

nalyzing Curvature Properties and Geometric Solitons of the wisted Sasaki Metric on the Tangent Bundle over a tatistical Manifold

u Yan [1], Yanlin Li [2,*], Lokman Bilen [3] and Aydın Gezer [4]

1 Department of Mathematics, Northeast Forestry University, Harbin 150040, China; yanlxmath@nefu.edu.cn
2 School of Mathematics, Hangzhou Normal University, Hangzhou 311121, China
3 Faculty of Science and Art, Department of Mathematics, Iğdır University, Iğdır 76100, Turkey; lokman.bilen@igdir.edu.tr
4 Faculty of Science, Department of Mathematics, Ataturk University, Erzurum 25240, Turkey; agezer@atauni.edu.tr
* Correspondence: liyl@hznu.edu.cn

Abstract: Let (M, ∇, g) be a statistical manifold and TM be its tangent bundle endowed with a twisted Sasaki metric G. This paper serves two primary objectives. The first objective is to investigate the curvature properties of the tangent bundle TM. The second objective is to explore conformal vector fields and Ricci, Yamabe, and gradient Ricci–Yamabe solitons on the tangent bundle TM according to the twisted Sasaki metric G.

Keywords: conformal vector field; Ricci and Yamabe solitons; statistical manifold; twisted Sasaki metric; tangent bundle

MSC: 53C07; 53B05; 53B12; 53A45

1. Introduction

Geometric flows are a mathematical concept that explore the evolution of metric structures on Riemannian manifolds. These flows enable the analysis of how the metric structure of a manifold changes over time and its consequent impact on the geometry of the manifold. One significant type of geometric flow is the Ricci flow, developed by Richard S. Hamilton [1], famously used by Grigori Perelman to prove the Poincaré Conjecture. The Ricci flow serves as a fundamental mathematical tool for studying the changes in metric structures on Riemannian manifolds over time. This flow systematically modifies the metric structure of a Riemannian manifold with the objective of achieving a constant Ricci curvature. The Ricci curvature is defined as the trace of the Ricci tensor, quantifying the curvature of the manifold at a specific point. Ricci flow is instrumental in comprehending the evolution of the metric structure of a manifold and deriving topological results.

Yamabe flow, introduced by Hidehiko Yamabe, is another type of geometric flow. It seeks to maintain the Yamabe curvature of a Riemannian manifold at a constant level while simultaneously altering the metric structure. The Yamabe curvature represents a specific aspect of the metric structure, and this flow aims to highlight these characteristics. Yamabe flow plays a crucial role in geometric analysis and understanding the evolution of metric structures on manifolds.

Ricci solitons are closely related to the Ricci flow and soliton solutions defined by a specific vector field on a Riemannian manifold. They represent particular solutions of the Ricci flow, describing how the manifold evolves in a specific manner. There are two main categories of Ricci solitons: gradient solitons, which correspond to metric structures with certain properties, and Ricci flow solitons, which are solutions of the Ricci flow that exhibit specific behavior.

Yamabe solitons are mathematical concepts used to characterize and study metric structures on Riemannian manifolds. These solitons are a special type that characterize the metric structure of a Riemannian manifold (M, g) through a specific vector field X. This vector field X characterizes a soliton that alters the Yamabe curvature of the manifold, a curvature measure reflecting the properties of the metric structure. A negative Yamabe curvature indicates specific curvature and contraction properties on the manifold.

The significance of Ricci solitons arises from their correspondence to self-similar solutions of the Ricci flow [1] and their natural generalization of Einstein metrics. Recent years have witnessed growing interest in the study of Ricci solitons and their various generalizations in the realm of Riemannian geometry.

Yamabe flow was introduced by Hamilton [1], coinciding with the development of Ricci flow. Both the Ricci soliton and Yamabe soliton are special solitons associated with Ricci flow and Yamabe flow, respectively. In soliton structures, if the potential vector field is the gradient of a function, it is termed a gradient soliton (such as gradient Ricci solitons or gradient Yamabe solitons).

These geometric flows and solitons play pivotal roles in addressing a wide range of problems related to the topology, geometry, and differential equations of manifolds. Regarding different geometric flows, some researchers have obtained different estimates and inequalities, such as, Li–Yau-type estimates, Perelman-type differential Harnack inequalities, etc. [2–5]. The studies on gradient estimates and differential Harnack inequalities were presented in [6–11]. We can find some papers regarding different manifolds of different connections in a tangent bundle [12–18]. All of these papers provide powerful mathematical tools for comprehending manifold evolution and changes, contributing to the derivation of numerous mathematical theorems and results.

Several authors have explored soliton structures in different contexts. For instance, Abbassi and Amri [19] investigated natural Ricci soliton structures on the tangent and unit tangent bundles of Riemannian manifolds. Chen and Deshmukh [20] introduced the concept of quasi-Yamabe solitons on Riemannian manifolds. Güler and Crasmareanu [21] introduced the notion of Ricci–Yamabe flow by considering a scalar combination of the Ricci flow and Yamabe flow. Dey and Majhi [22] studied generalized gradient Ricci–Yamabe solitons on complete Sasakian three-manifolds. Nurowski and Randall [23] introduced generalized Ricci solitons, and Kumara et al. [24] demonstrated that a Riemannian concurrent–recurrent manifold is Einstein when its metric is a generalized Ricci-type soliton. Gezer, Bilen, and De [25] explored almost Ricci and almost Yamabe soliton structures on the tangent bundle using the ciconia metric. Recently, Li and Khan et al. studied solitons, inequalities, and submanifolds using soliton theory, submanifold theory, and other related theories [26–31]. They obtained a number of interesting results and inspired the idea of this paper. De and Gezer [32] studied k-almost Yamabe solitons for the perfect fluid spacetime of general relativity. In particular, they constructed two examples to prove the existence of k-almost Yamabe solitons.

In the 1980s, the concept of a statistical structure emerged, playing a pivotal role in the development of an effective branch known as information geometry, which combines differential geometry and statistics. This field found applications across various scientific domains, including image processing, data analysis, physics, computer science, and machine learning (see [33–38]). A comprehensive survey of information geometry is available in [39]. Researchers have conducted numerous studies related to statistical manifolds. For example, Gezer, Peyghan, and Nourmohammadifar explored Kähler–Norden structures on statistical manifolds in [40]. Matsuzoe provided an overview of the geometry of statistical manifolds and discussed the connections between information geometry and affine differential geometry [41]. In [42], Peyghan, Seifipour, and Gezer investigated statistical structures on the tangent bundle TM equipped with two Riemannian g-natural metrics and lift connections.

In our present study, we introduce a novel natural metric for the tangent bundle TM over a statistical manifold, known as the twisted Sasaki metric. In this paper, firstly, we delve into the geometry of twisted Sasaki metrics, scrutinizing their properties and inherent geometrical aspects in depth. Our exploration aims to provide a comprehensive understanding of the behavior of these metrics, setting the stage for further investigations. Secondly, we shift our focus to the study of soliton structures associated with the twisted Sasaki metric when it is employed on the tangent bundle TM over a statistical manifold. Solitons, as mathematical constructs, reveal specific patterns of evolution within a given metric space. In our analysis, we aimed to elucidate and characterize the soliton structures that emerge from the utilization of the twisted Sasaki metric on TM within the context of a statistical manifold. This investigation sheds light on the dynamic interplay between the metric structure and the statistical properties of the underlying manifold.

Our research endeavors to enhance our comprehension of the geometry and curvature characteristics of the twisted Sasaki metric applied to TM, while also uncovering the intriguing soliton structures that manifest in this scenario. By doing so, we contribute to the broader fields of differential geometry and metric analysis, introducing a novel metric that challenges traditional rigidity when applied to tangent bundles and statistical manifolds.

Throughout this paper, we consistently assume that all manifolds, tensor fields, and connections are differentiable of class C^∞.

2. The Twisted Sasaki Metric on the Tangent Bundle over a Statistical Manifold

Statistical manifolds have found applications in various domains, including information science, information theory, neural networks, and statistical mechanics (as demonstrated in references [33–35]). Essentially, a statistical manifold is a mathematical space where the points represent probability distributions, providing a geometric model for understanding these distributions. A statistical structure on a differentiable manifold M is defined by a pair (∇, g), where g represents a (pseudo-)Riemannian metric, and ∇ denotes a torsion-free linear connection with the property that ∇g is totally symmetric. When a manifold possesses such a statistical structure, it is referred to as a statistical manifold. It is worth noting that a typical example of a statistical manifold is a (pseudo-)Riemannian manifold (M, g) paired with a Levi-Civita connection ∇ for g. In essence, statistical manifolds serve as generalizations of (pseudo-)Riemannian manifolds, providing a broader geometric framework for probabilistic modeling.

Definition 1. *Consider an arbitrary linear connection ∇ defined on a (pseudo-)Riemannian manifold (M, g). With the given pair (∇, g), we construct the $(0,3)$-tensor field denoted as F through the expression*

$$F(X, Y, Z) := (\nabla_Z g)(X, Y).$$

Clearly, $F(X, Y, Z) = F(Y, X, Z)$, due to the symmetry of g. This tensor field F is occasionally referred to as the cubic form associated with the pair (∇, g) [43]. Now, when we have a symmetric bilinear form ρ defined on a manifold M, we designate (∇, ρ) as a Codazzi pair if the covariant derivative $(\nabla\rho)$ is (totally) symmetric concerning vector fields X, Y, and Z [44]:

$$(\nabla_Z \rho)(X, Y) = (\nabla_X \rho)(Z, Y) = (\nabla_Y \rho)(Z, X).$$

Expressed in terms of the cubic form F, this condition can be rephrased as

$$F(X, Y, Z) = F(Z, Y, X) = F(Z, X, Y),$$

which means that the condition for (∇, g) to form a Codazzi pair is equivalent to F being entirely symmetric with respect to all of its indices.

Now, let us consider a torsion-free linear connection ∇ defined on a (pseudo-)Riemannian manifold (M, g). In the case where the pair (∇, g) forms a Codazzi pair, a concept well known to information geometers as characterizing statistical structures, the manifold M, when coupled with this statistical structure (∇, g), is termed a statistical manifold. It is important to note that the notion of a statistical manifold was originally introduced by Lauritzen [45]. These statistical manifolds are extensively explored in the realm of affine differential geometry, as evidenced in [45,46], and they play a pivotal role in the field of information geometry.

Consider an n-dimensional statistical manifold denoted as (M_n, ∇, g). In this article, we employ the C^∞-category to comprehensively elucidate various concepts. We focus on connected manifolds with dimension of $n > 1$. To facilitate our analysis, we introduce the tangent bundle of M_n, denoted as TM. We define the natural projection as $\pi : TM \to M_n$. When we employ a system of local coordinates (U, x^i) in M_n, it induces a corresponding system of local coordinates on TM, denoted as $\left(\pi^{-1}(U), x^i, x^{\bar{i}} = u^i\right)$, where $\bar{i}$ ranges from $n+1$ to $2n$. Here, (u^i) represents the cartesian coordinates within each tangent space T_pM for all $p \in U$. It is important to note that p is any arbitrary point within U.

Consider the linear connection ∇ on the statistical manifold (M_n, ∇, g). We can decompose the tangent space of the tangent bundle TM into two distributions: the horizontal distribution determined by ∇ and the vertical distribution defined by ker π_*. In this context, the local frame is given by

$$E_i = \frac{\partial}{\partial x^i} - u^s \Gamma_{is}^h \frac{\partial}{\partial u^h}; \; i = 1, ..., n,$$

and

$$E_{\bar{i}} = \frac{\partial}{\partial u^i}; \; \bar{i} = n+1, ..., 2n.$$

Here, Γ_{is}^h represents the Christoffel symbols of the linear connection ∇. The local frame $\{E_\beta\} = (E_i, E_{\bar{i}})$ is commonly referred to as the adapted frame. Let $A = A^i \frac{\partial}{\partial x^i}$ be a vector field. We can obtain the horizontal and vertical lifts of A with respect to the adapted frame as follows [47]:

$$\begin{aligned} {}^H A &= A^i E_i, \\ {}^V A &= A^i E_{\bar{i}}. \end{aligned}$$

Within TM, the local $1-$form system $(dx^i, \delta u^i)$ forms the dual frame of the adapted frame $\{E_\beta\}$, where

$$\delta u^i = {}^H\left(dx^i\right) = du^i + u^s \Gamma_{hs}^i dx^h.$$

From the Riemannian manifold (M_n, g) to its tangent bundle TM, a variety of Riemannian or pseudo-Riemannian metrics have been devised. These metrics are constructed by naturally lifting the Riemannian metric g to the tangent bundle TM and are commonly referred to as g-natural metrics. In [48], the authors systematically derived a family of Riemannian g-natural metrics, which depend on six arbitrary functions characterizing the norm of a vector $u \in TM$. The study of natural metrics on tangent bundles arises from the need to understand the geometric and physical properties of objects moving on a Riemannian manifold. These metrics provide a way to extend the geometry of the base manifold to the tangent bundle, which is essential in various fields, including physics, differential geometry, and mechanics. Now, let us introduce the twisted Sasaki metric on the tangent bundle of a statistical manifold.

Let (M_n, ∇, g) be a statistical manifold and $a, b \in \mathbb{R}$. On the tangent bundle TM, the twisted Sasaki metric G is defined by

$$\begin{aligned} (i) \; G({}^H X, {}^H Y) &= a g(X, Y), \\ (ii) \; G({}^V X, {}^H Y) &= 0, \\ (iii) \; G({}^V X, {}^V Y) &= b g(X, Y) \end{aligned}$$

for all vector fields X and Y.

In the adapted frame $\{E_\beta\}$, we can express the twisted Sasaki metric and its inverse as follows:

$$(G) = \begin{pmatrix} ag_{ij} & 0 \\ 0 & bg_{ij} \end{pmatrix}$$

and

$$\left(G^{-1}\right) = \begin{pmatrix} \frac{1}{a}g^{ij} & 0 \\ 0 & \frac{1}{b}g^{ij} \end{pmatrix}.$$

The horizontal lift ${}^H\nabla$ of a linear connection ∇ on the manifold M_n to the tangent bundle TM is a unique linear connection defined on the TM and is characterized by the following conditions:

$$\begin{aligned} {}^H\nabla_{{}^HX}{}^HY &= {}^H(\nabla_X Y),\ {}^H\nabla_{{}^HX}{}^VY = {}^V(\nabla_X Y), \\ {}^H\nabla_{{}^VX}{}^HY &= 0,\ {}^H\nabla_{{}^VX}{}^VY = 0 \end{aligned}$$

for all vector fields X and Y on M_n. The torsion tensor ${}^HT$ of ${}^H\nabla$ satisfies the conditions

$$\begin{aligned} {}^HT({}^VX, {}^VY) &= 0,\ {}^HT({}^VX, {}^HY) = {}^V(T(X,Y)), \\ {}^HT({}^HX, {}^HY) &= {}^H(T(X,Y)) - \gamma R(X,Y). \end{aligned}$$

In these expressions, T and R represent the torsion and curvature tensor fields, respectively, of the linear connection ∇ on the manifold M_n. Importantly, the last set of identities implies that the connection ${}^H\nabla$ can have a non-zero torsion tensor even if ∇ is selected as a torsion-free linear connection. For a more detailed explanation, we can refer to [47]. Now, our objective is to determine the conditions under which the pair $({}^H\nabla, G)$ forms a Codazzi pair on the tangent bundle TM. To facilitate future use, we provide the following.

$$\begin{aligned} & ({}^H\nabla_{E_i}G)(E_j, E_k) \\ = {} & E_iG(E_j, E_k) - G({}^H\nabla_{E_i}E_j, E_k) - G(E_j, {}^H\nabla_{E_i}E_k) \\ = {} & E_i(ag_{jk}) - G({}^H\nabla_{{}^H(\partial_i)}{}^H(\partial_j), {}^H(\partial_k)) - G({}^H(\partial_j), {}^H\nabla_{{}^H(\partial_i)}{}^H(\partial_k)) \\ = {} & a(\partial_i g_{jk}) - G({}^H(\nabla_{\partial_i}\partial_j), {}^H(\partial_k)) - G({}^H(\partial_j), {}^H(\nabla_{\partial_i}\partial_k)) \\ = {} & a(\partial_i g_{jk}) - ag(\nabla_{\partial_i}\partial_j, \partial_k) - ag(\partial_j, \nabla_{\partial_i}\partial_k) \\ = {} & a(\partial_i g_{jk}) - ag(\Gamma^h_{ij}\partial_h, \partial_k) - ag(\partial_j, \Gamma^h_{ik}\partial_h) \\ = {} & a(\partial_i g_{jk} - \Gamma^h_{ij}g_{hk} - \Gamma^h_{ik}g_{jh}) \\ = {} & a\nabla_i g_{jk}, \end{aligned} \tag{1}$$

$$\begin{aligned} & ({}^H\nabla_{E_i}G)(E_j, E_{\bar{k}}) \\ = {} & E_iG(E_j, E_{\bar{k}}) - G({}^H\nabla_{E_i}E_j, E_{\bar{k}}) - G(E_j, {}^H\nabla_{E_i}E_{\bar{k}}) \\ = {} & E_iG({}^H(\partial_j), {}^V(\partial_k)) - G({}^H\nabla_{{}^H(\partial_i)}{}^H(\partial_j), {}^V(\partial_k)) - G({}^H(\partial_j), {}^H\nabla_{{}^H(\partial_i)}{}^V\partial_k) \\ = {} & -G({}^H(\nabla_{\partial_i}\partial_j), {}^V(\partial_k)) - G({}^H(\partial_j), {}^V(\nabla_{\partial_i}\partial_k)) \\ = {} & 0, \end{aligned} \tag{2}$$

$$\begin{aligned} ({}^H\nabla_{E_i}G)(E_{\bar{j}}, E_{\bar{k}}) &= E_iG(E_{\bar{j}}, E_{\bar{k}}) - G({}^H\nabla_{E_i}E_{\bar{j}}, E_{\bar{k}}) - G(E_{\bar{j}}, {}^H\nabla_{E_i}E_{\bar{k}}) \\ &= E_i bg(\partial_j, \partial_k) - G({}^V(\nabla_{\partial_i}\partial_j), {}^V(\partial_k)) - G({}^V(\partial_j), {}^V(\nabla_{\partial_i}\partial_k)) \\ &= b(\partial_i g_{jk} - \Gamma^h_{ij}g_{hk} - \Gamma^h_{jk}g_{jh}) \\ &= b\nabla_i g_{jk}, \end{aligned} \tag{3}$$

$$
\begin{aligned}
({}^H\nabla_{E_{\bar{i}}}G)(E_j,E_k) &= E_{\bar{i}}G(E_j,E_k) - G({}^H\nabla_{E_{\bar{i}}}E_j,E_k) - G(E_j,{}^H\nabla_{E_{\bar{i}}}E_k) \\
&= E_{\bar{i}}G({}^H(\partial_j),{}^H(\partial_k)) - G({}^H\nabla_{V(\partial_i)}{}^H(\partial_j),{}^H(\partial_k)) \\
&\quad -G({}^H(\partial_j),{}^H\nabla_{V(\partial_i)}{}^H(\partial_k)) \\
&= 0, \qquad (4)
\end{aligned}
$$

$$({}^H\nabla_{E_{\bar{i}}}G)(E_j,E_{\bar{k}}) = E_{\bar{i}}G(E_j,E_{\bar{k}}) - G({}^H\nabla_{E_{\bar{i}}}E_j,E_{\bar{k}}) - G(E_j,{}^H\nabla_{E_{\bar{i}}}E_{\bar{k}}) = 0, \qquad (5)$$

$$({}^H\nabla_{E_{\bar{i}}}G)(E_{\bar{j}},E_{\bar{k}}) = E_{\bar{i}}G(E_{\bar{j}},E_{\bar{k}}) - G({}^H\nabla_{E_{\bar{i}}}E_{\bar{j}},E_{\bar{k}}) - G(E_{\bar{j}},{}^H\nabla_{E_{\bar{i}}}E_{\bar{k}}) = 0.$$

Theorem 1. *Given a statistical manifold (M_n, g, ∇), and the tangent bundle TM equipped with the twisted Sasaki metric G, the pair $({}^H\nabla, G)$ can be classified as a Codazzi pair if and only if ∇ is a metric connection with respect to g, where ${}^H\nabla$ represents the horizontal lift of the linear connection ∇.*

Proof. Consider a statistical manifold (M_n, g, ∇) with a (pseudo-)Riemannian metric g and a linear connection ∇. Then, the following relationships hold:

$$\nabla_i g_{jk} = \nabla_j g_{ik} = \nabla_k g_{ij}.$$

Now, let $(TM, G, {}^H\nabla)$ be a statistical manifold derived from (M_n, g, ∇), where G is the twisted Sasaki metric, and ${}^H\nabla$ is the associated horizontal lift connection. From the above relationships (1), we obtain

$$
\begin{aligned}
({}^H\nabla_{E_i}G)(E_j,E_k) &= ({}^H\nabla_{E_j}G)(E_i,E_k) = ({}^H\nabla_{E_k}G)(E_i,E_j) \\
a\nabla_i g_{jk} &= a\nabla_j g_{ik} = a\nabla_k g_{ij}.
\end{aligned}
$$

Additionally, from Equations (2) and (4), we find that

$$({}^H\nabla_{E_i}G)(E_j,E_{\bar{k}}) = ({}^H\nabla_{E_{\bar{k}}}G)(E_j,E_i) = ({}^H\nabla_{E_j}G)(E_{\bar{k}},E_i) = 0.$$

Furthermore, from Equations (3) and (5), we have

$$
\begin{aligned}
({}^H\nabla_{E_i}G)(E_{\bar{j}},E_{\bar{k}}) &= ({}^H\nabla_{E_{\bar{j}}}G)(E_i,E_{\bar{k}}) = ({}^H\nabla_{E_{\bar{k}}}G)(E_i,E_{\bar{j}}) \\
b\nabla_i g_{jk} &= 0,
\end{aligned}
$$

which means that ∇ is a metric connection with respect to g. If the above equations are performed in reverse, the sufficient condition of the theorem is easily reached. □

To calculate the Levi-Civita connection $\tilde{\nabla}$ of the twisted Sasaki metric G, we need to find the Christoffel symbols $\tilde{\Gamma}^{\alpha}_{\gamma\beta}$ associated with this connection. The Levi-Civita connection ensures that the metric is compatible with the connection, meaning that the metric is parallel with respect to $\tilde{\nabla}$. To find the connection coefficients, you can use the following formula for the Christoffel symbols:

$$\tilde{\Gamma}^{\alpha}_{\gamma\beta} = \frac{1}{2}G^{\alpha\varepsilon}(E_\gamma G_{\varepsilon\beta} + E_\beta G_{\gamma\varepsilon} - E_\varepsilon G_{\gamma\beta}) + \frac{1}{2}(\Omega_{\gamma\beta}{}^{\alpha} + \Omega^{\alpha}_{\gamma\beta} + \Omega^{\alpha}_{\beta\gamma}),$$

where

$$
\begin{cases}
\Omega^{\alpha}_{\gamma\beta} = G^{\alpha\varepsilon}G_{\delta\beta}\Omega_{\varepsilon\gamma}{}^{\delta}, \\
\Omega_{j\bar{i}}{}^{\bar{h}} = -\Omega_{\bar{i}j}{}^{\bar{h}} = -R_{jis}{}^{h}y^s, \\
\Omega_{j\bar{i}}{}^{\bar{h}} = -\Omega_{\bar{i}j}{}^{\bar{h}} = \Gamma^h_{ji}
\end{cases}
$$

and it will be used as $\gamma = j;\bar{j}$ $\beta = i;\bar{i}$ $\alpha = h;\bar{h}$ $\varepsilon = k;\bar{k}$ $\delta = m;\bar{m}$.

We present the following proposition regarding the Levi-Civita connection $\tilde{\nabla}$ associated with the twisted Sasaki metric G.

Proposition 1. *In the context of a statistical manifold (M_n, g, ∇) and its tangent bundle (TM, G) equipped with the twisted Sasaki metric G, the local expression for the Levi-Civita connection $\tilde{\nabla}$ associated with the twisted Sasaki metric G on TM can be stated as follows:*

$$
\begin{aligned}
\tilde{\nabla}_{E_i}E_j &= \left(\Gamma^k_{ij}\right)E_k + \left(\frac{1}{2}y^s R_{jis}{}^k\right)E_{\bar{k}},\\
\tilde{\nabla}_{E_i}E_{\bar{j}} &= \left(\frac{b}{2\alpha}y^s R_{sji}{}^k\right)E_k + \left(\Gamma^k_{ij} + \frac{1}{2}g^{km}(\nabla_i g_{mj})\right)E_{\bar{k}},\\
\tilde{\nabla}_{E_{\bar{i}}}E_j &= \left(\frac{b}{2\alpha}y^s R_{sij}{}^k\right)E_k + \left(\frac{1}{2}g^{km}(\nabla_j g_{mi})\right)E_{\bar{k}},\\
\tilde{\nabla}_{E_{\bar{i}}}E_{\bar{j}} &= \left(-\frac{b}{2a}g^{kh}(\nabla_h g_{ij})\right)E_k,
\end{aligned}
$$

where R is the curvature tensor of the linear connection ∇.

3. Curvature Properties of the Twisted Sasaki Metric

The components of the Riemannian curvature tensor for the twisted Sasaki metric G are calculated using the following expression:

$$
\tilde{R}_{\delta\gamma\beta}{}^{\alpha} = E_\delta\tilde{\Gamma}^\alpha_{\gamma\beta} - E_\gamma\tilde{\Gamma}^\alpha_{\delta\beta} + \tilde{\Gamma}^\alpha_{\delta\varepsilon}\tilde{\Gamma}^\varepsilon_{\gamma\beta} - \tilde{\Gamma}^\alpha_{\gamma\varepsilon}\tilde{\Gamma}^\varepsilon_{\delta\beta} - \Omega_{\delta\gamma}{}^{\varepsilon}\tilde{\Gamma}^\alpha_{\varepsilon\beta}.
$$

In this context, the indices are denoted as follows: $\gamma = i;\bar{i}$ $\beta = j;\bar{j}$ $\alpha = k;\bar{k}$ $\varepsilon = h;\bar{h}$ $\delta = m;\bar{m}$. In the forthcoming proposition, we provide the components of the Riemannian curvature tensor associated with the twisted Sasaki metric G.

Proposition 2. *In the context of a statistical manifold (M_n, g, ∇) and its tangent bundle (TM, G) equipped with the twisted Sasaki metric G, the local expression for the corresponding Riemannian curvature tensor $\tilde{R}$ can be stated as follows:*

$$
\begin{aligned}
\tilde{R}_{mij}{}^k &= R_{mij}{}^k + \frac{b}{4a}y^s y^p\left[R_{shm}{}^k R_{jip}{}^h - R_{shi}{}^k R_{jmp}{}^h + 2R_{mis}{}^h R_{phj}{}^k\right],\\
\tilde{R}_{mij}{}^{\bar{k}} &= -\frac{1}{4}y^s\left[R_{jis}{}^h A^k_{mh} - R_{jms}{}^h A^k_{ih} + 2R_{mis}{}^h A^k_{jh}\right]\\
&\quad + \frac{1}{2}y^s\left[\nabla_m R_{jis}{}^k - \nabla_i R_{jms}{}^k\right],\\
\tilde{R}_{\bar{m}ij}{}^k &= \frac{b}{2a}y^s\nabla_i R_{msj}{}^k - \frac{b}{4a}y^s\left[R_{jis}{}^h A^k_{mh} + R_{shi}{}^k A^h_{jm}\right],\\
\tilde{R}_{\bar{m}ij}{}^{\bar{k}} &= \frac{1}{2}R_{jim}{}^k - \frac{b}{4a}y^s y^p R_{his}{}^k R_{pmj}{}^h - \frac{1}{2}\nabla_i A^k_{jm} - \frac{1}{4}A^k_{ih}A^h_{jm},\\
\tilde{R}_{m\bar{i}j}{}^k &= \frac{b}{2a}y^s\nabla_m R_{sij}{}^k + \frac{b}{4a}y^s\left[R_{shm}{}^k A^h_{ji} + R_{jms}{}^h A^k_{ih}\right],\\
\tilde{R}_{m\bar{i}j}{}^{\bar{k}} &= \frac{1}{2}R_{mji}{}^h + \frac{b}{4a}y^s y^p R_{hms}{}^k R_{pij}{}^h + \frac{1}{2}\nabla_m A^k_{ji} + \frac{1}{4}A^k_{mh}A^h_{ji},\\
\tilde{R}_{mi\bar{j}}{}^k &= \frac{b}{4a}y^s\left[R_{shm}{}^k A^h_{ij} - R_{shi}{}^k A^h_{mj} - 2R_{mis}{}^h A^k_{hj}\right]\\
&\quad + \frac{b}{2a}y^s\left[\nabla_m R_{sji}{}^k - \nabla_i R_{sjm}{}^k\right],\\
\tilde{R}_{mi\bar{j}}{}^{\bar{k}} &= R_{mij}{}^k + \frac{1}{2}\left[\nabla_m A^k_{ij} - \nabla_i A^k_{mj}\right] + \frac{1}{4}\left[A^k_{mh}A^h_{ij} - A^k_{ih}A^h_{mj}\right]\\
&\quad + \frac{b}{4a}y^s y^p\left[R_{hms}{}^k R_{pji}{}^h - R_{his}{}^k R_{pjm}{}^h\right],
\end{aligned}
$$

$$
\begin{aligned}
\tilde{R}_{\bar{m}\bar{i}j}^{\ \ \ \ k} &= \frac{b}{4a}\left[A_{ih}^{k}A_{jm}^{h} - A_{mh}^{k}A_{ji}^{h}\right] + \frac{b^2}{4a^2}y^s y^p\left[R_{smh}^{\ \ \ \ k}R_{pij}^{\ \ \ h} - R_{sih}^{\ \ \ k}R_{pmj}^{\ \ \ \ h}\right] \\
&\quad + \frac{b}{a}R_{mij}^{\ \ \ \ k}, \\
\tilde{R}_{\bar{m}\bar{i}j}^{\ \ \ \ \bar{k}} &= \frac{b}{4a}y^s\left[R_{sij}^{\ \ \ h}A_{hm}^{k} - R_{smj}^{\ \ \ \ h}A_{hi}^{k}\right], \\
\tilde{R}_{\bar{m}i\bar{j}}^{\ \ \ \ k} &= \frac{b}{2a}R_{mji}^{\ \ \ \ k} + \frac{b}{2a}\left(\nabla_i A_{mj}^{k}\right) - \frac{b}{4a}A_{mh}^{k}A_{ij}^{h} + \frac{b^2}{4a^2}y^s y^p R_{smh}^{\ \ \ \ k}R_{pji}^{\ \ \ h}, \\
\tilde{R}_{\bar{m}i\bar{j}}^{\ \ \ \ \bar{k}} &= \frac{b}{4a}y^s\left[R_{sji}^{\ \ \ h}A_{hm}^{k} + R_{his}^{\ \ \ k}A_{mj}^{h}\right], \\
\tilde{R}_{m\bar{i}\bar{j}}^{\ \ \ \ k} &= \frac{b}{2a}R_{jim}^{\ \ \ \ k} - \frac{b}{2a}\left(\nabla_m A_{ij}^{k}\right) + \frac{b}{4a}A_{ih}^{k}A_{mj}^{h} - \frac{b^2}{4a^2}y^s y^p R_{sih}^{\ \ \ k}R_{pjm}^{\ \ \ \ h}, \\
\tilde{R}_{m\bar{i}\bar{j}}^{\ \ \ \ \bar{k}} &= \frac{b}{4a}y^s\left[R_{mhs}^{\ \ \ \ k}A_{ij}^{h} + R_{jsm}^{\ \ \ \ h}A_{hi}^{k}\right], \\
\tilde{R}_{\bar{m}\bar{i}\bar{j}}^{\ \ \ \ k} &= \frac{b^2}{4a^2}y^s\left[R_{sih}^{\ \ \ k}A_{mj}^{h} - R_{smh}^{\ \ \ \ k}A_{ij}^{h}\right], \\
\tilde{R}_{\bar{m}\bar{i}\bar{j}}^{\ \ \ \ \bar{k}} &= \frac{b}{4a}\left[A_{hi}^{k}A_{mj}^{h} - A_{hm}^{k}A_{ij}^{h}\right],
\end{aligned}
$$

where R is the curvature tensor of the linear connection ∇ and $A_{ij}^{k} = g^{kl}\left(\nabla_i g_{jl}\right)$.

Continuing with our analysis, we now turn our attention to the Ricci and scalar curvature tensors. Utilizing the results from Proposition 2 and performing standard calculations, we obtain the following outcomes.

Proposition 3. *In the context of a statistical manifold (M_n, g, ∇) and its tangent bundle (TM, G) equipped with the twisted Sasaki metric G, the local expression for the corresponding Ricci curvature tensor $\tilde{R}_{IJ} = \tilde{R}_{MIJ}^{\ \ \ \ \ M}$ can be stated as follows:*

$$
\begin{aligned}
\tilde{R}_{ij} &= R_{ij} + \frac{b}{4a}y^s y^p\left[R_{mis}^{\ \ \ \ h}R_{phj}^{\ \ \ \ m} + R_{msi}^{\ \ \ \ h}R_{jhp}^{\ \ \ \ m}\right] - \frac{1}{4}\left(\nabla_i g^{ml}\right)\left(\nabla_j g_{ml}\right) \\
&\quad - \frac{1}{2}g^{ml}\left(\nabla_i \nabla_j g_{ml}\right), \\
\tilde{R}_{\bar{i}j} &= \frac{b}{2a}y^s \nabla_m R_{sij}^{\ \ \ m} + \frac{b}{4a}y^s\left[R_{sij}^{\ \ \ h}A_{hm}^{m} + R_{jsm}^{\ \ \ \ h}A_{ih}^{m}\right], \\
\tilde{R}_{i\bar{j}} &= \frac{b}{2a}y^s \nabla_m R_{sji}^{\ \ \ m} + \frac{b}{4a}y^s\left[R_{sji}^{\ \ \ h}A_{hm}^{m} + R_{ism}^{\ \ \ \ h}A_{jh}^{m}\right], \\
\tilde{R}_{\bar{i}\bar{j}} &= \frac{b}{4a}\left[2A_{hi}^{m}A_{mj}^{h} - A_{hm}^{m}A_{ij}^{h}\right] - \frac{b}{2a}\left(\nabla_m A_{ij}^{m}\right) - \frac{b^2}{4a^2}y^s y^p R_{sih}^{\ \ \ m}R_{pjm}^{\ \ \ \ h},
\end{aligned}
$$

where R is the curvature tensor of the linear connection ∇, and $A_{ij}^{k} = g^{kl}\left(\nabla_i g_{jl}\right)$.

Proposition 4. *In the context of a statistical manifold (M_n, g, ∇) and its tangent bundle (TM, G) equipped with the twisted Sasaki metric G, the local expression for the corresponding scalar curvature tensor $\tilde{r}$ can be stated as follows:*

$$
\tilde{r} = \frac{1}{a}r + \frac{b}{4a^2}\|R\| + \frac{1}{4a}g^{ij}\left[A_{hi}^{m}A_{mj}^{h} - A_{hm}^{m}A_{ij}^{h}\right] - \frac{1}{2a}g^{ij}\left[\nabla_i A_{jm}^{m} + \nabla_m A_{ij}^{m}\right],
$$

where $A_{ij}^{k} = g^{kl}\left(\nabla_i g_{jl}\right)$, and r and R are the scalar curvature and curvature tensor of the linear connection ∇, respectively.

4. Conformal Vector Fields according to the Twisted Sasaki Metric

Let $L_{\tilde{X}}$ be the Lie derivation with respect to the vector field $\tilde{X}$. A vector field $\tilde{X}$ with components $(v^h, v^{\bar{h}})$ is fiber-preserving if and only if v^h depends only on the variables (x^h). Therefore, every fiber-preserving vector field $\tilde{X}$ on TM induces a vector field $V = v^h \frac{\partial}{\partial x^h}$ on M_n. We shall first state the following lemmas, which are needed later on.

Lemma 1. *Consider a statistical manifold (M_n, g, ∇) and its tangent bundle TM. Let $\tilde{X}$ be a fiber-preserving vector field on TM with the components $(v^h, v^{\bar{h}})$. Then, the Lie derivates of the adapted frame and the dual basis are given as follows (see also [49]):*

$$\begin{aligned}
(i)\ L_{\tilde{X}}E_i &= -(E_i v^h)E_h + \left[y^b v^c R_{icb}{}^h - v^{\bar{b}}\Gamma^h_{bi} - (E_i v^{\bar{h}})\right]E_{\bar{h}},\\
(ii)\ L_{\tilde{X}}E_{\bar{i}} &= \left[v^b \Gamma^h_{bi} - (E_{\bar{i}} v^{\bar{h}})\right]E_{\bar{h}},\\
(iii)\ L_{\tilde{X}}dx^i &= (E_h v^i)dx^h,\\
(iv)\ L_{\tilde{X}}\delta y^i &= -\left[y^c v^b R_{bhc}{}^i + v^{\bar{b}}\Gamma^i_{bh} + (E_h v^{\bar{i}})\right]dx^h - \left[v^b\Gamma^i_{bh} - (E_{\bar{h}} v^{\bar{i}})\right]\delta y^h.
\end{aligned}$$

Here, R denotes the curvature tensor of the linear connection ∇, and Γ^h_{is} represents the Christoffel symbols of the linear connection ∇.

Through the Lemma given above, we provide the following lemma that we will use later.

Lemma 2. *In the context of a statistical manifold (M_n, g, ∇) and its tangent bundle (TM, G) equipped with the twisted Sasaki metric G, the Lie derivative of twisted Sasaki metric G with respect to the fiber-preserving vector field $\tilde{X}$ is given as follows:*

$$\begin{aligned}
L_{\tilde{X}}G &= a\left[L_V g_{ij} + 2\left(E_i v^h\right)g_{hj}\right]dx^i dx^j\\
&\quad -2b g_{hj}\left[y^s v^b R_{bis}{}^h + v^{\bar{b}}\Gamma^h_{bi} + \left(E_i v^{\bar{h}}\right)\right]dx^i \delta y^j\\
&\quad +b\left[L_V g_{ij} - 2v^b\Gamma^h_{bi} g_{hj} + 2g_{hj}\left(E_{\bar{i}} v^{\bar{h}}\right)\right]\delta y^i \delta y^j,
\end{aligned}$$

where $L_V g_{ij}$ denotes the components of the Lie derivative of $L_V g$, and $\tilde{X} = v^h E_h + v^{\bar{h}} E_{\bar{h}}$. $V = v^h \frac{\partial}{\partial x^h}$ is a vector field on M_n.

If we have the tangent bundle TM over a statistical manifold (M_n, g, ∇) equipped with the twisted Sasaki metric G, and there exists a scalar function Ω such that the equation

$$L_{\tilde{X}}G = 2\Omega G \tag{6}$$

is satisfied, then we refer to $\tilde{X}$ as an infinitesimal fiber-preserving conformal vector field on (TM, G).

Theorem 2. *In the context of a statistical manifold (M_n, g, ∇) of dimension $n \geq 2$ and its tangent bundle (TM, G) equipped with the twisted Sasaki metric G, the fiber-preserving vector field $\tilde{X}$ is an infinitesimal fiber-preserving conformal vector field on (TM, G) if and only if the scalar function Ω on TM depends only on the variables $\left(x^h\right)$ with respect to the induced coordinates $\left(x^h, x^{\bar{h}}\right)$ and the following conditions are satisfied:*

$$
\begin{aligned}
&(i) \quad \tilde{X} = \left(v^h, v^{\bar{h}}\right) = \left(v^h, y^s A_s^h + B^h\right) = {}^H V + {}^V B + \gamma A, \\
&(ii) \quad \Omega = \frac{1}{2n} g^{ij}(L_V g_{ij}) + \frac{1}{n}\left(E_i v^i\right), \\
&(iii) \quad v^l R_{lis}{}^h + \nabla_i \nabla_s v^h = 0, \\
&(iv) \quad \nabla_i v^h = A_i^h, \\
&(v) \quad \nabla_i B^h = 0,
\end{aligned}
$$

where $\tilde{X} = v^h E_h + v^{\bar{h}} E_{\bar{h}}$ is a fiber-preserving vector field on TM, $V = v^h \frac{\partial}{\partial x^h}$ is a vector field on M_n, Ω is a scalar function, ${}^H V$ is the horizontal lift of V, ${}^V B$ is the vertical lift of B, and $A = \left(A_i^h\right)$ and $B = \left(B^h\right)$ are $(1,1)$ and $(1,0)$ tensor fields on M_n, respectively.

Proof. Starting from the expressions given in Lemma 2 and Lemma 6, we obtain the following equations:

$$L_{\tilde{X}} G_{ij} = 2\Omega G_{ij}$$

$$L_V g_{ij} + 2(E_i v^h) g_{hj} = 2\Omega g_{ij}, \tag{7}$$

$$L_{\tilde{X}} G_{\bar{i}j} = 2\Omega G_{\bar{i}j}$$

$$2b g_{hj}\left[y^s v^l R_{lis}{}^h + v^{\bar{l}} \Gamma_{li}^h + \left(E_i v^{\bar{h}}\right)\right] = 0 \tag{8}$$

and

$$L_{\tilde{X}} G_{\bar{i}\bar{j}} = 2\Omega G_{\bar{i}\bar{j}}$$

$$L_V g_{ij} - 2v^l \Gamma_{li}^h g_{hj} + 2g_{hj}\left(E_{\bar{i}} v^{\bar{h}}\right) = 2\Omega g_{ij}. \tag{9}$$

Now, let us apply $E_{\bar{k}}$ to both sides of Equation (9):

$$g_{hj} E_{\bar{k}}\left(E_{\bar{i}} v^{\bar{h}}\right) = (E_{\bar{k}}\Omega) g_{ij}. \tag{10}$$

By interchanging i with k in the last equation, we obtain

$$g_{hj} E_{\bar{i}}\left(E_{\bar{k}} v^{\bar{h}}\right) = (E_{\bar{i}}\Omega) g_{kj}. \tag{11}$$

From the equalities of (10) and (11), we can write

$$(E_{\bar{k}}\Omega) g_{ij} = (E_{\bar{i}}\Omega) g_{kj}.$$

By contracting with g^{ij} in the above equation, we obtain

$$
\begin{aligned}
n(E_{\bar{k}}\Omega) &= (E_{\bar{k}}\Omega) \\
(E_{\bar{k}}\Omega) &= 0.
\end{aligned} \tag{12}
$$

This means that the scalar function Ω depends only on the variables $\left(x^h\right)$. Thus, we can see Ω as a function on M_n. Substituting Equation (12) into Equation (11), we can write

$$E_{\bar{i}}\left(E_{\bar{k}} v^{\bar{h}}\right) = 0$$

$$v^{\bar{h}} = y^s A_s^h + B^s, \tag{13}$$

where $A = \left(A_i^h\right)$ and $B = \left(B^h\right)$ are $(1,1)$ and $(1,0)$ tensor fields on M_n, respectively. Through substituting Equation (13) into Equation (8), we have

$$v^l R_{lis}{}^h + \nabla_i A_s^h = 0$$

and

$$\nabla_i B^h = 0.$$

By substituting Equation (13) into Equation (9), we have

$$L_V g_{ij} - 2v^l \Gamma_{li}^h g_{hj} + 2g_{hj} A_i^h = 2\Omega g_{ij}. \tag{14}$$

When Equations (7) and (14) are evaluated together, we can write

$$\nabla_i v^h = A_i^h.$$

If the last equation is used in equality (14), the following result is obtained

$$v^l R_{lis}{}^h + \nabla_i \nabla_s v^h = 0.$$

By contracting Equation (7) with $\frac{1}{2n} g^{ij}$, we have

$$\Omega = \frac{1}{2n} g^{ij}(L_V g_{ij}) + \frac{1}{n}\left(E_i v^i\right).$$

This completes the proof. □

5. The Ricci Solitons according to the Twisted Sasaki Metric

A Ricci soliton on a smooth manifold M_n (where $n \geq 2$) is defined as a triple (g, X, λ), where g represents a pseudo-Riemannian metric on M_n, Ric is the associated Ricci tensor, X is a vector field, and λ is a real constant. This triple can satisfy the following equation:

$$Ric + \frac{1}{2} L_X g = \lambda g,$$

where L_X represents the Lie derivative with respect to the vector field X. A Ricci soliton can be classified into three categories, shrinking, steady, or expanding, depending on the sign of the constant λ. It is called "shrinking" when λ is positive, "steady" when λ is zero, and "expanding" when λ is negative. These classifications are important because Ricci solitons correspond to self-similar solutions of the Ricci flow [1] and serve as natural generalizations of Einstein metrics.

In the case of a gradient Ricci soliton, the soliton vector field X can be expressed as the gradient of a smooth function f on M_n. For gradient Ricci solitons, the following equation holds:

$$\left(\nabla^2 f\right) + Ric = \lambda g,$$

where ∇^2 denotes the Hessian operator.

A Ricci soliton defined on the tangent bundle TM equipped with the twisted Sasaki metric G, which is defined over a statistical manifold (M_n, g), is characterized by the equation

$$\widetilde{Ric} + \frac{1}{2} L_{\tilde{X}} G = \lambda G, \tag{15}$$

where $\widetilde{Ric}$ is the Ricci tensor of G, $\tilde{X}$ is a vector field on TM, and λ is a smooth function on TM.

Theorem 3. *In the context of a statistical manifold* (M_n, g, ∇) *of dimension* $n \geq 2$ *and its tangent bundle* (TM, G) *equipped with the twisted Sasaki metric* G*,* (TM, G, λ) *constitutes a Ricci soliton if and only if the base manifold is flat, and the following conditions are satisfied:*

$$(i) \quad \tilde{X} = (v^h, v^{\bar{h}}) = (v^h, y^s M_s^h + N^h),$$
$$(ii) \quad \lambda = \frac{1}{n}\left[\frac{1}{2}g^{ij}(L_V g_{ij}) + (E_i v^i) - \frac{1}{2a}\nabla^j \nabla^m g_{jm} - \frac{1}{4a} g^{ij}(\nabla^m g_{ih})\left(\nabla^h g_{jm}\right)\right],$$
$$(iii) \quad \nabla M = \nabla N = 0,$$

where $\tilde{X} = v^h E_h + v^{\bar{h}} E_{\bar{h}}$ *is a vector field on* TM*,* $V = v^h \frac{\partial}{\partial x^h}$ *is a vector field on* M_n*,* $\lambda \in \mathbb{R}$*, and* $M = (M_s^h)$ *and* $N = (N^h)$ *are* $(1,1)$ *and* $(1,0)$ *tensor fields on* M_n*, respectively.*

Proof. With the help of Equation (15), we have

$$\tilde{R}_{ij} + \frac{1}{2} L_{\tilde{X}} G_{ij} = \lambda G_{ij},$$

$$\begin{aligned} \lambda a g_{ij} &= R_{ij} + \frac{b}{4a} y^s y^p [R_{mis}{}^h R_{phj}{}^m + R_{msi}{}^h R_{jhp}{}^m] - \frac{1}{4} A_{ih}^m A_{jm}^h \\ &\quad - \frac{1}{2}\nabla_i A_{jm}^m + \frac{a}{2}[L_V g_{ij} + 2(E_i v^h) g_{hj}], \end{aligned} \tag{16}$$

$$\tilde{R}_{i\bar{j}} + \frac{1}{2} L_{\tilde{X}} G_{i\bar{j}} = \lambda G_{i\bar{j}},$$

$$\frac{1}{2a}\nabla_m R_{sij}{}^m + \frac{1}{4a}(R_{sij}{}^h A_{hm}^m + R_{jsm}{}^h A_{ih}^m) - g_{hj} v^l R_{lis}{}^h = 0$$

and

$$v^{\bar{l}} \Gamma_{li}^h + (E_i v^{\bar{h}}) = 0. \tag{17}$$

Also,

$$\tilde{R}_{\bar{i}\bar{j}} + \frac{1}{2} L_{\tilde{X}} G_{\bar{i}\bar{j}} = \lambda G_{\bar{i}\bar{j}},$$

$$\begin{aligned} \lambda g_{ij} &= \frac{1}{4a}[2A_{hi}^m A_{mj}^h - A_{hm}^m A_{ij}^h] - \frac{1}{2a}(\nabla_m A_{ij}^m) + \frac{1}{2} L_V g_{ij} \\ &\quad + g_{hj}(E_{\bar{i}} v^{\bar{h}}) - v^l \Gamma_{li}^h g_{hj} \end{aligned} \tag{18}$$

and

$$\frac{b^2}{4a^2} y^s y^p R_{sih}{}^m R_{pjm}{}^h = 0 \Rightarrow \frac{b^2}{4a^2} y^s y^p \|R\| = 0.$$

This means that the base manifold is flat. By contracting with g^{ij} in Equation (18), we obtain

$$\begin{aligned} \lambda n &= \frac{1}{4a} g^{ij}[2A_{hi}^m A_{mj}^h - A_{hm}^m A_{ij}^h] - \frac{1}{2a} g^{ij}(\nabla_m A_{ij}^m) + \frac{1}{2} g^{ij} L_V g_{ij} \\ &\quad + (E_{\bar{i}} v^{\bar{i}}) - v^l \Gamma_{li}^i. \end{aligned}$$

Applying $E_{\bar{k}}$ to both sides of the last equation, we have

$$E_{\bar{k}}\left(E_{\bar{i}} v^{\bar{i}}\right) = 0$$

$$v^{\bar{i}} = y^s M_s^i + N^i, \tag{19}$$

where $M = (M_s^i)$ and $N = (N^i)$ are $(1,1)$ and $(1,0)$ tensor fields on M_n. Substituting Equation (19) in Equation (17), we obtain

$$y^s \nabla_i M_s^h + \nabla_i N^h = 0.$$

By contracting with $\frac{1}{na} g^{ij}$ in Equation (16) and considering that the base manifold is flat, we have the following equation

$$\lambda = \frac{1}{n}\left[\frac{1}{2} g^{ij}(L_V g_{ij}) + (E_i v^i) - \frac{1}{2a}\nabla^j \nabla^m g_{jm} - \frac{1}{4a} g^{ij}(\nabla^m g_{ih})\left(\nabla^h g_{jm}\right)\right],$$

which completes the proof. □

6. The Yamabe Solitons according to the Twisted Sasaki Metric

In the context of a complete Riemannian manifold (M_n, g), where $n \geq 2$, the Riemannian metric g is considered to have a Yamabe soliton if it satisfies the following equation:

$$\frac{1}{2} L_X g = (r - \lambda) g,$$

where λ is a scalar constant, X is a differentiable vector field known as the soliton vector field, and r represents the scalar curvature of the Riemannian manifold (M_n, g). If λ is a smooth function, then the Riemannian metric g is said to admit an almost Yamabe soliton, which is a natural generalization of Yamabe solitons [50]. Additionally, almost Yamabe solitons are classified as steady, expanding, or shrinking, depending on the value of λ. Specifically, they are categorized as follows: (1) steady if $\lambda = 0$, (2) expanding if $\lambda < 0$, and (3) shrinking if $\lambda > 0$. It is important to note that Einstein manifolds are a special case of almost Yamabe solitons.

Now, when considering a Yamabe soliton defined on the tangent bundle TM with the twisted Sasaki metric G over a statistical manifold (M_n, g, ∇), we can describe it with the following equation:

$$\frac{1}{2} L_{\tilde{X}} G = (\tilde{r} - \lambda) G, \tag{20}$$

where $\tilde{r}$ represents the scalar curvature of (TM, G), $\tilde{X}$ is a vector field on TM, and λ is a smooth function on TM.

In the case of a gradient Yamabe soliton, the soliton vector field X can be expressed as the gradient of a C^∞ function f on M_n, and this leads to the following equation:

$$\left(\nabla^2 f\right) = (r - \lambda) g.$$

Theorem 4. *In the context of a statistical manifold (M_n, g, ∇) of dimension $n \geq 2$ and its tangent bundle (TM, G) equipped with the twisted Sasaki metric G, (TM, G, λ) is a Yamabe soliton if and only if the following conditions are satisfied:*

$$\begin{aligned}
&(i) \quad \tilde{X} = (v^h, v^{\bar{h}}) = (v^h, y^s M_s^h + N^h),\\
&(ii) \quad \lambda = \tilde{r} - \frac{1}{2n} g^{ij}(L_V g_{ij}) - \frac{1}{n}(E_i v^i),\\
&(iii) \quad \nabla_i M_s^h = v^l R_{lis}{}^h,\\
&(iv) \quad \nabla_i N^h = 0,
\end{aligned}$$

where $\tilde{X} = v^h E_h + v^{\bar{h}} E_{\bar{h}}$ is a vector field on TM, $V = v^h \frac{\partial}{\partial x^h}$ is a vector field on M_n, $\lambda \in \mathbb{R}$, and $M = (M_s^h)$ and $N = (N^h)$ are $(1,1)$ and $(1,0)$ tensor fields on M_n, respectively.

Proof. Starting from Equation (20), we have

$$\frac{1}{2}L_{\tilde{X}}G_{ij} = (\tilde{r} - \lambda)G_{ij},$$

$$\lambda g_{ij} = \tilde{r} g_{ij} - \frac{1}{2}L_V g_{ij} - (E_i v^h) g_{hj}.$$

By contracting the last equation with $\frac{1}{n}g^{ij}$, we obtain

$$\lambda = \tilde{r} - \frac{1}{2n}(L_V g_{ij}) - \frac{1}{n}(E_i v^i). \tag{21}$$

Now, from Equation (20), we also have

$$\frac{1}{2}L_{\tilde{X}}G_{\bar{i}j} = (\tilde{r} - \lambda)G_{\bar{i}j} = 0,$$

which implies that

$$y^s v^l R_{lis}{}^h + \Gamma^h_{li} v^{\bar{l}} + E_i v^{\bar{h}} = 0 \tag{22}$$

$$\frac{1}{2}L_{\tilde{X}}G_{\bar{i}\bar{j}} = (\tilde{r} - \lambda)G_{\bar{i}\bar{j}}$$

$$\frac{1}{2}[L_V g_{ij} - 2v^l \Gamma^h_{li} g_{hj} + 2g_{hj}(E_{\bar{i}} v^{\bar{h}})] = (\tilde{r} - \lambda) g_{ij}.$$

By contracting the last equation with $\frac{1}{n}g^{ij}$, we find that

$$\lambda = \tilde{r} - \frac{1}{2n}g^{ij}(L_V g_{ij}) + \frac{1}{n}v^l \Gamma^i_{li} - \frac{1}{n}(E_{\bar{i}} v^{\bar{i}}). \tag{23}$$

When Equations (21) and (23) are evaluated together, we obtain

$$-(E_i v^i) = v^l \Gamma^i_{li} - (E_{\bar{i}} v^{\bar{i}}).$$

Applying $E_{\bar{k}}$ to both sides of the last equation, we have

$$E_{\bar{k}}\left(E_{\bar{i}} v^{\bar{i}}\right) = 0,$$

$$v^{\bar{i}} = y^s M^i_s + N^i. \tag{24}$$

Substituting Equation (24) into Equation (22), we obtain

$$y^s v^l R_{lis}{}^h + \Gamma^h_{li}(y^s M^l_s + N^l) + E_i(y^s M^h_s + N^h) = 0$$

and

$$y^s(v^l R_{lis}{}^h + \nabla_i M^h_s) + \nabla_i N^h = 0.$$

Form the above equations, we obtain

$$\nabla_i M^h_s = -v^l R_{ils}{}^h$$

and

$$\nabla_i N^h = 0.$$

Thus, the proof is complete. □

7. The Generalized Ricci–Yamabe Solitons according to the Twisted Sasaki Metric

Güler and Crasmareanu [21] introduced the concept of a Ricci–Yamabe flow on a Riemannian manifold (M_n, g), where $n \geq 2$, by combining the Ricci flow and Yamabe flow into a scalar equation:

$$\frac{\partial g}{\partial t}(t) + 2\alpha R(t) + \beta r(t) g(t) = 0, \tag{25}$$

where g is a Riemannian metric, R is the Ricci tensor, r is the scalar curvature tensor, and $\alpha, \beta \in \mathbb{R}$.

A Riemannian manifold (M_n, g) with $n > 2$ is said to have a generalized Ricci–Yamabe soliton $(g, X, \lambda, \alpha, \beta, \gamma)$ if the following equation holds:

$$L_X g + 2\alpha R = (2\lambda - \beta r) g + 2\gamma X^{\#} \otimes X^{\#}, \tag{26}$$

where $\lambda, \alpha, \beta, \gamma \in \mathbb{R}$, and $X^{\#}$ is the associated $1-$form with X.

If X is the gradient of a smooth function f on M_n, then this concept is referred to as a generalized gradient Ricci–Yamabe soliton, and Equation (26) simplifies to

$$\nabla^2 f + \alpha R = \left(\lambda - \frac{1}{2}\beta r\right) g + \gamma df \otimes df.$$

The generalized (gradient) Ricci–Yamabe soliton is categorized as expanding if $\lambda < 0$, steady if $\lambda = 0$, or shrinking if $\lambda > 0$.

Theorem 5. *In the context of a statistical manifold (M_n, g, ∇) of dimension $n \geq 2$ and its tangent bundle (TM, G) equipped with the twisted Sasaki metric G, $(TM, G, {}^C V, \lambda)$ is a generalised Ricci–Yamabe soliton if and only if the following conditions are satisfied:*

$$\begin{aligned}
(i) \quad & \lambda = \frac{1}{2n}\left[g^{ij}(L_V g_{ij}) + 2(E_i v^i) - g^{ij}\partial_t g_{ij} - 2\gamma a g_{kt} v^k v^t\right], \\
(ii) \quad & L_V g_{ij} = 2\lambda g_{ij} + 2g_{hj}(\nabla_i v^h) + \partial_t g_{ij}, \\
(iii) \quad & g_{kt}(\partial_s v^k)(\partial_p v^t) = 0,
\end{aligned}$$

where the potential vector field ${}^C V$ is the complete lift of a vector field V on M_n to the tangent bundle TM. This lift is given by ${}^C V = (v^m, v^{\overline{m}}) = (v^m, y^s \partial_s v^m)$, where $V = v^h \frac{\partial}{\partial x^h}$ is a vector field on M_n, and $\lambda \in \mathbb{R}$.

Proof. We will demonstrate the existence of the scalar λ. Starting from Equation (25), we obtain

$$2\alpha R + \beta r G = -\partial_t G. \tag{27}$$

When we substitute the expressions for $L_{{}^C X} G$ and Equation (27) into Equation (26), we obtain

$$L_{{}^C X} G_{ij} + 2\alpha \tilde{R}_{ij} = (2\lambda - \beta \tilde{r}) G_{ij} + 2\gamma\, {}^C X^{\#}_i \otimes {}^C X^{\#}_j, \tag{28}$$

$$L_{{}^C X} G_{\overline{i}j} + 2\alpha \tilde{R}_{\overline{i}j} = (2\lambda - \beta \tilde{r}) G_{\overline{i}j} + 2\gamma\, {}^C X^{\#}_{\overline{i}} \otimes {}^C X^{\#}_j$$

and

$$L_{{}^C X} G_{\overline{i}\overline{j}} + 2\alpha \tilde{R}_{\overline{i}\overline{j}} = (2\lambda - \beta \tilde{r}) G_{\overline{i}\overline{j}} + 2\gamma\, {}^C X^{\#}_{\overline{i}} \otimes {}^C X^{\#}_{\overline{j}}. \tag{29}$$

Here, the potential vector field ${}^C X$ and its associated $1-$form ${}^C X^{\#}$ are defined as follows:

$${}^C X = \begin{pmatrix} v^m \\ v^{\overline{m}} \end{pmatrix} = \begin{pmatrix} v^m \\ y^s \partial_s v^m \end{pmatrix}$$

and

$$
\begin{aligned}
{}^{C}X^{\#}_{j} &= {}^{C}X^{I}G_{Ij} = a g_{ij} v^{i}, \\
{}^{C}X^{\#}_{\bar{j}} &= {}^{C}X^{I}G_{I\bar{j}} = b g_{ij} y^{s} \partial_s v^{i}.
\end{aligned}
$$

From (28), we deduce that

$$L_V g_{ij} + 2(E_i v^h) g_{hj} - \partial_t g_{ij} = 2\lambda g_{ij} + 2\gamma a g_{ki} v^k g_{tj} v^t.$$

By contracting this equation with $\frac{1}{2n} g^{ij}$, we obtain

$$\lambda = \frac{1}{2n}\left[g^{ij}(L_V g_{ij}) + 2(E_i v^i) - g^{ij}\partial_t g_{ij} - 2\gamma a g_{kt} v^k v^t\right].$$

From Equation (29), we have

$$b\left[L_V g_{ij} - 2v^l \Gamma^h_{li} g_{hj} + 2h_{hj}(E_{\bar{i}} v^{\bar{h}})\right] + 2\alpha \tilde{R}_{\bar{i}\bar{j}} + \beta \tilde{r} G_{\bar{i}\bar{j}} = 2\lambda G_{\bar{i}\bar{j}} + 2\gamma\, {}^{C}X^{\#}_{\bar{i}} \otimes {}^{C}X^{\#}_{\bar{j}}.$$

Using Equation (27) in the last equation, we can rewrite it as

$$
\begin{aligned}
0 &= L_V g_{ij} - 2\lambda g_{ij} - 2g_{hj} v^l \Gamma^h_{li} + 2g_{hj}(\partial_i v^h) - \partial_t g_{ij} \\
L_V g_{ij} &= 2\lambda g_{ij} + 2g_{hj}(\nabla_i v^h) + \partial_t g_{ij}.
\end{aligned}
$$

Finally, from the contracted equation

$$2\gamma b y^s y^p g_{ki}(\partial_s v^k) g_{tj}(\partial_p v^t) = 0,$$

we obtain

$$g_{kt}(\partial_s v^k)(\partial_p v^t) = 0.$$

Therefore, the proof is complete. □

8. Conclusions

This paper endeavors to examine the geometric characteristics of the tangent bundles of statistical manifolds by employing the twisted Sasaki metric on their tangent bundles. Our inquiry was structured into two principal phases, each delineating specific yet interconnected objectives.

Firstly, we delved into the intricacies of the twisted Sasaki metrics, meticulously scrutinizing their properties and inherent geometrical aspects. Our comprehensive analysis has provided a deeper understanding of these metrics, elucidating their behavior and implications for differential geometry within the context of statistical manifolds.

Secondly, we shifted our focus toward the study of soliton structures arising from the application of the twisted Sasaki metric to the tangent bundle. Solitons, as mathematical constructs, offer invaluable insights into the dynamic interplay between metric structures and statistical properties. By characterizing these soliton structures, we uncovered novel patterns of evolution within the manifold, further enriching our comprehension of its geometry.

Our research makes a significant contribution to the fields of differential geometry and metric analysis by introducing a novel metric that challenges traditional rigidity when applied to tangent bundles and statistical manifolds. By elucidating the curvature properties of the tangent bundle and uncovering intriguing soliton structures, we have expanded the horizons of geometric understanding in this domain. In the future, our findings will facilitate further investigations into the manifold implications of the twisted Sasaki metric and its role in shaping the geometric and statistical landscapes of diverse mathematical structures. It is our hope that this paper will inspire future endeavors aimed at unraveling the profound mysteries of geometric structures in statistical manifolds.

Author Contributions: Conceptualization, Y.L., L.B., and A.G.; methodology, Y.L., L.B., and A.G.; investigation, Y.L., L.B., and A.G.; writing—original draft preparation, Y.L., L.B., and A.G.; writing—review and editing, L.Y., Y.L., L.B., and A.G. All authors have read and agreed to the published version of the manuscript.

Funding: This research received no external funding.

Data Availability Statement: No new data were created or analyzed in this study. Data sharing is not applicable to this article.

Acknowledgments: The authors would like to express their sincere thanks to the editor and the anonymous reviewers for their helpful comments and suggestions.

Conflicts of Interest: The authors declare no conflicts of interest.

ferences

Hamilton, R.S. The Ricci flow on surfaces. *Contemp. Math.* **1988**, *71*, 237–262.
Guo, H.; He, T. Harnack estimates for geometric flows, applications to Ricci flow coupled with harmonic map flow. *Geom. Dedicata* **2014**, *169*, 411–418. [CrossRef]
Hui, S.K.; Azami, S.; Bhattacharyya, S. Hamilton and Souplet-Zhang type estimations on semilinear parabolic system along geometric flow. *arXiv* **2022**, arXiv:2208.12582.
Guo, H.; Ishida, M. Harnack estimates for nonlinear backward heat equations in geometric flows. *J. Func. Anal.* **2014**, *267*, 2638–2662. [CrossRef]
Abolarinwa, A. Harnack estimates for heat equations with potentials on evolving manifolds. *Mediterr. J. Math.* **2016**, *13*, 3185–3204. [CrossRef]
Abolarinwa, A. Differential Harnack inequalities for nonlinear parabolic equation on time-dependent metrics. *Adv. Theor. Appl. Math.* **2014**, *9*, 155–166.
Cao, X.; Hamilton, R. Differential Harnack estimates for time-dependent heat equations with potentials. *Geom. Funct. Anal.* **2009**, *19*, 989–1000. [CrossRef]
Fang, S. Differential Harnack inequalities for heat equations with potentials under the geometric flow. *Arch. Math.* **2013**, *100*, 179–189. [CrossRef]
Ma, L. Gradient estimates for a simple elliptic equation on non-compact Riemannian manifolds. *J. Funct. Anal.* **2006**, *241*, 374–382.
Abolarinwa, A. Gradient estimates for heat-type equations on evolving manifolds. *J. Nonlinear Evol. Equ. Appl.* **2015**, *1*, 1–19.
Azami, S. Gradient estimates for a weighted parabolic equation under geometric flow. *arXiv* **2021**, arXiv:2112.01271v1.
Kumar, R.; Colney, L.; Khan, M.N.I. Proposed theorems on the lifts of Kenmotsu manifolds admitting a non-symmetric non-metric connection (NSNMC) in the tangent bundle. *Symmetry* **2023**, *15*, 2037. [CrossRef]
Kumar, R.; Colney, L.; Shenawy, S.; Turki, N.B. Tangent bundles endowed with quarter-symmetric non-metric connection (QSNMC) in a Lorentzian Para-Sasakian manifold. *Mathematics* **2023**, *11*, 4163. [CrossRef]
Khan, M.N.I.; Mofarreh, F.; Haseeb, A.; Saxena, M. Certain results on the lifts from an LP-Sasakian manifold to its tangent bundles associated with a quarter-symmetric metric connection. *Symmetry* **2023**, *15*, 1553. [CrossRef]
Khan, M.N.I. Liftings from a para-sasakian manifold to its tangent bundles. *Filomat* **2023**, *37*, 6727–6740.
Khan, M.N.I.; Mofarreh, F.; Haseeb, A. Tangent bundles of P-Sasakian manifolds endowed with a quarter-symmetric metric connection. *Symmetry* **2023**, *15*, 753. [CrossRef]
Khan, M.N.I.; Bahadur, O. Tangent bundles of LP-Sasakian manifold endowed with generalized symmetric metric connection. *Facta Univ. Ser. Math. Inform.* **2023**, *38*, 125–139.
Khan, M.N.I.; De, U.C.; Velimirovic, L.S. Lifts of a quarter-symmetric metric connection from a Sasakian manifold to its tangent bundle. *Mathematics* **2023**, *11*, 53. [CrossRef]
Abbassi, M.T.K.; Amri, N. Natural Ricci solitons on tangent and unit tangent bundles. *J. Math. Phys. Anal. Geom.* **2021**, *17*, 3–29.
Chen, B.Y.; Deshmukh, S. Yamabe and quasi-Yamabe solitons on Euclidean submanifolds. *Mediterr. J. Math.* **2018**, *15*, 194. [CrossRef]
Guler, S.; Crasmareanu, M. Ricci-Yamabe maps for Riemannian flows and their volume variation and volume entropy. *Turk. J. Math.* **2019**, *43*, 2631–2641. [CrossRef]
Dey, D.; Majhi, P. Sasakian 3-metric as a generalized Ricci-Yamabe soliton. *Quaest. Math.* **2022**, *45*, 409–421. [CrossRef]
Nurowski, P.; Randall, M. Generalized Ricci solitons. *J. Geom. Anal.* **2016**, *26*, 1280–1345. [CrossRef]
Aruna Kumara, H.; Mallesha Naik D.; Venkatesha, V. Geometry of generalized Ricci-type solitons on a class of Riemannian manifolds. *J. Geom. Phys.* **2022** *176*, 104506. [CrossRef]
Gezer, A.; Bilen, L.; De, U.C. Conformal vector fields and geometric solitons on the tangent bundle with the ciconia metric. *Filomat* **2023**, *37*, 8193–8204.
Li, Y.; Mofarreh, F.; Abolarinwa, A.; Alshehri, N.; Ali, A. Bounds for Eigenvalues of q-Laplacian on Contact Submanifolds of Sasakian Space Forms. *Mathematics* **2023**, *11*, 4717. [CrossRef]

27. Li, Y.; Aquib, M.; Khan, M.A.; Al-Dayel, I.; Youssef, M.Z. Chen-Ricci Inequality for Isotropic Submanifolds in Locally Metal Product Space Forms. *Axioms* **2024**, *13*, 183. [CrossRef]
28. Li, Y.; Siddiqi, M.; Khan, M.; Al-Dayel, I.; Youssef, M. Solitonic effect on relativistic string cloud spacetime attached with strang quark matter. *AIMS Math.* **2024**, *9*, 14487–14503. [CrossRef]
29. Li, J.; Yang, Z.; Li, Y.; Abdel-Baky, R.A.; Saad, M.K. On the Curvatures of Timelike Circular Surfaces in Lorentz-Minkowski Spa *Filomat* **2024**, *38*, 1–15.
30. Li, Y.; Güler, E. Twisted Hypersurfaces in Euclidean 5-Space. *Mathematics* **2023**, *11*, 4612. [CrossRef]
31. Li, Y.; Jiang, X.; Wang, Z. Singularity properties of Lorentzian Darboux surfaces in Lorentz–Minkowski spacetime. *Res. Math. S* **2024**, *11*, 7. [CrossRef]
32. De, K.; De, U.C.; Gezer, A. Perfect fluid spacetimes and *k*-almost Yamabe solitons. *Turk. J. Math.* **2023**, *47*, 1236–1246. [CrossRe
33. Amari, S. Information geometry of the EM and em algorithms for neural networks. *Neural Netw.* **1995**, *8*, 1379–1408. [CrossRe
34. Belkin, M.; Niyogi, P.; Sindhwani, V. Manifold regularization: A geometric framework for learning from labeled and unlabel examples. *J. Mach. Learn. Res.* **2006**, *7*, 2399–2434.
35. Sun, K.; Marchand-Maillet, S. An information geometry of statistical manifold learning. In Proceedings of the 31st Internatior Conference on Machine Learning (ICML-14), Beijing, China, 21–26 June 2014; pp. 1–9.
36. Bernstein, A.; Kuleshov, A.; Yanovich, Y. Statistical learning via manifold learning. In Proceedings of the 2015 IEEE 14 International Conference on Machine Learning and Applications (ICMLA), Miami, FL, USA, 9–11 December 2015; IEE Piscateville, NJ, USA, 2015; pp. 64–69.
37. Vert, J.; Yamanishi, Y. Supervised graph inference. *Adv. Neural Inf. Process. Syst.* **2005**, *17*, 1433–1440.
38. Tsang, I.; Kwok, J. Very large scale manifold regularization using core vector machines. In Proceedings of the NIPS 2005 Worksh on Large Scale Kernel Machines, Brussels, Belgium, 10 December 2005.
39. Amari, S.; Nagaoka, H. *Methods of Information Geometry*; American Mathematical Society: Providence, RI, USA, 2000.
40. Peyghan, E.; Gezer, A.; Nourmohammadifar, L. Kähler-Norden structures on statistical manifolds. *Filomat* **2022**, *36*, 5691–57 [CrossRef]
41. Matsuzoe, H. Statistical manifolds and affine differential geometry. In *Probabilistic Approach to Geometry*; Advanced Studies Pure Mathematics 57; Mathematical Society of Japan: Tokyo, Japan, 2010; pp. 303–321.
42. Peyghan, E.; Seifipour, D.; Gezer, A. Statistical structures on tangent bundles and tangent Lie groups. *Hacet. J. Math. Stat.* **2021**, 1140–1154. [CrossRef]
43. Fei, T.; Zhang, J. Interaction of Codazzi couplings with (Para-)Kähler geometry. *Result Math.* **2017**, *72*, 2037–2056. [CrossRef]
44. Schwenk-Schellschmidt, A.; Simon, U. Codazzi-equivalent affine connections. *Result Math.* **2009**, *56*, 211–229. [CrossRef]
45. Lauritzen, S.L. Statistical manifolds. In *Differential Geometry in Statistical Inferences*; IMS Lecture Notes Monographu Series Institute of Mathematical Statistics: Hayward, CA, USA, 1987; pp. 96–163.
46. Nomizu, K.; Sasaki, T. *Affine Differential Geometry: Geometry of Affine Immersions*; Vol. 111 of Cambridge Tracts in Mathemati Cambridge University Press: Cambridge, UK, 1994.
47. Yano, K.; Ishihara, S. *Tangent and Cotangent Bundles*; Marcel Dekker, Inc.: New York, NY, USA, 1973.
48. Abbassi, M.T.K.; Sarih, M. On some hereditary properties of Riemannian g-natural metrics on tangent bundles of Riemanni manifolds. *Difer. Geom. Appl.* **2005**, *22*, 19–47. [CrossRef]
49. Yamauchi, K. On infinitesimal conformal transformations of the tangent bundles over Riemannian manifolds. *Ann. Rep. Asahika Med. Coll.* **1994**, *15*, 1–10.
50. Blaga, A.M.; Perktas, S.Y. Remarks on almost-Ricci solitons in para Sasakian manifolds. *Commun. Fac. Sci. Univ. Ank. Ser. Math. Stat.* **2019**, *68*, 1621–1628. [CrossRef]

mathematics

MDPI

icle

icci–Bourguignon Almost Solitons with Special Potential on asaki-Like Almost Contact Complex Riemannian Manifolds

ncho Manev [1,2]

[1] Department of Algebra and Geometry, Faculty of Mathematics and Informatics, University of Plovdiv Paisii Hilendarski, 24 Tzar Asen St., 4000 Plovdiv, Bulgaria; mmanev@uni-plovdiv.bg; Tel.: +359-889-521-244

[2] Department of Medical Physics and Biophysics, Faculty of Pharmacy, Medical University of Plovdiv, 15A Vasil Aprilov Blvd., 4002 Plovdiv, Bulgaria

Abstract: Almost contact complex Riemannian manifolds, also known as almost contact B-metric manifolds, are equipped with a pair of pseudo-Riemannian metrics that are mutually associated with each other using the tensor structure. Here, we consider a special class of these manifolds, namely those of the Sasaki-like type. They have an interesting geometric interpretation: the complex cone of such a manifold is a holomorphic complex Riemannian manifold (also called a Kähler–Norden manifold). The basic metric on the considered manifold is specialized here as a soliton, i.e., has an additional curvature property such that the metric is a self-similar solution to an intrinsic geometric flow. Almost solitons are more general objects than solitons because they use functions rather than constants as coefficients in the defining condition. A β-Ricci–Bourguignon-like almost soliton (β is a real constant) is defined using the pair of metrics. The introduced soliton is a generalization of some well-known (almost) solitons (such as those of Ricci, Schouten, and Einstein) which, in principle, arise from a single metric rather than a pair of metrics. The soliton potential is chosen to be pointwise collinear to the Reeb vector field, or the Lie derivative of any B-metric along the potential to be the same metric multiplied by a function. The resulting manifolds equipped with the introduced almost solitons are characterized geometrically. Suitable examples for both types of almost solitons are constructed, and the properties obtained in the theoretical part are confirmed.

Keywords: Ricci–Bourguignon; almost contact B-metric manifold; almost contact complex Riemannian manifold; Sasaki-like manifold; vertical potential; conformal potential

MSC: 53C25; 53D15; 53C50; 53C44; 53D35; 70G45

1. Introduction

The notion of *Ricci–Bourguignon flow* was introduced by J. P. Bourguignon in 1981 [1]. A time-dependent family of (pseudo-)Riemannian metrics $g(t)$ considered on a smooth manifold $\mathcal{M}$ is said to evolve through Ricci–Bourguignon flow if $g(t)$ satisfies the following evolution equation:

$$\frac{\partial}{\partial t}g = -2(\rho - \beta\tau g), \qquad g(0) = g_0,$$

where β is a real constant, $\rho(t)$ and $\tau(t)$ are the Ricci tensor and the scalar curvature regarding $g(t)$, respectively.

This flow is an intrinsic geometric flow on $\mathcal{M}$ whose fixed points or self-similar solutions are its solitons. The *Ricci–Bourguignon soliton* (RB soliton for short) is described by the following equation [2,3]:

$$\rho + \frac{1}{2}\mathcal{L}_\vartheta g + (\lambda + \beta\tau)g = 0, \tag{1}$$

where $\mathcal{L}_\vartheta g$ denotes the Lie derivative of g along the vector field ϑ, called the soliton potential, and λ is the soliton constant. Briefly, we denote this soliton by $(g, \beta; \vartheta, \lambda)$. In the case where λ is a differential function on $\mathcal{M}$, the solution is called the *RB almost soliton* [3]. The RB soliton is called expanding if $\lambda > 0$, steady if $\lambda = 0$, and shrinking if $\lambda < 0$. In the case where the soliton potential ϑ is a Killing vector field, i.e., $\mathcal{L}_\vartheta g = 0$, the RB soliton is called trivial.

This family of geometric flows contains the famous Ricci flow for $\beta = 0$, the Einstein flow for $\beta = \frac{1}{2}$, the traceless Ricci flow for $\beta = \frac{1}{m}$, and the Schouten flow for $\beta = \frac{1}{2(m-1)}$, where m is the dimension of the manifold [4,5]. For this reason, we consider it more correct to say *β-RB solitons* and *β-RB almost solitons*, respectively.

Many authors have recently investigated various aspects of these solitons and almost solitons considered on manifolds with different additional structures. For example, some results for conformal Ricci–Bourguignon flow were obtained in [6]. Also, for suitable values of the scalar parameter involved in these flows, the short-time existence is proven in [2], and curvature estimates are provided.

Other recent studies on (almost) solitons of this type have been performed in the last four years. In particular, some integral formulas were derived in [3] for compact gradient Ricci–Bourguignon (almost) solitons. The scalar curvatures of Ricci–Bourguignon solitons of various special types were studied in [7,8]. Ricci–Bourguignon solitons and almost solitons with a concurrent potential vector field on Riemannian manifolds were studied in [9], where such (almost) solitons were first classified and then their geometric properties were obtained. The geometrical characteristics of Ricci–Bourguignon almost solitons with a potential parallel to the Reeb vector field on Kenmotsu manifolds were given in [10]. In [11], some properties of Ricci–Bourguignon almost solitons with a gradient and a torse-forming potential vector field were first derived, then some of these results were generalized to η-Ricci–Bourguignon almost solitons using a structure 1-form. Furthermore, gradient Ricci–Bourguignon almost solitons on a doubly warped product manifold were investigated. In [12], properties of η-Ricci–Bourguignon solitons on a Riemannian manifold equipped with a semi-symmetric metric (or non-metric) connection are studied. Moreover, the potential vector field is torse-forming with respect to the corresponding metric. Almost η-Ricci–Bourguignon solitons in compact and non-compact cases were investigated in [13]. The object of study in [14] was (gradient) η-Ricci–Bourguignon almost solitons on almost Kenmotsu manifolds and (κ, μ) almost Kenmotsu manifolds.

These studies inspire us to study almost solitons of this type generated by the pair of metrics and the contact 1-form on almost contact complex Riemannian manifolds, which we have been working on for a long time.

2. Almost Contact Complex Riemannian Manifolds

Let us consider a $(2n+1)$-dimensional smooth manifold $\mathcal{M}$, equipped with an almost contact structure (φ, ξ, η) and a B-metric g. It is called an *almost contact B-metric manifold* or *almost contact complex Riemannian* (abbreviated *accR*) *manifold*, and is denoted by $(\mathcal{M}, \varphi, \xi, \eta, g)$. In detail, φ is an endomorphism of the tangent bundle $T\mathcal{M}$, ξ is a Reeb vector field, η is its dual contact 1-form, and g is a pseudo-Riemannian metric of signature $(n+1, n)$, such that

$$\begin{gathered}\varphi\xi = 0, \qquad \varphi^2 = -\iota + \eta \otimes \xi, \qquad \eta \circ \varphi = 0, \qquad \eta(\xi) = 1, \\ g(\varphi x, \varphi y) = -g(x, y) + \eta(x)\eta(y),\end{gathered} \tag{2}$$

where ι denotes the identity on $\Gamma(T\mathcal{M})$ [15].

In the last equality and further, x, y, and z will denote arbitrary elements of $\Gamma(T\mathcal{M})$ or vectors in the tangent space $T_p\mathcal{M}$ of $\mathcal{M}$ at an arbitrary point p in $\mathcal{M}$.

The following equations are immediate consequences of Equation (2):

$$g(\varphi x, y) = g(x, \varphi y), \qquad g(x, \xi) = \eta(x), \qquad g(\xi, \xi) = 1, \qquad \eta(\nabla_x \xi) = 0, \tag{3}$$

where ∇ denotes the Levi-Civita connection of g.

The associated metric $\tilde{g}$ of g on $\mathcal{M}$ is also a B-metric, and is defined by

$$\tilde{g}(x, y) = g(x, \varphi y) + \eta(x)\eta(y). \tag{4}$$

The Ganchev–Mihova–Gribachev classification of the investigated manifolds, given in [15], consists of eleven basic classes $\mathcal{F}_i$, $i \in \{1,2,\ldots,11\}$, determined by conditions for the (0,3)-tensor F defined by

$$F(x,y,z) = g((\nabla_x\varphi)y,z). \tag{5}$$

It has the following basic properties:

$$F(x,y,z) = F(x,z,y) = F(x,\varphi y,\varphi z) + \eta(y)F(x,\xi,z) + \eta(z)F(x,y,\xi), \tag{6}$$

$$F(x,\varphi y,\xi) = (\nabla_x\eta)y = g(\nabla_x\xi,y). \tag{7}$$

2.1. Sasaki-Like accR Manifolds

An interesting class of accR manifolds was introduced in [16] by the condition that the complex cone of such a manifold is a Kähler–Norden manifold. They are called *Sasaki-like manifolds*, and are defined by the condition

$$F(x,y,z) = g(\varphi x,\varphi y)\eta(z) + g(\varphi x,\varphi z)\eta(y). \tag{8}$$

The class of Sasaki-like manifolds is contained in the basic class $\mathcal{F}_4$ of the Ganchev–Mihova–Gribachev classification, not intersecting with the special class $\mathcal{F}_0$ of cosymplectic accR manifolds defined by $F = 0$.

Moreover, the following identities hold for this type of accR manifolds [16]:

$$\nabla_x\xi = -\varphi x, \qquad \rho(x,\xi) = 2n\,\eta(x), \qquad \rho(\xi,\xi) = 2n, \tag{9}$$

where ρ stands for the Ricci tensor for g.

Let τ and $\tilde{\tau}$ be the scalar curvatures with respect to g and $\tilde{g}$, respectively, and let τ^* be the associated quantity of τ regarding φ, defined by $\tau^* = g^{ij}\rho(e_i,\varphi e_j)$. Then, for a Sasaki-like manifold, we have

$$\tilde{\tau} = -\tau^* + 2n. \tag{10}$$

2.2. Einstein-Like accR Manifolds

The following concept was introduced in [17]. An accR manifold $(\mathcal{M},\varphi,\xi,\eta,g)$ is said to be *Einstein-like* if its Ricci tensor ρ satisfies

$$\rho = a\,g + b\,\tilde{g} + c\,\eta\otimes\eta \tag{11}$$

for some triplet of constants (a,b,c). In particular, when $b = 0$ and $b = c = 0$, the manifold is called an *η-Einstein manifold* and an *Einstein manifold*, respectively. If a, b, c in Equation (11) are functions on $\mathcal{M}$, then the manifold is called *almost Einstein-like, almost η-Einstein*, and *almost Einstein*, respectively [17].

The consequences of Equation (11) are the following: $\tau = (2n+1)a + b + c$ and $\tau^* = -2nb$.

3. β-RB Almost Solitons

A generalization of the known RB soliton on a manifold with additional 1-form η is the *η-Ricci–Bourguignon soliton*, defined following Equation (1) by

$$\rho + \frac{1}{2}\mathcal{L}_\vartheta g + (\lambda + \beta\tau)g + \mu\eta\otimes\eta = 0, \tag{12}$$

where μ is also a constant [11]. Obviously, the η-Ricci–Bourguignon soliton with $\mu = 0$ is an RB soliton. Again, in the case where λ and μ are functions on the manifold, almost solitons of the corresponding kind are said to be given.

In the present paper, we study the accR manifold. The presence of two B-metrics, g and $\tilde{g}$, related to each other with respect to the structure of such a manifold gives us reason to introduce a more natural generalization of the β-RB soliton than Equation (12). In addition, we also have the structure 1-form η so that $\eta\otimes\eta$ is included in both B-metrics g and $\tilde{g}$ as their restriction on the vertical distribution span(ξ).

Definition 1. *An accR manifold* $(\mathcal{M}, \varphi, \xi, \eta, g)$ *is called a* β*-Ricci–Bourguignon-like soliton (briefly* β*-RB-like soliton) with potential vector field* ϑ *if its Ricci tensor* ρ *satisfies the following condition for a pair of constants* $(\lambda, \tilde{\lambda})$*:*

$$\rho + \frac{1}{2}\mathcal{L}_{\vartheta}g + \frac{1}{2}\mathcal{L}_{\vartheta}\tilde{g} + (\lambda + \beta\tau)g + (\tilde{\lambda} + \beta\tilde{\tau})\tilde{g} = 0, \tag{13}$$

where $\tilde{\tau}$ *is the scalar curvature of the manifold with respect to* $\tilde{g}$ *and the corresponding Levi-Civita connection* $\tilde{\nabla}$*. If* $(\lambda, \tilde{\lambda})$ *is a pair of functions on* $\mathcal{M}$ *satisfying Equation* (13)*, then* $(\mathcal{M}, \varphi, \xi, \eta, g)$ *is called a* β*-Ricci–Bourguignon-like almost soliton (briefly* β*-RB-like almost soliton).*

Taking the trace in (13) with respect to g, we obtain

$$[1 + (2n+1)\beta]\tau + \beta\tilde{\tau} + \operatorname{div}_g \vartheta + \frac{1}{2}\operatorname{tr}_g(\mathcal{L}_{\vartheta}\tilde{g}) + (2n+1)\lambda + \tilde{\lambda} = 0$$

by means of the formula $\operatorname{div}_g \vartheta = \frac{1}{2}\operatorname{tr}_g(\mathcal{L}_{\vartheta}g) = g^{ij}g(\nabla_{e_i}\vartheta, e_j)$. The trace expression in the above equality is the following:

$$\frac{1}{2}\operatorname{tr}_g(\mathcal{L}_{\vartheta}\tilde{g}) = g^{ij}\tilde{g}(\tilde{\nabla}_{e_i}\vartheta, e_j).$$

3.1. The Potential Is a Conformal Vector Field

Recall that a vector field, e.g., the potential ϑ on $\mathcal{M}$, is called a *conformal vector field with respect to* g if there exists a function ψ on $\mathcal{M}$ such that [3]

$$\mathcal{L}_{\vartheta}g = 2\psi g.$$

The conformal vector field is nontrivial if $\psi \neq 0$. If $\psi = 0$, then ϑ is called a *Killing vector field* with respect to g.

Similarly, ϑ is called a *conformal vector field with respect to* $\tilde{g}$ if there exists a function $\tilde{\psi}$ on $\mathcal{M}$ such that

$$\mathcal{L}_{\vartheta}\tilde{g} = 2\tilde{\psi}\tilde{g}.$$

Depending on whether $\tilde{\psi}$ is nonzero or zero, we have a vector field ϑ that is nontrivial scalar or Killing, respectively.

Theorem 1. *Let* $(\mathcal{M}, \varphi, \xi, \eta, g)$ *be a* $(2n+1)$*-dimensional Sasaki-like accR manifold that is a* β*-RB-like almost soliton with a pair of soliton functions* $(\lambda, \tilde{\lambda})$ *and a conformal potential vector field* ϑ *with potential functions* ψ *and* $\tilde{\psi}$ *with respect to* g *and* $\tilde{g}$*, respectively. Then, the manifold is Einstein-like, and has the following Ricci tensor:*

$$\rho = \left(\frac{\tau}{2n} - 1\right)g + \left(\frac{\tilde{\tau}}{2n} - 1\right)\tilde{g} \tag{14}$$

and the following property is valid:

$$\tau + \tilde{\tau} = 4n(n+1). \tag{15}$$

If $\beta \neq -\frac{1}{2n}$*, then the scalar curvatures with respect to* g *and* $\tilde{g}$ *can be expressed separately as*

$$\tau = -\frac{2n}{1+2n\beta}(\psi + \lambda - 1), \qquad \tilde{\tau} = -\frac{2n}{1+2n\beta}(\tilde{\psi} + \tilde{\lambda} - 1) \tag{16}$$

and the following condition for the used functions is valid:

$$\psi + \lambda + \tilde{\psi} + \tilde{\lambda} + 2n[1 + 2(n+1)\beta] = 0. \tag{17}$$

If $\beta = -\frac{1}{2n}$*, then the following properties hold:*

$$\psi + \lambda = 1, \qquad \tilde{\psi} + \tilde{\lambda} = 1. \tag{18}$$

Proof. Under these circumstances, regarding the considered manifold, due to Equation (13), its Ricci tensor for g has the following form:

$$\rho = -(\psi + \lambda + \beta\tau)g - (\tilde{\psi} + \tilde{\lambda} + \beta\tilde{\tau})\tilde{g}. \tag{19}$$

Taking the trace of the last expression, we obtain that the scalar curvatures of $(\mathcal{M}, \varphi, \xi, \eta, g)$ with respect to g and $\tilde{g}$ are related as follows:

$$[1 + (2n+1)\beta]\tau + \beta\tilde{\tau} + (2n+1)(\psi + \lambda) + \tilde{\psi} + \tilde{\lambda} = 0. \tag{20}$$

Using Equation (19), we take the appropriate trace to obtain τ^* on the left-hand side, and the resulting relation to $\tilde{\tau}$ is

$$\tau^* = 2n(\tilde{\psi} + \tilde{\lambda} + \beta\tilde{\tau}). \tag{21}$$

By virtue of Equation (10) for a Sasaki-like accR manifold, and Equation (21), we obtain the expression of the scalar curvature with respect to $\tilde{g}$ as follows:

$$\tilde{\tau} = -2n\frac{\tilde{\psi} + \tilde{\lambda} - 1}{1 + 2n\beta}. \tag{22}$$

The last formula is true for the case $\beta \neq -\frac{1}{2n}$.

Combining Equations (20) and (22), we obtain the following form of the scalar curvature with respect to g:

$$\tau = -\frac{1}{1 + (2n+1)\beta}\left\{(2n+1)(\psi + \lambda) + 1 + \frac{\tilde{\psi} + \tilde{\lambda} - 1}{1 + 2n\beta}\right\}, \tag{23}$$

where $\beta \neq -\frac{1}{2n+1}$. Otherwise, for $\beta = -\frac{1}{2n+1}$, the identity in Equation (17) is valid for this value of β.

On the other hand, a consequence of Equation (19) and the value of $\rho(\xi, \xi)$ from Equation (9) for the Sasaki-like case gives the following relation:

$$\beta\tau + \beta\tilde{\tau} + \psi + \lambda + \tilde{\psi} + \tilde{\lambda} + 2n = 0. \tag{24}$$

Let us check what follows in the particular case where $\beta = 0$. The last relation implies

$$\psi + \lambda + \tilde{\psi} + \tilde{\lambda} + 2n = 0,$$

which we apply in Equations (22) and (23) to specialize them in the following form:

$$\tau = -2n(\psi + \lambda - 1), \qquad \tilde{\tau} = -2n(\tilde{\psi} + \tilde{\lambda} - 1). \tag{25}$$

Therefore, the identity in Equation (15) holds, as do the formulas in Equation (16). Thus, we find that no different results are obtained for $\beta = 0$ compared to the case $\beta \neq -\frac{1}{2n}$.

Let us return to Equation (24) and the solution of the system of equations in (20) and (24) in terms of τ and $\tilde{\tau}$ for $\beta \neq 0$ and $\beta \neq -\frac{1}{2n}$, which gives us

$$\tau = -2n\frac{\psi + \lambda - 1}{1 + 2n\beta}, \qquad \tilde{\tau} = -\frac{1}{\beta}\left\{\frac{\psi + \lambda - 1}{1 + 2n\beta} + \tilde{\psi} + \tilde{\lambda} + 2n + 1\right\}. \tag{26}$$

After comparing the equalities for τ in Equations (23) and (26), we obtain Equation (17). The same relation results from a similar comparison for $\tilde{\tau}$ in Equations (22) and (26). The identity in Equation (17) is a generalization of the corresponding result in the $\beta = 0$ case. Thus, we obtain the formulas in Equation (16).

Because of Equation (17), the sum of τ and $\tilde{\tau}$ is a constant that depends only on the dimension of the manifold, as in Equation (15).

For the case $\beta = -\frac{1}{2n}$, the equalities in Equations (15) and (17) imply the relations in Equation (18). Therefore, Equations (20) and (24) take the form in Equation (15).

In conclusion, using Equations (16) and (19) , we obtain the Ricci tensor expression in Equation (14), which shows that the manifold is Einstein-like, since Equation (11) is satisfied for $a = \frac{\tau}{2n} - 1$, $b = \frac{\tilde{\tau}}{2n} - 1$ and $c = 0$. $\square$

Example of a β-RB Almost Soliton with a Conformal Potential

As in [16], let us consider a Sasaki-like accR manifold of arbitrary dimension with Einstein metric $\tilde{g}$. The image of this manifold by contact homothetic transformation of the metric given by $g = p\tilde{g} + q\tilde{\tilde{g}} + (1 - p - q)\eta \otimes \eta$ for $p, q \in \mathbb{R}$, $(p, q) \neq (0, 0)$, is also a Sasaki-like accR manifold. The corresponding Ricci tensor has the form

$$\rho = \frac{2n}{p^2 + q^2}\Big\{pg - q\tilde{g} + \Big(p^2 + q^2 - p\Big)\eta \otimes \eta\Big\}.$$

We now calculate the scalar curvatures with respect to B-metrics g and $\tilde{g}$ as follows:

$$\tau = 2n\left\{1 + \frac{2np}{p^2 + q^2}\right\}, \qquad \tilde{\tau} = 2n\left\{1 - \frac{2nq}{p^2 + q^2}\right\}. \tag{27}$$

Let ϑ be a conformal vector field with respect to the two B-metrics, g and $\tilde{g}$, with functions ψ and $\tilde{\psi}$, respectively. We construct a β-RB-like almost soliton on the transformed manifold with potential ϑ and a pair of functions $(\lambda, \tilde{\lambda})$ satisfying Equation (13).

By virtue of Equations (15) and (27), we obtain the following condition $p^2 + q^2 - p + q = 0$, which has a solution $p = \frac{1}{2}(1 + \sqrt{2}\cos t)$, $q = -\frac{1}{2}(1 - \sqrt{2}\sin t)$ for $t \in \mathbb{R}$. Then, the scalar curvatures from Equation (27) specialize into the following form for $t \neq (8l + 3)\frac{\pi}{4}$, $l \in \mathbb{Z}$:

$$\begin{aligned} \tau &= 2n\frac{(n+1)\sqrt{2} + (2n+1)\cos t - \sin t}{\sqrt{2} + \cos t - \sin t}, \\ \tilde{\tau} &= 2n\frac{(n+1)\sqrt{2} + \cos t - (2n+1)\sin t}{\sqrt{2} + \cos t - \sin t}. \end{aligned} \tag{28}$$

We then determine the functions ψ, $\tilde{\psi}$, λ, $\tilde{\lambda}$ in the case $\beta \neq -\frac{1}{2n}$, as follows:

$$\begin{aligned} \psi + \lambda &= 1 - (1 + 2n\beta)\frac{(n+1)\sqrt{2} + (2n+1)\cos t - \sin t}{\sqrt{2} + \cos t - \sin t}, \\ \tilde{\psi} + \tilde{\lambda} &= 1 - (1 + 2n\beta)\frac{(n+1)\sqrt{2} + \cos t - (2n+1)\sin t}{\sqrt{2} + \cos t - \sin t}. \end{aligned} \tag{29}$$

In the case $\beta = -\frac{1}{2n}$, the equalities in Equation (29) obviously reduce to those in Equation (18).

We directly verify that Equations (28) and (29) satisfy the expressions in Equations (16) and (17).

In conclusion, we found that the constructed manifold satisfies the conditions of Theorem 1.

3.2. The Potential Is a Vertical Vector Field

Suppose that $(\mathcal{M}, \varphi, \xi, \eta, g)$ is a Sasaki-like accR manifold admitting a β-RB-like almost soliton, whose potential vector field ϑ is pointwise collinear with ξ, i.e., $\vartheta = k\xi$, where k is a differentiable function on $\mathcal{M}$. It is clear that $k = \eta(\vartheta)$ is true and, therefore, ϑ belongs to the vertical distribution $H^{\perp} = \mathrm{span}(\xi)$, which is orthogonal to the contact distribution $H = \ker(\eta)$ with respect to both g and $\tilde{g}$.

We easily obtain the expression $(\mathcal{L}_{\vartheta}g)(x, y) = h(x, y) - 2kg(x, \varphi y)$ for the vertical potential ϑ, where the first equality of Equation (9) is used and the symmetric tensor $h(x, y) = \mathrm{d}k(x)\eta(y) + \mathrm{d}k(y)\eta(x)$ is denoted for brevity. Then, by virtue of Equation (4), we obtain

$$\mathcal{L}_{\vartheta}g = h - 2k(\tilde{g} - \eta \otimes \eta). \tag{30}$$

Similarly, since one can obtain $\tilde{\nabla}_x\xi = -\varphi x$ for a Sasaki-like accR manifold, we have $(\mathcal{L}_{\vartheta}\tilde{g})(x, y) = h(x, y) - 2kg(\varphi x, \varphi y)$, and considering the last equality of Equation (2), the last expression takes the following form:

$$\mathcal{L}_{\vartheta}\tilde{g} = h + 2k(g - \eta \otimes \eta). \tag{31}$$

Theorem 2. *Let $(\mathcal{M}, \varphi, \xi, \eta, g)$ be a $(2n+1)$-dimensional Sasaki-like accR manifold that is a β-RB-like almost soliton with a pair of soliton functions $(\lambda, \tilde{\lambda})$ and a vertical potential vector field ϑ with potential function k. Then, the manifold is Einstein-like, and has the following Ricci tensor:*

$$\rho = \left(\frac{\tau}{2n} - 1\right)g + \left(\frac{\tilde{\tau}}{2n} - 1\right)\tilde{g} - \left\{\frac{\tau + \tilde{\tau}}{2n} - 2(n+1)\right\}\eta \otimes \eta. \tag{32}$$

In the case of $\beta \neq -\frac{1}{2n}$, the scalar curvatures with respect to g and $\tilde{g}$ are determined by:

$$\tau = -\frac{2n}{1+2n\beta}(\lambda + k - 1), \qquad \tilde{\tau} = -\frac{2n}{1+2n\beta}(\tilde{\lambda} - k - 1). \tag{33}$$

Moreover, the following condition is true for the functions used:

$$\mathrm{d}k = \mathrm{d}k(\xi)\eta, \qquad \mathrm{d}k(\xi) = -\frac{\lambda + \tilde{\lambda} - 2}{2(1+2n\beta)} - n - 1. \tag{34}$$

In the case of $\beta = -\frac{1}{2n}$, the two scalar curvatures satisfy the following relation:

$$\tau + \tilde{\tau} = 4n\{\mathrm{d}k(\xi) + n + 1\} \tag{35}$$

and the soliton functions are expressed in terms of k as follows:

$$\lambda = 1 - k, \qquad \tilde{\lambda} = 1 + k. \tag{36}$$

Proof. Substituting Equations (30) and (31) into Equation (13), we obtain

$$\rho = -(\lambda + \beta\tau + k)g - (\tilde{\lambda} + \beta\tilde{\tau} - k)\tilde{g} - h. \tag{37}$$

For a Sasaki-like accR manifold, we know the expression of $\rho(x, \xi)$ from Equation (9). Then, we compare it with the corresponding consequence of Equation (37), and obtain $\mathrm{d}k = \mathrm{d}k(\xi)\eta$, where

$$\mathrm{d}k(\xi) = -\frac{1}{2}\{\lambda + \tilde{\lambda} + \beta(\tau + \tilde{\tau}) + 2n\}. \tag{38}$$

Therefore, k is a horizontal constant, and h takes the following form:

$$h = -\{\lambda + \tilde{\lambda} + \beta(\tau + \tilde{\tau}) + 2n\}\eta \otimes \eta. \tag{39}$$

Then, applying the last equality and Equation (38) in (37), for the Ricci tensor ρ, we obtain

$$\rho = -(\lambda + \beta\tau + k)g - (\tilde{\lambda} + \beta\tilde{\tau} - k)\tilde{g} + \{\lambda + \tilde{\lambda} + \beta(\tau + \tilde{\tau}) + 2n\}\eta \otimes \eta. \tag{40}$$

Taking the appropriate traces in Equation (40), we obtain for $\beta \neq -\frac{1}{2n}$ the first equality in Equation (33), as well as

$$\tau^* = 2n(\tilde{\lambda} + \beta\tilde{\tau} - k). \tag{41}$$

Then, bearing in mind Equation (10), the equality in Equation (41) implies, for $\beta \neq -\frac{1}{2n}$, the second equality in Equation (33). Substituting the expressions of τ and $\tilde{\tau}$ from Equation (33) into Equation (40), we obtain Equation (32).

As a consequence of Equations (33) and (38), we express $\tau + \tilde{\tau}$ in two ways. One is given in Equation (35), and the other is as follows:

$$\tau + \tilde{\tau} = -\frac{2n}{1+2n\beta}(\lambda + \tilde{\lambda} - 2). \tag{42}$$

Comparing two expressions implies the relation in Equation (34) between the functions used and the constant β.

Comparing Equation (40) with Equation (11), it follows that $(\mathcal{M}, \varphi, \xi, \eta, g)$ is Einstein-like, and the coefficients in (11) are

$$a=\frac{\tau}{2n}-1, \qquad b=\frac{\tilde{\tau}}{2n}-1, \qquad c=-\frac{\tau+\tilde{\tau}}{2n}+2(n+1). \tag{43}$$

In the particular case where $\beta=-\frac{1}{2n}$, the dependencies in Equation (36) are valid due to Equation (33). Then, Equation (40) is specialized, as in Equation (32). In addition, according to Equation (38), the equality in Equation (35) also holds. Note that, in this case, the scalar curvatures with respect to each of the B-metrics cannot be expressed separately. □

Example of a β-RB Almost Soliton with a Vertical Potential

Let us consider an explicit example, given as Example 2 in [16]. It concerns a Sasaki-like accR manifold constructed on a Lie group G of dimension 5, i.e., for $n=2$, with a basis of left-invariant vector fields $\{e_0,\dots,e_4\}$. The corresponding Lie algebra is defined by the commutators

$$\begin{array}{ll}[e_0,e_1]=pe_2+e_3+qe_4, & [e_0,e_2]=-pe_1-qe_3+e_4,\\ [e_0,e_3]=-e_1-qe_2+pe_4, & [e_0,e_4]=qe_1-e_2-pe_3, \qquad p,q\in\mathbb{R}.\end{array} \tag{44}$$

The introduced accR structure is defined as follows:

$$\begin{array}{l}g(e_0,e_0)=g(e_1,e_1)=g(e_2,e_2)=-g(e_3,e_3)=-g(e_4,e_4)=1,\\ g(e_i,e_j)=0, \quad i,j\in\{0,1,\dots,4\},\ i\neq j,\\ \xi=e_0, \quad \varphi e_1=e_3, \quad \varphi e_2=e_4, \quad \varphi e_3=-e_1, \quad \varphi e_4=-e_2.\end{array} \tag{45}$$

Then, in [17], the components of the curvature tensor $R_{ijkl}=R(e_i,e_j,e_k,e_l)$ and those of the Ricci tensor $\rho_{ij}=\rho(e_i,e_j)$ are calculated. The non-zero ones are determined by the following equalities and the property $R_{ijkl}=-R_{jikl}=-R_{ijlk}$:

$$\begin{array}{l}R_{0110}=R_{0220}=-R_{0330}=-R_{0440}=1,\\ R_{1234}=R_{1432}=R_{2341}=R_{3412}=1,\\ R_{1331}=R_{2442}=1, \qquad \rho_{00}=4.\end{array} \tag{46}$$

Therefore, its Ricci tensor has the form $\rho=4\eta\otimes\eta$, and the manifold is η-Einstein. Hence, the scalar curvature of g is $\tau=4$, and the constructed manifold is $*$-scalar flat, i.e., $\tau^*=0$. Then, due to Equation (10), we obtain the value $\tilde{\tau}=4$ for the scalar curvature of $\tilde{g}$.

Using Equation (13), let us construct a β-RB-like almost soliton on (G,φ,ξ,η,g) with vertical potential $\vartheta=k\xi$ for a constant β and a pair of functions $(\lambda,\tilde{\lambda})$.

From Equation (33), we determine the following conditions in the case where $\beta\neq-\frac{1}{2n}$:

$$\lambda+k=-4\beta, \qquad \tilde{\lambda}-k=-4\beta. \tag{47}$$

Then, Equation (34) implies $\mathrm{d}k(\xi)=-2$ and $\mathrm{d}k=-2\eta$. A solution to the last equation is, e.g., $k=-2t$, assuming that $\eta=\mathrm{d}t$. This form of k also satisfies the condition in Equation (35) for the case $\beta=-\frac{1}{2n}$. This allows us to determine functions $(\lambda,\tilde{\lambda})$ from Equation (36) and Equation (47) for all values of β by

$$\lambda=2(t-2\beta), \qquad \tilde{\lambda}=-2(t+2\beta). \tag{48}$$

Hence, and from Equation (39), we obtain $h=-4\eta\otimes\eta$. Consequently, Equations (30) and (31) take the following form:

$$\mathcal{L}_\vartheta g=4t\tilde{g}-4(t+1)\eta\otimes\eta, \qquad \mathcal{L}_\vartheta\tilde{g}=-4tg+4(t+1)\eta\otimes\eta.$$

Finally, we found that all the findings in Theorem 2 are true for the given example.

Funding: The research is partially supported by project FP23-FMI-002 of the Scientific Research Fund, University of Plovdiv Paisii Hilendarski.

Data Availability Statement: Data are contained within the article.

Conflicts of Interest: The funders had no role in the design of the study; in the collection, analyses, or interpretation of data; in the writing of the manuscript; or in the decision to publish the results. The author declares no conflict of interest.

ferences

Bourguignon, J.-P. Ricci curvature and Einstein metrics. In *Global Differential Geometry and Global Analysis (Berlin, 1979)*; Lecture Notes in Math; Springer: Berlin/Heidelberg, Germany, 1981; Volume 838, pp. 42–63.
Catino, G.; Cremaschi, L.; Djadli, Z.; Mantegazza, C.; Mazzieri, L. The Ricci–Bourguignon Flow. *Pac. J. Math.* **2017**, *287*, 337–370. [CrossRef]
Dwivedi, S. Some results on Ricci–Bourguignon solitons and almost solitons. *Can. Math. Bull.* **2021**, *64*, 591–604. [CrossRef]
Siddiqui, A.N.; Siddiqi, M.D. Almost Ricci–Bourguignon solitons and geometrical structure in a relativistic perfect fluid spacetime. *Balk. J. Geom. Appl.* **2021**, *26*, 126–138.
Catino, G.; Mazzieri, L.; Mongodi, S. Rigidity of gradient Einstein shrinkers. *Commun. Contemp. Math.* **2015**, *17*, 1550046. [CrossRef]
Lu, P.; Qing, J.; Zheng, Y. A note on conformal Ricci flow. *Pac. J. Math.* **2014**, *268*, 413–434. [CrossRef]
Mi, R. Remarks on scalar curvature of gradient Ricci–Bourguignon sollitons. *Bull. Sci. Math.* **2021**, *171*, 103034. [CrossRef]
Cunha, A.W.; Lemos, R.; Roing, F. On Ricci–Bourguignon solitons: Triviality, uniqueness and scalar curvature estimates. *J. Math. Anal. Appl.* **2023**, *526*, 127333. [CrossRef]
Soylu, Y. Ricci–Bourguignon solitons and almost solitons with concurrent vector field. *Differ. Geom. Dyn. Syst.* **2022**, *24*, 191–200.
Prakasha, D.G.; Amruthalakshmi, M.R.; Suh, Y.J. Geometric characterizations of almost Ricci–Bourguignon solitons on Kenmotsu manifolds. *Filomat* **2024**, *38*, 861–871.
Blaga, A.M.; Taştan, H.M. Some results on almost η-Ricci–Bourguignon solitons. *J. Geom. Phys.* **2021**, *168*, 104316. [CrossRef]
Dogru, Y. η-Ricci–Bourguignon solitons with a semi-symmetric metric and semi-symmetric non-metric connection. *AIMS Math.* **2023**, *8*, 11943–11952. [CrossRef]
Traore, M.; Taştan, H.M.; Aydın, S.G. On almost η-Ricci–Bourguignon solitons. *Miskolc. Math. Notes* **2024**, *25*, 493–508. [CrossRef]
Dey, S. Certain results on gradient almost η-Ricci–Bourguignon soliton. *Quaest. Math.* **2023**, *47*, 1–22. [CrossRef]
Ganchev, G.; Mihova, V.; Gribachev, K. Almost contact manifolds with B-metric. *Math. Balk.* **1993**, *7*, 261–276.
Ivanov, S.; Manev, H.; Manev, M. Sasaki-like almost contact complex Riemannian manifolds. *J. Geom. Phys.* **2016**, *105*, 136–148. [CrossRef]
Manev, M. Ricci-like solitons on almost contact B-metric manifolds. *J. Geom. Phys.* **2020**, *154*, 103734. [CrossRef]

mathematics

MDPI

Article

Characterization of Bach and Cotton Tensors on a Class of Lorentzian Manifolds

Yanlin Li [1,*,†], M. S. Siddesha [2,†], H. Aruna Kumara [3,†] and M. M. Praveena [4,†]

[1] School of Mathematics, Hangzhou Normal University, Hangzhou 311121, China
[2] Department of Data Analytics and Mathematical Science, Jain (Deemed to be University), Global Campus, Bangalore 562112, India; m.siddesh@jainuniversity.ac.in
[3] Department of Mathematics, BMS Institute of Technology and Management, Yelahanka, Bangalore 560064, India; arunakumara@bmsit.in
[4] Department of Mathematics, M. S. Ramaiah Institute of Technology, Bangalore 560054, India; mmpmaths@msrit.in
* Correspondence: liyl@hznu.edu.cn
† These authors contributed equally to this work.

Abstract: In this work, we aim to investigate the characteristics of the Bach and Cotton tensors on Lorentzian manifolds, particularly those admitting a semi-symmetric metric ω-connection. First, we prove that a Lorentzian manifold admitting a semi-symmetric metric ω-connection with a parallel Cotton tensor is quasi-Einstein and Bach flat. Next, we show that any quasi-Einstein Lorentzian manifold admitting a semi-symmetric metric ω-connection is Bach flat.

Keywords: Bach tensor; Cotton tensor; Lorentzian manifolds; semi-symmetric metric connection

MSC: 53C15; 53C40; 53D10

1. Introduction

Consider an n-dimensional pseudo-Riemannian manifold denoted as M. When the torsion tensor $\widetilde{\mathcal{T}}$ associated with a linear connection $\widetilde{\nabla}$ vanishes, i.e., $\widetilde{\mathcal{T}}(\zeta_1,\zeta_2) = \widetilde{\nabla}_{\zeta_1}\zeta_2 - \widetilde{\nabla}_{\zeta_2}\zeta_1 - [\zeta_1,\zeta_2] = 0$, then $\widetilde{\nabla}$ is referred to as a symmetric connection. Conversely, if the torsion tensor does not vanish, the connection is termed non-symmetric. Many geometers classify linear connection $\widetilde{\nabla}$ into different categories based on distinct forms. For example, a connection is referred to as semi-symmetric (abbreviated as $\mathcal{SS}$) if it satisfies the following relation:

$$\widetilde{\mathcal{T}}(\zeta_1,\zeta_2) = \omega^\sharp(\zeta_2)\zeta_1 - \omega^\sharp(\zeta_1)\zeta_2, \ \forall \zeta_1,\zeta_2 \in \Gamma(TM) \tag{1}$$

where the one-form $\omega^\sharp$ and the associated vector field ω are linked through a pseudo-Riemannian metric g by $g(\cdot,\omega) = \omega^\sharp(\cdot)$. If $\widetilde{\nabla} g = 0$, the connection $\widetilde{\nabla}$ is referred to as a metric connection. Otherwise, it is categorized as a non-metric connection [1]. A connection is symmetric and metric if and only if it is a Levi–Civita connection. Hayden [2] investigated a metric connection $\widetilde{\nabla}$ with a non-vanishing torsion on a Riemannian manifold, which later became known as the Hayden connection. Following Pak's work [3], which showed that it is a semi-symmetric metric connection (abbreviated as $\mathcal{SSM}$-connection), numerous questions emerged about this type of connection. Yano [4] subsequently initiated an investigation into Riemannian manifolds equipped with an $\mathcal{SSM}$-connection, finding that such manifolds become conformally flat when the curvature tensor vanishes. Recently, Chaubey et al. [5] commenced the investigation of the concept of semi-symmetric metric ω-connection (briefly, $\mathcal{SSM}\omega$-connection) on a Riemannian manifold and explored its geometric properties. This concept was later extended to Lorentzian manifolds by Chaubey et al. [6], where they studied its geometric and physical properties within specific classifications. Duggal studied a class of almost Ricci solitons (ARSs), he showed three mathematical models of conformally flat almost-Ricci-soliton manifolds, leading to several significant findings [7]. In recent years, Kumar and Colney et al. considered the tangent bundles with NSNMC and QSNMC [8–10]. The properties, theorems, and results of the curvature tensor and Ricci tensor relevant to the connection on the tangent bundles were obtained in [11–13]. Moreover, Li and Mihai et al. conducted research relevant to inequalities [14,15], solitons [16,17], submanifolds [18–20], classical differential geometry [21–23], non-Euclidean geomgery [24–27], etc., under the viewpoint of soliton theory, singularity theory, submanifold theory, and other theories [28–30]. The results and methods of those papers motivated us to write this paper and helped us in future research.

In Bach's work [31], a novel type of conformally invariant tensor was introduced as a tool for investigating conformal relativity. The analysis involves considering a specific function based on the squared norm of the Weyl conformal curvature denoted by $\mathcal{W}(g)$, which is evaluated over a Riemannian compact manifold of dimension 4. In the context of critical points of functionals, the Bach tensor (briefly, $\mathcal{B}_g$-tensor) plays a crucial role in the study of Einstein metrics, which are Riemannian metrics that satisfy the Einstein field equations. The $\mathcal{B}_g$-tensor is used to characterize the critical points of functionals that arise from the Einstein–Hilbert action, a functional that defines the action of a physical system. For any pseudo-Riemannian manifold (M,g) of dimension n, the $\mathcal{B}_g$-tensor is defined as

$$\begin{aligned}\mathcal{B}_g(\zeta_1,\zeta_2) &= \tfrac{1}{n-3}\textstyle\sum_{i,j=1}^{n}(\nabla_{e_i}\nabla_{e_j}\mathcal{W})(\zeta_1,e_i,e_j,\zeta_2)\\ &+\tfrac{1}{n-2}\textstyle\sum_{i,j=1}^{n}\mathrm{Ric}_g(e_i,e_j)\mathcal{W}(\zeta_1,e_i,e_j,\zeta_2),\end{aligned} \tag{2}$$

where $e_i,\ i=1,2,\cdots,n$ represents a local orthonormal frame on (M,g), Ric_g is the Ricci tensor of type $(0,2)$, and $\mathcal{W}$ is the Weyl tensor of type $(0,4)$. The Weyl conformal tensor of type $(1,3)$ is defined by

$$\begin{aligned}\mathcal{W}(\zeta_1,\zeta_2)\zeta_3 &= R(\zeta_1,\zeta_2)\zeta_3-\frac{1}{n-2}\{\mathrm{Ric}_g(\zeta_2,\zeta_3)\zeta_1-\mathrm{Ric}_g(\zeta_1,\zeta_3)\zeta_2+g(\zeta_2,\zeta_3)Q\zeta_1\\ &\quad -g(\zeta_1,\zeta_3)Q\zeta_2\}+\frac{r}{(n-1)(n-2)}\{g(\zeta_2,\zeta_3)\zeta_1-g(\zeta_1,\zeta_3)\zeta_2\},\end{aligned} \tag{3}$$

where R is the Riemannian curvature tensor, Q is the Ricci operator defined by $\mathrm{Ric}_g=g(Q\cdot,\cdot)$, and r is the scalar curvature of g. Utilizing the properties of the Weyl conformal tensor $\mathcal{W}$ and the contraction of the Bianchi second identity, it can be inferred that $div\mathcal{W}=(\frac{n-3}{n-2})\mathcal{C}$, where $\mathcal{C}$ is the Cotton tensor (briefly, $\mathcal{C}_g$-tensor) of type $(0,3)$ defined by (see [32])

$$\begin{aligned}\mathcal{C}(\zeta_1,\zeta_2)\zeta_3 &= (\nabla_{\zeta_1}\mathrm{Ric}_g)(\zeta_2,\zeta_3)-(\nabla_{\zeta_2}\mathrm{Ric}_g)(\zeta_1,\zeta_3)\\ &\quad-\frac{1}{2(n-1)}[(\zeta_1 r)g(\zeta_2,\zeta_3)-(\zeta_2 r)g(\zeta_1,\zeta_3)].\end{aligned} \tag{4}$$

A pseudo-Riemannian manifold (M,g) is called conformally flat if its conformally transformed pseudo-Riemannian metric is locally Euclidean. For $n>3$, this implies $\mathcal{W}=0$, and for $n=3$, that $\mathcal{C}=0$. Furthermore, the vanishing of the Weyl tensor also implies that the divergence of $\mathcal{W}$ is zero, which is equivalent to $\mathcal{C}=0$. Thus, the condition $\mathcal{C}=0$ generalizes both Einstein metrics and conformally flat pseudo-Riemannian metrics, encompassing a broader class of geometric structures. For instance, a Riemannian manifold that satisfies the condition $\mathcal{C}=0$ is characterized by a harmonic curvature tensor, denoted as $divR=0$. Interestingly, this condition implies that the scalar curvature r is constant. As a result, such a manifold generalizes Einstein manifolds and locally symmetric manifolds. Furthermore, the converse holds: the system of equations $divR=0$ implies that the Levi–Civita connection ∇ of the metric g constitutes a Yang–Mills connection while the metric is fixed [33].

Utilizing Equations (3) and (4), the expression for the $\mathcal{B}_g$-tensor (2) can be expressed as shown in Chen and He [34]:

$$\mathcal{B}_g(\zeta_1, \zeta_2) = \frac{1}{(n-2)}\left[\sum_{i=1}^{n}(\nabla_{e_i}\mathcal{C})(e_i, \zeta_1)\zeta_2 + \sum_{i,j=1}^{n} \mathrm{Ric}_g(e_i, e_j)\mathcal{W}(\zeta_1, e_i, e_j, \zeta_2)\right]. \quad (5)$$

A closer examination of the $\mathcal{B}_g$-tensor in (5) has led us to realize that its final term can be expressed as

$$g(Qe_i, e_j)g(\mathcal{W}(\zeta_1, e_i)e_j, \zeta_2) = -g(\mathcal{W}(\zeta_1, e_i)\zeta_2, Qe_i) = -g(Q\mathcal{W}(\zeta_1, e_i)\zeta_2, e_i).$$

As a consequence, expression (5) takes the following form:

$$\mathcal{B}_g(\zeta_1, \zeta_2) = \frac{1}{(n-2)}\left[\sum_{i}(\nabla_{e_i}\mathcal{C})(e_i, \zeta_1)\zeta_2 - \sum_{i,j} g(QW(\zeta_1, e_i)\zeta_2, e_i)\right]. \quad (6)$$

As the Weyl tensor W becomes zero in three dimensions, the Bach tensor can be formulated as

$$\mathcal{B}_g(\zeta_1, \zeta_2) = \sum_{i=1}^{3}(\nabla_{e_i}\mathcal{C})(e_i, \zeta_1)\zeta_2. \quad (7)$$

A Riemannian metric g is called Bach flat when its associated $\mathcal{B}_g$-tensor becomes zero. Notably, a Bach-flat metric encompasses Einstein metrics and conformally flat metrics. Consequently, a Riemannian manifold endowed with a vanishing $\mathcal{B}_g$-tensor is intriguing, as it accommodates both Einstein metrics and conformally flat metrics. Given that 4-dimensional Bach-flat metrics are conformally invariant, and this property is obtained by conformally related Einstein metrics, we know that these metrics are also Bach flat. Additionally, in 4-dimensional space, the $\mathcal{B}_g$-tensor plays a special role as a symmetric, divergence-free tensor of type $(0, 2)$ that is both quadratic in Riemann curvature (representing a linear combination of the products of two Riemann curvature tensors) and exhibits desirable conformal properties (for detailed insights, refer to [35]). It is well known that Einstein metrics are Bach flat, which naturally leads to the question of whether there exist Bach-flat metrics that are not Einstein metrics. Initially, this question was addressed by the authors Leistner and Nurwoski [36]. In their work, they presented 4-dimensional pp-waves that demonstrate Bach flatness. Furthermore, they provided numerous examples exhibiting Bach flatness without being Einstein metrics. Later, Ghosh [37] validated this concept within the context of an almost contact Riemannian manifold and obtained many fruitful results. Recently, Naik et al. [38] studied the $\mathcal{B}_g$-tensor and $\mathcal{C}_g$-tensor on a cosymplectic manifold and obtained many fruitful results. These works on the Bach and $\mathcal{C}_g$-tensors motivate us to investigate the validity of the converse statement in the context of Lorentzian manifolds.

The paper is structured as follows. In Section 2, we revisit the foundational definitions and formulas pertinent to Lorentzian manifolds with $\mathcal{SSM}\omega$-connections. Section 3 encompasses the derivation of auxiliary results, and subsequently, establishes the theorem that asserts the quasi-Einstein and Bach-flat nature of a Lorentzian manifold admitting $\mathcal{SSM}\omega$-connection with a parallel $\mathcal{C}_g$-tensor. Continuing, we explore the scenario of a Lorentzian manifold admitting $\mathcal{SSM}\omega$-connection with a purely transversal $\mathcal{B}_g$-tensor. Concretely, we prove that it is a quasi-Einstein manifold. Furthermore, we consider specific conditions on the $\mathcal{B}_g$-tensor, which are addressed in Section 4.

2. Notes on Lorentzian Manifolds

Lorentzian manifolds are one of the most important subclasses of pseudo-Riemannian manifolds, playing a vital role in mathematical physics (especially in the development of the theory of general relativity and cosmology). Let M be a connected paracompact Hausdorff smooth manifold of dimension n. Then, M is said to be a Lorentzian manifold if it admits a $(0,2)$-type non-degenerate smooth symmetric tensor $g_p : T_pM \times T_pM \to R$ that is $\forall p \in M$ of signature $(-,+,+,\cdots,+)$, where T_pM represents the tangent vector space on M at point p. A non-zero vector field $\omega \in T_pM$ is said to be time-like (resp. non-space-like, null, and space-like) if $g_p(\omega,\omega) < 0$ (resp. $\leq, =, > 0$) [39].

A one-form $\omega^\sharp$ and an associated vector field ω are connected through a pseudo-Riemannian metric g by $g(\zeta_1,\omega) = \omega^\sharp(\zeta_1)$. A linear connection $\widetilde{\nabla}$ on M is defined by

$$\widetilde{\nabla}_{\zeta_1}\zeta_2 = \nabla_{\zeta_1}\zeta_2 + \omega^\sharp(\zeta_2)\zeta_1 - g(\zeta_1,\zeta_2)\omega, \ \forall\zeta_1,\zeta_2 \in \aleph(M). \tag{8}$$

Mishra et al. [40] and Chaubey et al. [41] comprehensively studied an almost contact metric manifold admitting $\mathcal{SSM}\omega$-connection, characterizing situations where ω is a Reeb vector field ζ and $\widetilde{\nabla}\omega = 0$, and this characterization led to several significant geometric findings. Inspired by this study, Chaubey et al. [5,6] extended this concept to both Riemannian and Lorentzian manifolds. A linear connection $\widetilde{\nabla}$ defined on a Lorentzian manifold M is called a semi-symmetric metric ω-connection if it satisfies Equations (1) and (8), $\widetilde{\nabla}g = 0$, and $\widetilde{\nabla}\omega$. Consider $\widetilde{\nabla}\omega = 0$, then from Equation (8) we have

$$\nabla_{\zeta_1}\omega = \zeta_1 + \omega^\sharp(\zeta_1)\omega, \tag{9}$$

where ω is a unit-time-like vector field, that is, $g(\omega,\omega) = -1$. Now, we state the following restricted curvature with respect to the Levi–Civita connection ∇ as follows.

Lemma 1 ([6]). *Let M be an n-dimensional Lorentzian manifold with $\mathcal{SSM}\omega$-connection $\widetilde{\nabla}$, then*

$$R(\zeta_1,\zeta_2)\omega = \omega^\sharp(\zeta_2)\zeta_1 - \omega^\sharp(\zeta_1)\zeta_2, \tag{10}$$
$$R(\omega,\zeta_1)\zeta_2 = g(\zeta_1,\zeta_2)\omega - \omega^\sharp(\zeta_2)\zeta_1, \tag{11}$$
$$\omega^\sharp(R(\zeta_1,\zeta_2)\zeta_3) = \omega^\sharp(\zeta_1)g(\zeta_2,\zeta_3) - \omega^\sharp(\zeta_2)g(\zeta_1,\zeta_3), \tag{12}$$

for $\zeta_1,\zeta_2,\zeta_3 \in \chi(M)$.

Taking the g-trace of Equation (10) yields

$$\mathrm{Ric}_g(\zeta_1,\omega) = (n-1)\omega^\sharp(\zeta_1), \tag{13}$$

which is equivalent to

$$Q\omega = (n-1)\omega, \tag{14}$$

Taking the differentiation of $\omega^\sharp(\zeta_1) = g(\zeta_1,\omega)$ along V, together with Equation (9), we infer

$$(\nabla_V\omega^\sharp)\zeta_1 = g(\zeta_1,V) + \omega^\sharp(\zeta_1)\omega^\sharp(V). \tag{15}$$

Now, we recall the following results which are used to prove our main outcomes.

Lemma 2. *A Lorentzian manifold of dimension n equipped with $\mathcal{SSM}\omega$-connection satisfies*

$$(\nabla_{\zeta_1}Q)\omega = (n-1)\zeta_1 - Q\zeta_1, \tag{16}$$
$$(\nabla_\omega Q)\zeta_1 = -2Q\zeta_1 + 2(n-1)\zeta_1. \tag{17}$$

If a Lorentzian manifold M of dimension n with non-vanishing Ricci tensor Ric_g satisfies

$$\mathrm{Ric}_g = \alpha g + \beta\omega^\sharp \otimes \omega^\sharp, \tag{18}$$

for smooth functions α and β, where $\omega^\sharp$ is a non-zero one-form, and the vector field corresponding to the one-form $\omega^\sharp$ is a unit-time-like vector field, then it is referred to as a perfect fluid spacetime. However, some geometers also call M quasi-Einstein [42]. Particularly, if $\beta = 0$ and $\alpha = constant$, then M is called Einstein.

3. Characterization of Lorentzian Manifolds Admitting $\mathcal{SSM}\omega$-Connection with Parallel $\mathcal{C}_g$-Tensor

Before proceeding to explore our main results, we first establish the following key outcome.

Lemma 3. *On a Lorentzian manifold with $\mathcal{SSM}\omega$-connection the following relations are valid:*

$$\sum_{i}^{n}(\nabla_{e_i}\mathcal{C})(e_i,\zeta_2)\omega = \frac{1}{2}\zeta_2(r) - \frac{1}{2(n-1)}[(divDr)\omega^{\sharp}(\zeta_2) - g(\nabla_{\omega}Dr,\zeta_2)], \quad (19)$$

$$\sum_{i}^{n}g(QW(\zeta_1,e_i)\omega,e_i) = \frac{(r-(n-1))^2}{(n-1)(n-2)}\omega^{\sharp}(\zeta_1) + \left[\frac{|Q|^2-(n-1)^2}{(n-2)}\right]\omega^{\sharp}(\zeta_1), \quad (20)$$

where Dr denotes the gradient of r and div denotes divergence.

Proof. Substituting ζ_3 with ω in Equation (4), we obtain

$$\mathcal{C}(\zeta_1,\zeta_2)\omega = g((\nabla_{\zeta_1}Q)\zeta_2,\omega) - g((\nabla_{\zeta_2}Q)\zeta_1,\omega) - \frac{1}{2(n-1)}\{\omega^{\sharp}(\zeta_2)(\zeta_1 r) - \omega^{\sharp}(\zeta_1)(\zeta_2 r)\}. \quad (21)$$

Utilizing Lemma 2 in Equation (21), one can easily obtain

$$\mathcal{C}(\zeta_1,\zeta_2)\omega = -\frac{1}{2(n-1)}[\omega^{\sharp}(\zeta_2)(\zeta_1 r) - \omega^{\sharp}(\zeta_1)(\zeta_2 r)]. \quad (22)$$

Differentiating Equation (22) covariantly along ζ_3 and employing Equation (9), we deduce

$$(\nabla_{\zeta_3}\mathcal{C})(\zeta_1,\zeta_2)\omega + \mathcal{C}(\zeta_1,\zeta_2)\zeta_3 + \omega^{\sharp}(\zeta_3)\mathcal{C}(\zeta_1,\zeta_2)\omega = -\frac{1}{2(n-1)}\{g(\zeta_2,\zeta_3)g(Dr,\zeta_1)$$
$$-g(\zeta_1,\zeta_3)g(Dr,\zeta_2) + \omega^{\sharp}(\zeta_3)\omega^{\sharp}(\zeta_2)g(Dr,\zeta_1) - \omega^{\sharp}(\zeta_1)\omega^{\sharp}(\zeta_3)g(Dr,\zeta_2)$$
$$+\omega^{\sharp}(\zeta_2)g(\zeta_1,\nabla_{\zeta_3}Dr) - \omega^{\sharp}(\zeta_1)g(\zeta_1,\nabla_{\zeta_3}Dr)\}. \quad (23)$$

Consider an orthonormal frame $e_i : i = 1,2,\cdots,n$ for the tangent space of M. Setting $\zeta_1 = \zeta_3 = e_i$ in Equation (23), and subsequently, summing over i, we obtain

$$\sum_{i}^{n}[(\nabla_{e_i}\mathcal{C})(e_i,\zeta_2)\omega + \mathcal{C}(e_i,\zeta_2)e_i] + \mathcal{C}(\omega,\zeta_2)\omega = -\frac{1}{2(n-1)}\{(2-n)(\zeta_2 r)$$
$$+\omega^{\sharp}(\zeta_2)(\omega r) + (divDr)\omega^{\sharp}(\zeta_2) - g(\nabla_{\omega}Dr,\zeta_2)\}. \quad (24)$$

Switching ζ_1 with ω in Equation (23), a straightforward computation gives

$$\mathcal{C}(\omega,\zeta_2)\omega = -\frac{1}{2(n-1)}\{\omega^{\sharp}(\zeta_2)(\omega r) + \zeta_2 r\}. \quad (25)$$

As we know, $\mathcal{C}_g$-tensor is trace-free, so employing Equation (25) in Equation (24) leads to the proof of (i).

Replacing ζ_3 with ω in Equation (3) and calling back Equations (10) and (14), we arrive at

$$QW(\zeta_1,\zeta_2)\omega = \frac{r-(n-1)}{(n-1)(n-2)}[\omega^{\sharp}(\zeta_2)Q\zeta_1 - \omega^{\sharp}(\zeta_1)Q\zeta_2]$$
$$-\frac{1}{(n-2)}[\omega^{\sharp}(\zeta_2)Q^2\zeta_1 - \omega^{\sharp}(\zeta_1)Q^2\zeta_2]. \quad (26)$$

Substituting ζ_2 with e_i in the above equation, taking the inner product with e_i, summing over i, and utilizing Equation (14), we derive (ii). □

We know that the parallelism of the $\mathcal{C}_g$-tensor encompasses both conformal flatness and Cotton flatness. So, we were intrigued to study Lorentzian manifolds admitting $\mathcal{SSM}\omega$-connection M with a parallel $\mathcal{C}_g$-tensor and proceed to prove the following outcome.

Theorem 1. *Let M be a Lorentzian manifold admitting $\mathcal{SSM}\omega$-connection. If M has a parallel $\mathcal{C}_g$-tensor, then M is quasi-Einstein and Bach flat.*

Proof. The assumption of parallelism of the $\mathcal{C}_g$-tensor ensures its divergence-free nature. So that, from Equation (19), one can obtain

$$\frac{1}{2}(\zeta_2 r) - \frac{1}{2(n-1)}[(divDr)\omega^\sharp(\zeta_2) - g(\nabla_\omega Dr, \zeta_2)] = 0. \tag{27}$$

On the other hand,

$$Dr = -(\omega r)\omega, \qquad \text{and} \qquad (\omega r) = 2(r - n(n-1)). \tag{28}$$

Employing the foregoing equation in Equation (22) leads to $\mathcal{C}(\zeta_1, \zeta_2)\omega = 0$. Then, by taking the covariant derivative of this equation and utilizing Equation (9), we obtain

$$(\nabla_{\zeta_3}\mathcal{C})(\zeta_1, \zeta_2)\omega + \mathcal{C}(\zeta_1, \zeta_2)\zeta_3 + \omega^\sharp(\zeta_3)\mathcal{C}(\zeta_1, \zeta_2)\omega = 0.$$

Because of the parallelism of the $\mathcal{C}_g$-tensor and the fact that $\mathcal{C}(\zeta_1, \zeta_2)\omega = 0$, the preceding equation indicates that the manifold possesses a vanishing $\mathcal{C}_g$-tensor. Consequently, referring to Equation (4), we have

$$(\nabla_{\zeta_1}\mathrm{Ric}_g)(\zeta_2, \zeta_3) - (\nabla_{\zeta_2}\mathrm{Ric}_g)(\zeta_1, \zeta_3) - \frac{1}{2(n-1)}[(\zeta_1 r)g(\zeta_2, \zeta_3) - (\zeta_2 r)g(\zeta_1, \zeta_3)] = 0.$$

Replacing ζ_1 with ω in the previous equation and utilizing Lemma 2 and Equation (17), we conclude that

$$(n-1)g(\zeta_2, \zeta_3) - g(Q\zeta_2, \zeta_3) = \frac{1}{2(n-1)}[(\omega r)g(\zeta_2, \zeta_3) - (\zeta_2 r)\omega^\sharp(\zeta_3)]. \tag{29}$$

Utilizing Equation (28) in Equation (29), followed by simplification, reveals that the Ricci operator Q can be represented as

$$Q\zeta_2 = \left[\frac{r}{(n-1)} - 1\right]\zeta_2 + \left[\frac{r}{(n-1)} - n\right]\omega^\sharp(\zeta_2)\omega. \tag{30}$$

Therefore, M is quasi-Einstein. Continuing, we employ Equations(10), (14) and (30), in Equation (3), resulting in $\mathcal{W}(\zeta_1, \zeta_2)\omega = 0$. Given the parallelism of the $\mathcal{C}_g$-tensor, the first term of the $\mathcal{B}_g$-tensor (6) becomes zero. Consequently, utilizing Equation (30), the $\mathcal{B}_g$-tensor is simplified to

$$\begin{aligned} \mathcal{B}_g(\zeta_1, \zeta_2) &= -\frac{1}{(n-2)}\sum\{\left(\frac{r}{(n-1)} - 1\right)g(\mathcal{W}(\zeta_1, e_i)\zeta_2, e_i) \\ &\quad + \left(\frac{r}{(n-1)} - n\right)g(\mathcal{W}(\zeta_1, \omega)\zeta_2, \omega)\}. \end{aligned} \tag{31}$$

As the Weyl tensor is trace-free and $\mathcal{W}(\zeta_1, \zeta_2)\omega = 0$, the aforementioned equation demonstrates the vanishing of $\mathcal{B}_g$ on M. This completes the proof. □

Remark 1. *It is well known that the Weyl tensor vanishes in three dimensions; however, this assertion does not hold true for higher dimensions [43]. Consequently, substituting* $\zeta_2 = \zeta_3 = \omega$, $n = 3$, *and* $\mathcal{W} = 0$ *into Equation (3) yields*

$$R(\zeta_1, \omega)\omega = -[2\zeta_1 + 4\omega^\sharp(\zeta_1)\omega + Q\zeta_1] + \frac{r}{2}[\zeta_1 + \omega^\sharp(\zeta_1)\omega].$$

Utilizing of (10) in the preceding equation, we obtain

$$Q\zeta_1 = \left(\frac{r}{2} - 1\right)\zeta_1 + \left(\frac{r}{2} - 3\right)\omega^\sharp(\zeta_1)\omega.$$

Under the same assumptions as stated in the preceding theorem and guided by Equation (7), the conclusion trivially holds in three dimensions.

If the unit-time-like vector field ω leaves the scalar curvature r invariant, then from the second term of Equation (28) we obtain $r = n(n-1)$. Applying this result to Equation (30), we observe that $Q\zeta_2 = (n-1)\zeta_2$. As a result of Theorem 1, the following conclusion can be drawn.

Corollary 1. *Let M be a Lorentzian manifold admitting $\mathcal{SSM}\omega$-connection with parallel Cotton tensor. If a unit-vector field ω leaves the scalar curvature invariant, then M is Einstein.*

4. $\mathcal{B}_g$-Tensor on Lorentzian Manifolds with $\mathcal{SSM}\omega$-Connection

In this section, we consider a purely transversal $\mathcal{B}_g$-tensor on a class of Lorentzian manifolds and obtain the following outcome.

Theorem 2. *A Lorentzian manifold admitting $\mathcal{SSM}\omega$–connection has a purely transversal $\mathcal{B}_g$-tensor if and only if it is quasi-Einstein.*

Proof. Firstly, we derive the expression for the transversal $\mathcal{B}_g$-tensor. To achieve this, we substitute ω into Equation (6) and with the help of Equations (19) and (20), we obtain

$$\begin{aligned}\mathcal{B}_g(\zeta_1, \omega) &= \frac{1}{(n-2)}\left\{\frac{(\zeta_1 r)}{2} - \frac{1}{2(n-1)}[div(Dr)\omega^\sharp(\zeta_1) - g(\nabla_\omega Dr, \zeta_1)]\right. \\ &\quad \left. + \frac{[r-(n-1)]^2}{(n-1)(n-2)}\omega^\sharp(\zeta_1) - \left[\frac{|Q|^2 - (n-1)^2}{(n-2)}\right]\omega^\sharp(\zeta_1)\right\}. \end{aligned} \tag{32}$$

Suppose the $\mathcal{B}_g$-tensor is purely transversal, then the foregoing equation transforms into

$$\begin{aligned}&\frac{(\zeta_1 r)}{2} - \frac{1}{2(n-1)}[div(Dr)\omega^\sharp(\zeta_1) - g(\nabla_\omega Dr, \zeta_1)] \\ &+ \frac{[r-(n-1)]^2}{(n-1)(n-2)}\omega^\sharp(\zeta_1) - \left[\frac{|Q|^2 - (n-1)^2}{(n-2)}\right]\omega^\sharp(\zeta_1) = 0. \end{aligned} \tag{33}$$

Covariant differentiate the first term of Equation (28) along ζ_1, and consider Equation (9) to obtain $\nabla_{\zeta_1} Dr = -[\zeta_1(\omega r)\omega + (\omega r)(\zeta_1 + \omega^\sharp(\zeta_1)\omega)]$. Contracting this and recalling Equation (28) gives us $div(Dr) = -(n-3)(\omega r)$. Putting this together with Equation (33), we arrive at

$$\begin{aligned}&\frac{(\zeta_1 r)}{2} + \frac{1}{2(n-1)}[(n-3)(\omega r)\omega^\sharp(\zeta_1) + 2(\omega r)\omega] \\ &+ \frac{[r-(n-1)]^2}{(n-1)(n-2)}\omega^\sharp(\zeta_1) - \left[\frac{|Q|^2 - (n-1)^2}{(n-2)}\right]\omega^\sharp(\zeta_1) = 0. \end{aligned}$$

Plugging $\zeta_1 = \omega$ into the previous equation yields

$$(n-1)[|Q|^2 - (n-1)^2] - [r-(n-1)]^2 = 0. \tag{34}$$

Utilizing Equations (14), calculating the magnitude of the tensor T becomes

$$|T|^2 = \frac{1}{(n-1)}\{(n-1)(|Q|^2 - (n-1)^2) - [r - (n-1)]^2\}. \tag{35}$$

Employing Equation (34) in Equation (35) demonstrates that $T = 0$, consequently establishing that M is quasi-Einstein. On the other hand, if we assume M is quasi-Einstein, then consequently, $\mathcal{W}(\zeta_1, \zeta_2)\omega = 0$ (as proved in Theorem 1). Differentiating this along ζ_3 and recalling Equation (10) gives

$$(\nabla_{\zeta_3}\mathcal{W})(\zeta_1, \zeta_2)\omega = -\mathcal{W}(\zeta_1, \zeta_2)\zeta_3 + \omega^\sharp(\zeta_3)\mathcal{W}(\zeta_1, \zeta_2)\omega = -\mathcal{W}(\zeta_1, \zeta_2)\zeta_3.$$

By contracting this equation over ζ_3 and taking into account that the Weyl tensor is trace-free, we deduce $(divW)(\zeta_1, \zeta_2)\omega = \mathcal{C}(\zeta_1, \zeta_2)\omega = 0$. Considering the quasi-Einstein, Equation (35) implies $(n-1)[|Q|^2 - (n-1)^2] - [r - (n-1)]^2 = 0$. By incorporating all of these results into Equation (32), we can deduce that $\mathcal{B}_g(\zeta_1, \omega) = 0$. □

As a consequence of Theorem 2, we demonstrate the following.

Corollary 2. *A quasi-Einstein Lorentzian manifold admitting $\mathcal{SSM}\omega$-connection is Bach flat.*

Proof. As a result of Theorem 2, we know that a quasi-Einstein Lorentzian manifold admitting $\mathcal{SSM}\omega$-connection possesses a purely transversal $\mathcal{B}_g$-tensor. Now, covariant differentiate Equation (30) along ζ_1 and invoke Equation (8) to obtain

$$\begin{aligned}(\nabla_{\zeta_1}Q)\zeta_2 &= \frac{\zeta_1 r}{(n-1)}[\zeta_2 + \omega^\sharp(\zeta_2)\omega] + \left(\frac{r}{(n-1)} - n\right) \\ &\quad \{g(\zeta_1, \zeta_2)\omega + \omega^\sharp(\zeta_2)\zeta_1 + 2\omega^\sharp(\zeta_1)\omega^\sharp(\zeta_2)\omega\}.\end{aligned} \tag{36}$$

By virtue of Equation (36), the $\mathcal{C}_g$-tensor Equation (4) can be expressed as

$$\begin{aligned}\mathcal{C}(\zeta_1, \zeta_2)\zeta_3 &= \frac{\zeta_1 r}{2(n-1)}g(\zeta_2, \zeta_3) - \frac{\zeta_2 r}{2(n-1)}g(\zeta_1, \zeta_3) + \frac{\zeta_1 r}{(n-1)}\omega^\sharp(\zeta_2)\omega^\sharp(\zeta_3) \\ &- \frac{\zeta_2 r}{(n-1)}\omega^\sharp(\zeta_1)\omega^\sharp(\zeta_3) + \left(\frac{r}{(n-1)} - n\right)\{\omega^\sharp(\zeta_2)g(\zeta_1, \zeta_3) - \omega^\sharp(\zeta_1)g(\zeta_2, \zeta_3)\}.\end{aligned}$$

Applying Equation (28) to the previous equation, we can deduce that $\mathcal{C}(\zeta_1, \zeta_2)\zeta_3 = 0$. As a result, the first term of the Bach tensor Equation (6) becomes zero. Conversely, considering M to be a quasi-Einstein manifold, we have $\mathcal{W}(\zeta_1, \zeta_2)\omega = 0$ for all vector fields ζ_1 and ζ_2 in M (a consequence of Theorem 1). The subsequent steps of the proof align with the implications arising from Equation (31). □

Combining the results of Corollary 2 with Theorem 1, we conclude the following.

Corollary 3. *In a Lorentzian manifold admitting $\mathcal{SSM}\omega$-connection, the following conditions are equivalent:*

- *M is quasi-Einstein;*
- *M has a vanishing $\mathcal{C}_g$-tensor;*
- *M has a vanishing $\mathcal{B}_g$-tensor.*

Finally, by extending the implications of Theorem 2, we establish the following.

Theorem 3. *Let M be a Lorentzian manifold admitting $\mathcal{SSM}\omega$-connection. If a unit-vector field ω leaves the scalar curvature invariant, then M is Einstein if one of the following conditions holds:*

- *The $\mathcal{B}_g$-tensor containing the unit-vector field ω is parallel along ω;*
- *The curvature transformation annihilates the $\mathcal{B}_g$-tensor.*

Proof. Suppose the vector field ω leaves the scalar curvature invariant, that is, $\omega(r) = 0$. Consequently, referring to Equation (28), it becomes evident that $r = n(n-1)$. As a consequence of Equation (32), it can be deduced that

$$\mathcal{B}_g(\zeta_1, \omega) = f\omega^\sharp(\zeta_1), \tag{37}$$

where $f = \frac{1}{(n-2)}[(n-1)^3 + (n-1)^2 - |Q|^2]$. Covariant differentiating Equation (37) along ω and applying the hypothesis $(\nabla_\omega \mathcal{B}_g)(\zeta_1, \omega) = 0$, we can deduce $\nabla_\omega |Q|^2 = 0$. Utilizing Equation (17), this inference leads to

$$0 = \sum_{i=1}^{n} g((\nabla_\omega Q)e_i, Qe_i) = -2\sum_{i=1}^{n} g(Qe_i - (n-1)e_i, Qe_i).$$

Therefore, we have $|Q|^2 = (n-1)r = n(n-1)^2$. Considering this outcome, we can now proceed with the computation:

$$\begin{aligned} |Q - (n-1)I|^2 &= |Q|^2 - 2(n-1)r + n(n-1)^2 \\ &= n(n-1)^2 - 2n(n-1)^2 + n(n-1)^2 = 0. \end{aligned} \tag{38}$$

Hence, we conclude M is an Einstein manifold. This finishes the verification of (i).

Now, we assume that $R(\zeta_1, \zeta_2) \cdot \mathcal{B}_g = 0$. This equivalence can be expressed as follows: $\mathcal{B}_g(R(\zeta_1, \zeta_2)\zeta_3, U) + \mathcal{B}_g(R(\zeta_1, \zeta_2)U, \zeta_3) = 0$. By setting $\zeta_2 = \zeta_3 = \omega$, we obtain

$$\mathcal{B}_g(R(\zeta_1, \omega)\omega, U) + \mathcal{B}_g(R(\zeta_1, \omega)U, \omega) = 0. \tag{39}$$

Utilizing Equation (10), we deduce that $R(\zeta_1, \omega)\omega = -\zeta_1 - \omega^\sharp(\zeta_1)\omega$ and $R(\zeta_1, \omega)U = \omega^\sharp(U)\zeta_1 - g(\zeta_1, U)\omega$. Putting these results into Equation (39) and recalling $\mathcal{B}_g(\zeta_1, \omega) = f\omega^\sharp(\zeta_1)$, we derive $\mathcal{B}_g(\zeta_1, U) = fg(\zeta_1, U)$. Since the $\mathcal{B}_g$-tensor is trace-free, we immediately arrive at $f = 0$. This implies that

$$(n-1)^3 + (n-1)^2 - |Q|^2 = 0.$$

Consequently, by following Equation (38), we can deduce that M is an Einstein manifold. This concludes the proof of (ii). □

5. Conclusions

Rudolf Bach's introduction of the Bach tensor was a major milestone in the development of differential geometry and general relativity, and it has had far-reaching implications for the understanding of spacetime geometry and gravitational phenomena. Conformally, Einstein spaces and the Bach tensor are fundamental aspects of differential geometry and general relativity. The Bach tensor provides a powerful tool for studying the curvature properties of spacetime and has important implications for our understanding of gravitational phenomena. In the context of geometry and relativity, both the Bach and Cotton tensors play an important role in understanding spacetime geometry and its relation to matter and energy. The Bach tensor helps us to understand how matter and energy affect the curvature of spacetime, while the Cotton tensor describes how spacetime curvature responds to matter distribution. The Bach and Cotton tensors have been used in various modified gravity theories, such as Brans–Dicke theory, Jordan–Brans–Dicke theory, and $f(R)$-gravity, to describe deviations from general relativity. Our main results reveal that a Lorentzian manifold admitting $\mathcal{SSM}\omega$-connection with a parallel Cotton tensor is quasi-Einstein and Bach flat, which provides understanding in many aspects of black holes, dark energy, cosmology, and so on.

Author Contributions: Conceptualization, Y.L., H.A.K., M.S.S. and M.M.P.; methodology, Y.L., H.A.K. and M.S.S.; investigation, Y.L., H.A.K., M.S.S. and M.M.P.; writing—original draft preparation, Y.L., H.A.K., M.S.S. and M.M.P.; writing— review and editing, Y.L., H.A.K., M.S.S. and M.M.P. All authors have read and agreed to the published version of the manuscript.

Funding: This research received no external funding.

Data Availability Statement: The original contributions presented in the study are included in the article, further inquiries can be directed to the corresponding author.

Acknowledgments: We gratefully acknowledge the constructive comments from the editor and the anonymous referees.

Conflicts of Interest: The authors declare no conflicts of interest.

ferences

Yano, K.; Kon, M. *Structures on Manifolds*; Series in Pure Math; World Scientific: Singapore, 1985.
Hayden, H.A. Subspace of a Space with Torsion. *Proc. Am. Math. Soc.* **1957**, *34*, 294–298.
Pak, E. On the pseudo-Riemannian spaces. *J. Korean Math. Soc.* **1969**, *6*, 23–31.
Yano, K. On semi-symmetric metric connections. *Pures Appl. Rev. Roumaine Math.* **1970**, *15*, 1579–1586.
Chaubey, S.K.; Lee, J.; Yadav, S. Riemannian manifolds with a semi-symmetric metric P-connection. *Korean Math. Soc.* **2019**, *56*, 1113–1129.
Chaubey, S.K.; Suh, Y.J.; De, U.C. Characterizations of the Lorentzian manifolds admitting a type of semi-symmetric metric connection. *Anal. Math. Phys.* **2020**, *10*, 61. [CrossRef]
Duggal, K.L. Almost Ricci solitons and physical applications. *Int. Electron. J. Geom.* **2017**, *10*, 1–10.
Kumar, R.; Colney, L.; Shenawy, S.; Turki, N.B. Tangent bundles endowed with quarter-symmetric non-metric connection (QSNMC) in a Lorentzian Para-Sasakian manifold. *Mathematics* **2023**, *11*, 4163. [CrossRef]
Kumar, R.; Colney, L.; Khan, M.N.I. Proposed theorems on the lifts of Kenmotsu manifolds admitting a non-symmetric non-metric connection (NSNMC) in the tangent bundle. *Symmetry* **2023**, *15*, 2037. [CrossRef]
Khan, M.N.I. Liftings from a para-sasakian manifold to its tangent bundles. *Filomat* **2023**, *37*, 6727–6740. [CrossRef]
Khan, M.N.I.; De, U.C.; Velimirovic, L.S. Lifts of a quarter-symmetric metric connection from a Sasakian manifold to its tangent bundle. *Mathematics* **2023**, *11*, 53. [CrossRef]
Khan, M.N.I.; Mofarreh, F.; Haseeb, A.; Saxena, M. Certain results on the lifts from an LP-Sasakian manifold to its tangent bundles associated with a quarter-symmetric metric connection. *Symmetry* **2023**, *15*, 1553. [CrossRef]
Khan, M.N.I.; Mofarreh, F.; Haseeb, A. Tangent bundles of P-Sasakian manifolds endowed with a quarter-symmetric metric connection. *Symmetry* **2023**, *15*, 753. [CrossRef]
Li, Y.; Aquib, M.; Khan, M.A.; Al-Dayel, I.; Masood, K. Analyzing the Ricci Tensor for Slant Submanifolds in Locally Metallic Product Space Forms with a Semi-Symmetric Metric Connection. *Axioms* **2024**, *13*, 454. [CrossRef]
Li, Y.; Aquib, M.; Khan, M.A.; Al-Dayel, I.; Youssef, M.Z. Geometric Inequalities of Slant Submanifolds in Locally Metallic Product Space Forms. *Axioms* **2024**, *13*, 486. [CrossRef]
De, K.; De, U.C.; Gezer, A. Perfect Fluid Spacetimes and k-Almost Yamabe Solitons. *Turk. J. Math.* **2023**, *47*, 1236–1246. [CrossRef]
Li, Y.; Gezer, A.; Karakas, E. Exploring Conformal Soliton Structures in Tangent Bundles with Ricci-Quarter Symmetric Metric Connections. *Mathematics* **2024**, *12*, 2101. [CrossRef]
Mihai, I.; Mohammed, M. Optimal inequalities for submanifolds in trans-Sasakian manifolds endowed with a semi-symmetric metric connection. *Symmetry* **2023**, *15*, 877. [CrossRef]
Mihai, I.; Mihai, R.I. General Chen inequalities for statistical submanifolds in Hessian manifolds of constant Hessian curvature. *Mathematics* **2022**, *10*, 3061. [CrossRef]
Mihai, A.; Mihai, I. Curvature invariants for statistical submanifolds of Hessian manifolds of constant Hessian curvature. *Mathematics* **2018**, *6*, 44. [CrossRef]
Canlı, D.; Şenyurt, S.; Kaya, F.E.; Grilli, L. The Pedal Curves Generated by Alternative Frame Vectors and Their Smarandache Curves. *Symmetry* **2024**, *16*, 1012. [CrossRef]
Kaya, E.F.; Şenyurt, S. Curve-Surface Pairs on Embedded Surfaces and Involute D-Scroll of the Curve-Surface Pair in E3. *Symmetry* **2024**, *16*, 323. [CrossRef]
Gür, M.; Şenyurt, S.; Grilli, L. The invariants of Dual Parallel equidistant ruled surfaces. *Symmetry* **2023**, *15*, 206. [CrossRef]
Li, Y.; Güler, E.; Toda, M. Family of right conoid hypersurfaces with light-like axis in Minkowski four-space. *AIMS Math.* **2024**, *9*, 18732–18745. [CrossRef]
Li, Y.; Güler, E. Right Conoids Demonstrating a Time-like Axis within Minkowski Four-Dimensional Space. *Mathematics* **2024**, *12*, 2421. [CrossRef]
Li, Y.; Abdel-Aziz, H.; Serry, H.; El-Adawy, F.; Saad, M. Geometric visualization of evolved ruled surfaces via alternative frame in Lorentz-Minkowski 3-space. *AIMS Math.* **2024**, *9*, 25619–25635. [CrossRef]
Gür, M. Geometric properties of timelike surfaces in Lorentz-Minkowski 3-space. *Filomat* **2023**, *37*, 5735–5749. [CrossRef]

28. De, K.; Khan, M.; De, U. Almost co-Kahler manifolds and (m, ρ)-quasi-Einstein solitons. *Chaos Solitons Fractals* **2023**, *167*, 1130 [CrossRef]
29. Sardar, A.; De, U. Almost Schouten solitons and almost cosymplectic manifolds. *J. Geom.* **2023**, *114*, 13. [CrossRef]
30. Li, Y.; Turki, N.; Deshmukh, S.; Belova, O. Euclidean hypersurfaces isometric to spheres. *AIMS Math.* **2024**, *9*, 28306–283 [CrossRef]
31. Bach, R. Zur Weylschen Relativitatstheorie und der Weylschen Erweiterung des Krummungstensorbegriffs. *Math. Z.* **1921**, 110–135. [CrossRef]
32. Kuhnel, W.; Rademacher, H.B. Conformal vector fields on pseudo-Riemannian spaces. *Differ. Geom. Appl.* **1997**, *7*, 237–2 [CrossRef]
33. Bourguignon, J.-P. Harmonic curvature for gravitational and yang mills fields. In *Harmonic Maps*; Springer: Berlin/Heidelbe Germany, 1982; pp. 35–47.
34. Chen, Q.; He, C. On bach flat warped product Einstein manifolds. *Pac. J. Math.* **2013**, *265*, 313–326. [CrossRef]
35. Bergman, J.; Edgar, S.B.; Herberthson, M. The bach tensor and other divergence-free tensors. *Int. J. Geom. Methods Mod. Ph* **2005**, *2*, 13–21. [CrossRef]
36. Leistner, T.; Nurowski, P. Ambient metrics for n-dimensional pp-waves. *Commun. Math. Phys.* **2010**, *296*, 881–898. [CrossRef]
37. Ghosh, A. Cotton tensor, Bach tensor and Kenmotsu manifolds. *Afr. Mat.* **2020**, *31*, 1193–1205. [CrossRef]
38. Naik, D.M.; Venkatesha, V.; Kumara, H.A. Certain types of metrics on almost coKähler manifolds. *Ann. Math. Quebec* **2023**, 331–347. [CrossRef]
39. Niell, B.O. *Semi-Riemannian Geometry with Applications to Relativity*; Academic Press, Inc.: New York, NY, USA , 1983.
40. Mishra, R.S.; Pandey, S.N. Semi-symmetric metric connections in an almost contact manifold. *Indian J. Pure Appl. Math.* **19** 9, 570–580.
41. Khan, M.A.; Al-Dayel, I.; Chaubey, S.K. Semi-Symmetric Metric Connections and Homology of CR-Warped Product Submanifol in a Complex Space Form Admitting a Concurrent Vector Field. *Symmetry* **2024**, *16*, 719. [CrossRef]
42. Chaki, M.C.; Maity, R.K. On quasi einstein manifolds. *Publ. Math. Debrecen.* **2000**, *57*, 297–306. [CrossRef]
43. Besse, A.L. *Einstein Manifolds*; Springer Science and Business Media: New York, NY, USA, 2007.

mathematics MDPI

icle

inching Results for Submanifolds in Lorentzian–Sasakian Ianifolds Endowed with a Semi-Symmetric on-Metric Connection

hammed Mohammed [1,2,†], Ion Mihai [3,*,†] and Andreea Olteanu [4,†]

1 School of Mathematics, Statistics and Computer Science, University of KwaZulu-Natal, Private Bag X01, Scottsville 3209, South Africa; 218075524@stu.ukzn.ac.za
2 Department of Mathematics, Faculty of Mathematical Sciences and Statistics, AL-Neelain University, Khartoum 11121, Sudan
3 Department of Mathematics, University of Bucharest, 010014 Bucharest, Romania
4 Department of Mathematics, Physics and Terrestrial Measurements, Faculty of Land Reclamation and Environmental Engineering, University of Agronomic Sciences and Veterinary Medicine of Bucharest, 011464 Bucharest, Romania; andreea.olteanu@fifim.ro
* Correspondence: imihai@fmi.unibuc.ro
† These authors contributed equally to this work.

Abstract: We establish an improved Chen inequality involving scalar curvature and mean curvature and geometric inequalities for Casorati curvatures, on slant submanifolds in a Lorentzian–Sasakian space form endowed with a semi-symmetric non-metric connection. Also, we present examples of slant submanifolds in a Lorentzian–Sasakian space form.

Keywords: Chen invariants; Chen inequalities; Casorati curvature; Lorentzian–Sasakian manifold; semi-symmetric connection; non-metric connection

MSC: 53C40; 53C25; 53C50; 53D15

1. Introduction

Friedmann and Schouten introduced the concept of semi-symmetric linear and metric connections on differentiable manifolds in [1], while H. A. Hayden independently proposed the same in [2]. K. Yano further dealt with the properties of Riemannian manifolds endowed with a semi-symmetric metric connection in [3]. Agashe and Chafle, in their works [4,5], introduced the notion of a semi-symmetric non-metric connection on a Riemannian manifold and explored submanifolds within this context.

Addressing Chen's open problem, which involves identifying optimal relationships between intrinsic and extrinsic invariants of a Riemannian submanifold—a key challenge in the geometry of submanifolds—Chen defined the δ-invariants, well known as Chen invariants, in [6,7]. In this respect, many authors have explored Chen's theory in different ambient spaces, concentrating on specific types of submanifolds. For additional details, we refer to [8–12].

The principal extrinsic invariant is the mean curvature. The Casorati curvature of a submanifold in a Riemannian manifold is also an extrinsic invariant. This curvature is defined as the normalized square of the length of the second fundamental form of the submanifold, extending the concept of the principal directions of a hypersurface in a Riemannian manifold. Therefore, obtaining optimal inequalities for the Casorati curvatures of submanifolds in different manifolds is of significant interest. Decu et al. [13] derived optimal inequalities for the Casorati curvature in terms of the scalar curvature on a submanifold in a Riemannian space form. Several authors have obtained geometric inequalities for the Casorati curvatures of submanifolds, as evidenced by works such as [13–16].

In particular, discussions on the Chen basic inequality and the inequality of Casorati invariants for submanifolds in different ambient spaces, equipped with a semi-symmetric metric connection or a semi-symmetric non-metric one, can be found in [17–25].

In the present paper, we establish an improved Chen inequality involving scalar curvature and mean curvature and geometric inequalities for Casorati curvatures, for slant submanifolds in a Lorentzian–Sasakian space form equipped with a semi-symmetric non-metric connection. Also, we prove an improved Chen basic inequality for Legendrian submanifolds in such spaces. Examples of slant submanifolds in a Lorentzian–Sasakian space form are presented.

2. Preliminaries

Let $(\tilde{M}, g)$ be a Riemannian manifold, $\dim \tilde{M} = n + q$. A linear connection $\tilde{\nabla}$ on $\tilde{M}$ is called a semi-symmetric connection if the torsion tensor field $\tilde{T}$, given by

$$\tilde{T}(X,Y) = \tilde{\nabla}_X Y - \tilde{\nabla}_Y X - [X,Y],$$

for any vector fields X and Y on $\tilde{M}$, satisfies

$$\tilde{T}(X,Y) = \omega(Y)X - \omega(X)Y, \tag{1}$$

where ω is the 1-form associated with the vector field P on $\tilde{M}$, i.e., $\omega(X) = g(X,P)$.

If $\tilde{\nabla} g = 0$, then $\tilde{\nabla}$ is said to be a semi-symmetric metric connection. Otherwise, i.e., if $\tilde{\nabla} g \neq 0$, $\tilde{\nabla}$ is said to be a semi-symmetric non-metric connection.

Let $\overset{\circ}{\tilde{\nabla}}$ be the Levi-Civita connection with respect to the Riemannian metric g. Following [4], a semi-symmetric non-metric connection $\tilde{\nabla}$ on M is given by

$$\tilde{\nabla}_X Y = \overset{\circ}{\tilde{\nabla}}_X Y + \omega(Y)X,$$

for any vector fields X and Y tangent to $\tilde{M}$.

In the following, we consider $(\tilde{M}, g)$ be a Riemannian manifold endowed with a semi-symmetric non-metric connection $\tilde{\nabla}$ and the Levi-Civita connection $\overset{\circ}{\tilde{\nabla}}$, and we consider M to be a submanifold in $\tilde{M}$.

Let $\tilde{R}$ and $\overset{\circ}{\tilde{R}}$ be the curvature tensor fields of the Riemannian manifold $\tilde{M}$ corresponding to $\tilde{\nabla}$ and $\overset{\circ}{\tilde{\nabla}}$, respectively. Then, $\tilde{R}$ is expressed by (see [4])

$$\tilde{R}(X,Y,Z,W) = \overset{\circ}{\tilde{R}}(X,Y,Z,W) + s(X,Z)g(Y,W) - s(Y,Z)g(X,W), \tag{2}$$

for any vector fields X, Y, Z, and W on $\tilde{M}$, where s is $(0,2)$ tensor given by

$$s(X,Y) = (\overset{\circ}{\tilde{\nabla}}_X \omega)Y - \omega(X)\omega(Y).$$

The Gauss formulae for the connections $\tilde{\nabla}$ and $\overset{\circ}{\tilde{\nabla}}$ are written as

$$\tilde{\nabla}_X Y = \nabla_X Y + h(X,Y), \quad \overset{\circ}{\tilde{\nabla}}_X Y = \overset{\circ}{\nabla}_X Y + \overset{\circ}{h}(X,Y),$$

for any vector fields X and Y on the submanifold M, where $\overset{\circ}{h}$ is the second fundamental form of M and h is a $(0,2)$-tensor on M. From [5], it is known that $\overset{\circ}{h} = h$.

Let $\tilde{M}$ be a differentiable manifold of dimension $(2m+1)$, where $m \geq 1$, admitting a tensor field ϕ of type $(1,1)$, a vector field ξ, and an l-form η such that

$$\phi^2(X) = -X + \eta(X)\xi, \quad \phi\xi = 0, \qquad \eta(\phi X) = 0, \qquad \eta(\xi) = 1, \tag{3}$$

for any vector field X on $\tilde{M}$.

We say that $\tilde{M}$ has an almost contact structure (ϕ, ξ, η) and call $\tilde{M}$ an almost contact manifold.

Since $\tilde{M}$ has a globally defined vector field ξ, it is able to admit a Lorentzian metric g such that $g(\xi, \xi) = -1$.

If M admits a normal almost contact structure (ϕ, ξ, η) and a Lorentzian metric g with

$$g(\phi X, \phi Y) = g(X, Y) + \eta(X)\eta(Y), \qquad (\overset{\circ}{\tilde{\nabla}}_X \eta)(Y) = g(\phi X, Y),$$

where $\overset{\circ}{\tilde{\nabla}}$ is the Levi-Civita connection of g, then M is called a Sasakian manifold with a Lorentzian metric or simply a Lorentzian–Sasakian manifold.

On such a manifold, one also has

$$\eta(X) = -g(X, \xi), \qquad \overset{\circ}{\tilde{\nabla}}_X \xi = -\phi X,$$
$$(\overset{\circ}{\tilde{\nabla}}_X \phi)Y = -g(X, Y)\xi - \eta(Y)X.$$

A Lorentzian–Sasakian manifold having a constant ϕ-sectional curvature c is called a Lorentzian–Sasakian space form and is denoted by $\tilde{M}(c)$.

Theorem 1 ([26]). *The curvature tensor field of a Lorentzian–Sasakian space form $\tilde{M}(c)$ has the expression*

$$\begin{aligned}\overset{\circ}{\tilde{R}}(X, Y)Z = & \frac{c-3}{4}(g(Y, Z)X - g(X, Z)Y) \\ & + \frac{c+1}{4}[g(X, \phi Z)\phi Y - g(Y, \phi Z)\phi X + 2g(X, \phi Y)\phi Z \\ & + \eta(Y)\eta(Z)X - \eta(X)\eta(Z)Y + g(X, Z)\eta(Y)\xi \\ & - g(Y, Z)\eta(X)\xi]. \end{aligned} \tag{4}$$

If $\tilde{M}$ is a $(2m+1)$-dimensional Lorentzian–Sasakian space form of constant ϕ-sectional curvature c endowed with a semi-symmetric non-metric connection, then, from expressions (2) and (4), it follows that the curvature tensor field $\tilde{R}$ becomes

$$\begin{aligned}\tilde{R}(X, Y, Z, W) = & \frac{c-3}{4}[g(Y, Z)g(X, W) - g(X, Z)g(Y, W)] \\ & + \frac{c+1}{4}[g(X, \phi Z)g(\phi Y, W) - g(Y, \phi Z)g(\phi X, W) \\ & + 2g(X, \phi Y)g(\phi Z, W) + \eta(Y)\eta(Z)g(X, W) \\ & - \eta(X)\eta(Z)g(Y, W) + g(X, Z)\eta(Y)\eta(W) \\ & - g(Y, Z)\eta(X)\eta(W)] \\ & + s(X, Z)g(Y, W) - s(Y, Z)g(X, W). \end{aligned} \tag{5}$$

Denoting the tangent bundle as TM and the normal bundle of M as $T^{\perp}M$, we can write

$$\phi X = PX + FX, \qquad \phi N = tN + fN,$$

where PX and FX denote the tangential component and the normal component, respectively, of ϕX, and tN and fN are the tangential and the normal components, respectively, of ϕN.

We decompose the vector field P on M into its tangent and normal components $P^{\top}$ and $P^{\perp}$, respectively; one has $P = P^{\top} + P^{\perp}$.

The Gauss equation for the semi-symmetric non-metric connection is (see [5])

$$\begin{aligned}\tilde{R}(X,Y,Z,W) =& R(X,Y,Z,W) + g(h(X,Z),h(Y,W)) \\ &- g(h(X,W),h(Y,Z)) + g(P^{\perp},h(Y,Z))g(X,W) \\ &- g(P^{\perp},h(X,Z))g(Y,W), \end{aligned} \tag{6}$$

for all vector fields X, Y, Z, and W on M.

Recently, one of the authors of the present work and A. Mihai [27] defined the sectional curvature of a non-metric connection. Let $p \in M$, $\pi \subset T_pM$ be a two-plane section at p, and $\{e_1, e_2\}$ be an orthonormal basis of π.

Since $R(X,Y,Z,W) \neq R(X,Y,W,Z)$, we cannot define the sectional curvature by the standard definition.

The sectional curvature $K(\pi)$, with respect to the induced connection ∇, is defined as follows [27]:

$$K(\pi) = \frac{1}{2}[R(e_1,e_2,e_2,e_1) + R(e_2,e_1,e_1,e_2)]. \tag{7}$$

For any orthonormal basis $\{e_1, \cdots, e_n\}$ of the tangent space T_pM, the scalar curvature τ at p with respect to the semi-symmetric non-metric connection is given by

$$\tau(p) = \sum_{1 \leq i < j \leq n} K_{ij}, \tag{8}$$

where K_{ij} is the sectional curvature of the two-plane section spanned by e_i and e_j.

In particular, if M is a submanifold tangent to ξ of a Lorentzian–Sasakian manifold equipped with a semi-symmetric non-metric connection, for an orthonormal basis $\{e_1, e_2, \cdots, e_n, e_{n+1} = \xi\}$ of T_pM, the scalar curvature τ of M at p, from (8), becomes

$$2\tau = \sum_{1 \leq i \neq j \leq n}^{n} K(e_i \wedge e_j) + 2\sum_{i=1}^{n} K(e_i \wedge \xi). \tag{9}$$

The normalized scalar curvature ρ is given by

$$\rho = \frac{2\tau}{n(n+1)}. \tag{10}$$

The mean curvature vector $H(p)$ at $p \in M$ is given by

$$H(p) = \frac{1}{n+1}\sum_{i=1}^{n+1} h(e_i, e_i). \tag{11}$$

We denote

$$||P||^2 = \sum_{i,j=1}^{n} g^2(e_i, Pe_j), \qquad ||F||^2 = \sum_{i=1}^{n} ||Fe_i||^2, \tag{12}$$

where $||P||^2$ and $||F||^2$ are independent of the choice of the above orthonormal basis.

Denote by $h_{ij}^r = g(h(e_i,e_j),e_r)$, $i,j = 1,...,n+1$, $r \in \{n+2,...,2m+1\}$ the components of the second fundamental form; then, the square norm of h is given by

$$||h||^2 = \sum_{r=n+2}^{2m+1}\sum_{i,j=1}^{n+1}\left(h_{ij}^r\right)^2.$$

Lemma 1. *Let $\tilde{M}$ be a Lorentzian–Sasakian manifold and M a submanifold tangent to ξ. Then,*

(i) $h(\xi,\xi) = 0$;
(ii) $h(X,\xi) = -FX$, *for any vector field X on M.*

Proof. Let $p \in M$ and $X \in T_pM$. Then, $\overset{\circ}{\tilde{\nabla}}_X \xi = -\phi X = -PX - FX$.
On the other hand, by the Gauss formula,

$$\overset{\circ}{\tilde{\nabla}}_X \xi = \overset{\circ}{\nabla}_X \xi + h(X,\xi).$$

Then, one has $h(X,\xi) = -FX$.
For $X = \xi$, we obtain (i), and for X orthogonal to ξ, one obtains (ii). □

In [28], A. Lotta defined the following class of submanifolds. Let $\tilde{M}$ be an almost contact metric manifold. A submanifold M tangent to ξ is called a contact slant submanifold if, for all $p \in M$ and $X \in T_pM$ linearly independent of ξ, ϕX makes a constant angle θ with T_pM. In this case, $\theta \in [0, \pi/2]$ is called the slant angle of M. As particular cases, we mention the invariant submanifolds ($\theta = 0$) and the anti-invariant submanifolds ($\theta = \pi/2$).
A contact slant submanifold is said to be proper if it is neither invariant nor anti-invariant.
A proper contact slant submanifold is said to be a special contact slant submanifold if

$$(\overset{\circ}{\nabla}_X P)Y = -\cos^2\theta[g(X,Y)\xi + \eta(Y)X],$$

for all vector fields X and Y on M.
This represents the contact analogue of a Kählerian slant submanifold defined by B.Y. Chen in [29]. Such a manifold is Kählerian with respect to a suitable complex structure.
On a special contact slant submanifold, $A_{FX}Y = A_{FY}X$, for all space-like vector fields X and Y (see [9]).

Definition 1. *A submanifold M is*
(a) *totally geodesic if the second fundamental form vanishes identically;*
(b) *totally umbilical if* $h(X,Y) = g(X,Y)H$, *for any tangent vectors X and Y on M;*
(c) *called minimal if* $H = 0$.

A point p in an n-dimensional submanifold M of an m-dimensional Riemannian manifold $\tilde{M}$ is said to be an invariantly quasi-umbilical point if there are $m-n$ mutually orthogonal unit vectors $\xi_{n+1}, \cdots, \xi_m$ normal to M such that the shape operators with respect to all directions ξ_α have an eigenvalue of multiplicity $n-1$, and for each ξ_γ, the distinguished eigenvalue direction is the same. The submanifold M is an invariantly quasi-umbilical submanifold if all its points are invariantly quasi-umbilical points.

3. Example

Our goal is to find interesting examples of special contact slant submanifolds in Lorentzian–Sasakian space forms.
First, we state the following result:

Theorem 2. *Let S be a proper slant surface of* $\mathbb{C}^2$, *defined by the equation*

$$x(u,v) = (f_1(u,v), f_2(u,v), f_3(u,v), f_4(u,v)), \tag{13}$$

with $\partial/\partial x$ *and* $\partial/\partial y$ *non-null and orthogonal. Then,*

$$y(u,v,t) = 2(f_1(u,v), f_2(u,v), f_3(u,v), f_4(u,v), t) \tag{14}$$

defines a 3-dimensional contact slant submanifold M of $(\mathbb{R}^5, \phi_0, \eta, \xi, g)$*, such that if we put*

$$e_1 = \frac{\partial}{\partial u} + (2f_3\frac{\partial f_1}{\partial u} + 2f_4\frac{\partial f_2}{\partial u})\frac{\partial}{\partial t},$$
$$e_2 = \frac{\partial}{\partial v} + (2f_3\frac{\partial f_1}{\partial v} + 2f_4\frac{\partial f_2}{\partial v})\frac{\partial}{\partial t},$$

then $\{e_1, e_2, \xi\}$ *is an orthonormal frame on M.*

We consider on $\mathbb{R}^{2m+1}$ the following Lorentzian–Sasakian structure $(\mathbb{R}^{2m+1}, \phi_0, \eta, \xi, g)$, given by

$$\eta = \frac{1}{2}\left(dz - \sum_{i=1}^{m} y^i dx^i\right), \quad \xi = 2\frac{\partial}{\partial z},$$
$$g = -\eta \otimes \eta + \frac{1}{4}\sum_{i=1}^{m}(dx^i \otimes dx^i + dy^i \otimes dy^i),$$

$$\phi_0\left(\sum_{i=1}^{m}\left(X_i\frac{\partial}{\partial x^i} + Y_i\frac{\partial}{\partial y^i}\right) + Z\frac{\partial}{\partial z}\right) = \sum_{i=1}^{m}\left(Y_i\frac{\partial}{\partial x^i} - X_i\frac{\partial}{\partial y^i}\right) + \sum_{i=1}^{m} Y_i y^i\frac{\partial}{\partial z},$$

where $\{x^i, y^i, z\}, i = 1, \cdots, m$, are the Cartesian coordinates.

Example 1. *The equation* $x(u,v,t) = 2(u+v, k\cos v, v-u, k\sin v, t)$ *defines a special contact slant submanifold with slant angle* $\cos^{-1}\sqrt{2+k^2}$.

Proof. We have

$$e_1 = \frac{1}{\sqrt{2}}(1, 0, -1, 0, 0), \tag{15}$$

$$e_2 = \frac{1}{\sqrt{k^2+2}}(1, -k\sin v, 1, k\cos v, 0). \tag{16}$$

We compute the slant angle

$$\cos\theta = g(\phi_0 e_2, e_1) = -g(e_2, \phi_0 e_1) = \sqrt{\frac{2}{2+k^2}}$$

Denote $e_i^* = \frac{Fe_i}{\|Fe_i\|}, i - 1, 2$. By Lemma 1, $h(e_1, \xi) = -\sin\theta e_2^*$ and $h(e_2, \xi) = 0$. A straightforward calculation gives $h(e_1, e_2) = -\sin\theta e_1^*$. Then, the matrix of the second fundamental form is

$$[h(e_i, e_j)] = \begin{pmatrix} 0 & -\sin\theta e_1^* & -\sin\theta e_2^* \\ -\sin\theta e_1^* & \frac{1}{2k^2+8}[2(0, -k\cos v, 0, -k\sin v, 0)] & 0 \\ -\sin\theta e_2^* & 0 & 0 \end{pmatrix}.$$

□

4. A Chen Inequality for the Scalar Curvature

Our main interest was motivated by certain inequalities obtained by B.Y. Chen.

Regarding the mean curvature H and the scalar curvature τ, B.Y. Chen [30] proved that, on an n-dimensional submanifold M in a Riemannian space form of constant sectional curvature c,

$$\|H\|^2 \geq \frac{2\tau}{n(n-1)} - c.$$

The equality holds identically if and only if M is totally umbilical.

In this section, we give an improved Chen inequality for the scalar curvature of a special contact slant submanifold in a Lorentzian–Sasakian space form equipped with a semi-symmetric non-metric connection.

Recall that an $(n+1)$-dimensional submanifold M tangent to ξ of a $(2n+1)$-dimensional Lorentzian–Sasakian manifold $\tilde{M}$ is a contact H-umbilical submanifold if the second fundamental form has the form

$$\begin{aligned} &h(e_1,e_1) = \alpha e_1^*, h(e_2,e_2) = \ldots = h(e_n,e_n) = \beta e_1^*, \\ &h(e_1,e_j) = \beta e_j^*, h(e_j,e_k) = 0, j \neq k, j,k = 2,..n, \end{aligned} \tag{17}$$

for some smooth functions α and β with respect to some suitable orthonormal local frame fields $\{e_1, ..., e_n, e_{n+1} = \xi\}$ on TM and $\{e_1^*, ..., e_n^*\}$ on $T^\perp M$.

Theorem 3. *Let $\tilde{M}(c)$ be a Lorentzian–Sasakian space form of dimension $2n+1$ $(n \geq 2)$ equipped with a semi-symmetric non-metric connection $\tilde{\nabla}$ and M an $(n+1)$-dimensional special contact slant submanifold. Then,*

$$\begin{aligned} \|H\|^2 \geq & \frac{2(n+2)}{(n-1)(n+1)^2}\tau - \frac{n(n+2)}{(n+1)^2}\frac{c-3}{4} \\ & - 3\frac{n^2(n+2)}{(n-1)(n+1)^2}\frac{c+1}{4}\cos^2\theta \\ & + \frac{(n+2)(n-2)}{(n-1)(n+1)^2}\lambda - \frac{n(n+2)}{(n-1)(n+1)^2}s(\xi,\xi) + \frac{(n+2)(n-2)}{n^2-1}\omega(H) \\ & - \frac{2n(n+2)}{(n-1)(n+1)^2}(1+\sin^2\theta)\}, \end{aligned} \tag{18}$$

where $\lambda = \sum_{j=1}^n s(e_j, e_j)$.

Moreover, the equality sign holds identically if and only if M is a contact H-umbilical submanifold where $\alpha = 3\beta$.

Proof. Let M be an $(n+1)$-dimensional special contact slant submanifold of a $(2n+1)$-dimensional Lorentzian–Sasakian space form $\tilde{M}(c)$ equipped with a semi-symmetric non-metric connection $\tilde{\nabla}$. Let $p \in M$ and $\{e_1, \cdots, e_n, e_{n+1} = \xi\}$ be an orthonormal basis of T_pM. An orthonormal basis of $T_p^\perp M$ is given by $\{e_1^*, \cdots, e_n^*\}$, where $e_k^* = \frac{1}{\sin\theta}Fe_k$, $k = 1, \cdots, n$.

We use the formula

$$2\tau = \sum_{1\leq i\neq j\leq n} K(e_i \wedge e_j) + 2\sum_{i=1}^n K(e_i \wedge \xi). \tag{19}$$

If we put $X = W = e_i$ and $Y = Z = e_j$, $i \neq j$, $i,j = 1, \cdots, n$ in the Gauss equation, then

$$\begin{aligned} R(e_i,e_j,e_j,e_i) =& \tilde{R}(e_i,e_j,e_j,e_i) - g(h(e_i,e_j),h(e_i,e_j)) \\ & + g(h(e_i,e_i),h(e_j,e_j)) - g(P^\perp, h(e_j,e_j)). \end{aligned}$$

It follows that the scalar curvature is given by

$$\begin{aligned} 2\tau(p) =& \sum_{1\leq i\neq j\leq n} \tilde{R}(e_i,e_j,e_j,e_i) + \sum_{1\leq i\neq j\leq n} [g(h(e_i,e_i),h(e_j,e_j)) - g(h(e_i,e_j),h(e_i,e_j))] \\ & - \sum_{1\leq i\neq j\leq n} g(P^\perp, h(e_j,e_j)) + 2\sum_{j=1}^n K(\xi \wedge e_j). \end{aligned} \tag{20}$$

We calculate $\tilde{R}(e_i, e_j, e_j, e_i)$ using Formula (5) of the curvature tensor, for $i, j = 1, \cdots n, i \neq j$.

$$\begin{aligned}\tilde{R}(e_i, e_j, e_j, e_i) =& \frac{c-3}{4}\{g(e_j, e_j)g(e_i, e_i) - g(e_i, e_j)g(e_j, e_i)\} \\ &+ \frac{c+1}{4}[g(e_i, \phi e_j)g(\phi e_j, e_i) - g(e_j, \phi e_j)g(\phi e_i, e_i) \\ &+ 2g(e_i, \phi e_j)g(\phi e_j, e_i) \\ &+ \eta(e_j)\eta(e_j)g(e_i, e_i) - \eta(e_i)\eta(e_j)g(e_j, e_i) \\ &+ g(e_i, e_j)\eta(e_j)\eta(e_i) - g(e_j, e_j)\eta(e_i)\eta(e_i)] \\ &+ s(e_i, e_j)g(e_j, e_i) - s(e_j, e_j)g(e_i, e_i). \end{aligned} \tag{21}$$

A straightforward calculation implies

$$\tilde{R}(e_i, e_j, e_j, e_i) = \frac{c-3}{4} + 3\frac{c+1}{4}g^2(\phi e_i, e_j) - s(e_j, e_j). \tag{22}$$

We denote $\lambda = \sum_{i=1}^{n} s(e_i, e_i)$.
Substituting (22) in (20), we obtain

$$\begin{aligned}2\tau(p) = n(n-1)\frac{c-3}{4} + 3\frac{c+1}{4}\sum_{i\neq j} g^2(\phi e_i, e_j) - (n-1)\lambda \\ - \|h\|^2 + (n+1)^2\|H\|^2 - (n+1)(n-1)\omega(H) + 2\sum_{j=1}^{n} K(\xi \wedge e_j). \end{aligned} \tag{23}$$

From the definition of the sectional curvature and because ξ is time-like, we have

$$K(\xi \wedge e_j) = -\frac{1}{2}[R(\xi, e_j, e_j, \xi) + R(e_j, \xi, \xi, e_j)]. \tag{24}$$

If we put $X = W = \xi$ and $Y = Z = e_j$, for $j = 1, \cdots n$ in the Gauss equation, then we obtain

$$\begin{aligned}R(\xi, e_j, e_j, \xi) = \tilde{R}(\xi, e_j, e_j, \xi) \\ + g(h(\xi, \xi), h(e_j, e_j)) - g(h(\xi, e_j), h(\xi, e_j)) \\ - g(P^{\perp}, h(e_j, e_j))g(\xi, \xi). \end{aligned} \tag{25}$$

We calculate $\tilde{R}(\xi, e_j, e_j, \xi)$ using Formula (5) of the curvature tensor.

$$\tilde{R}(\xi, e_j, e_j, \xi) = -1 - s(e_j, e_j). \tag{26}$$

By using Equation (26) and Lemma 1, we have $h(\xi, \xi) = 0$ and $h(\xi, e_j) = -Fe_j$. Obviously, $\|Fe_j\| = \sin\theta$. Then, Equation (25) becomes

$$R(\xi, e_j, e_j, \xi) = -1 - s(e_j, e_j) - \sin^2\theta + g(P^{\perp}, h(e_j, e_j)). \tag{27}$$

Similarly, we have

$$R(e_j, \xi, \xi, e_j) = -1 - s(\xi, \xi) - \sin^2\theta. \tag{28}$$

We substitute (27) and (28) in (24); taking the summation, we obtain

$$\sum_{j=1}^{n} K(\xi \wedge e_j) = n(1 + \sin^2\theta) + \frac{\lambda}{2} + \frac{n}{2}s(\xi, \xi) + \frac{1}{2}(n+1)\omega(H). \tag{29}$$

If we put (29) in (23), we obtain

$$
\begin{aligned}
2\tau(p) &= n(n-1)\frac{c-3}{4} + 3\frac{c+1}{4}\sum_{1\le i\ne j\le n} g^2(\phi e_i, e_j) \\
&- (n-2)\lambda + ns(\xi,\xi) - (n+1)(n-2)\omega(H) + 2n(1+\sin^2\theta) \\
&- \|h\|^2 + (n+1)^2\|H\|^2. \qquad (30)
\end{aligned}
$$

We denote by $h_{ij}^k = g(h(e_i, e_j), e_k^*)$, $1 \le i, j, k \le n$.

Since M is a special contact slant submanifold in $\tilde{M}(c)$, we know from [9] that the shape operator satisfies $A_{FX}Y = A_{FY}X$, for any space-like vector fields $X, Y \in \Gamma(TM)$.

The above relation implies the symmetry of the coefficients h_{ij}^k of the second fundamental form, i.e.,

$$h_{ij}^k = h_{jk}^i = h_{ik}^j, \ \forall i, j, k \in \{1, ..., n\}. \qquad (31)$$

By the definition of the mean curvature, we obtain

$$
\begin{aligned}
(n+1)^2\|H\|^2 &= \sum_i g(h(e_i,e_i), h(e_i,e_i)) + \sum_{i\ne j} g(h(e_i,e_i), h(e_j,e_j)) \\
&= \sum_{i=1}^n [\sum_{j=1}^n (h_{jj}^i)^2 + 2\sum_{1\le j<k\le n} h_{jj}^i h_{kk}^i]. \qquad (32)
\end{aligned}
$$

Also, $\|h\|^2 = \sum_{i,j,k=1}^n (h_{ij}^k)^2$.

From Equations (30)–(32),

$$
\begin{aligned}
2\tau(p) &= n(n-1)\frac{c-3}{4} + 3\frac{c+1}{4}\sum_{i\ne j} g^2(\phi e_i, e_j) \\
&- (n-2)\lambda + ns(\xi,\xi) - (n+1)(n-2)\omega(H) + 2n(1+\sin^2\theta) \\
&+ 2\sum_i^n \sum_{1\le j<k\le n} h_{jj}^i h_{kk}^i - 2\sum_{1\le i\ne j\le n} (h_{jj}^i)^2 - 6\sum_{1\le i<j<k\le n} (h_{ij}^k)^2. \qquad (33)
\end{aligned}
$$

Let us introduce a parameter, $m = \frac{n+2}{n-1}$, where $n \ge 2$. For studying the inequality of $\|H\|^2$, we follow the technique used in [31]. Then, we obtain

$$
\begin{aligned}
&(n+1)^2\|H\|^2 - m\{2\tau(p) - n(n-1)\frac{c-3}{4} - 3\frac{c+1}{4}\sum_{1\le i\ne j\le n} g^2(\phi e_i, e_j) \\
&+ (n-2)\lambda - ns(\xi,\xi) + (n+1)(n-2)\omega(H) - 2n(1+\sin^2\theta)\} \\
&= \sum_{i=1}^n (h_{ii}^i)^2 + (1+2m)\sum_{1\le i\ne j\le n} (h_{jj}^i)^2 + 6m\sum_{1\le i<j<k\le n} (h_{ij}^k)^2 \\
&- 2(m-1)\sum_i^n \sum_{1\le j<k\le n} h_{jj}^i h_{kk}^i \\
&= \sum_{i=1}^n (h_{ii}^i)^2 + 6m\sum_{1\le i<j<k\le n} (h_{ij}^k)^2 + (m-1)\sum_{i=1}^n \sum_{1\le j<k\le n} (h_{jj}^i - h_{kk}^i)^2 \\
&+ [1+2m-(n-2)(m-1)]\sum_{1\le i\ne j\le n} (h_{jj}^i)^2 - 2(m-1)\sum_{1\le i\ne j\le n} h_{ii}^i h_{jj}^i \\
&= 6m\sum_{1\le i<j<k\le n} (h_{ij}^k)^2 + (m-1)\sum_{i\ne j,k}\sum_{1\le j<k\le n} (h_{jj}^i - h_{kk}^i)^2 \\
&+ \frac{1}{n-1}\sum_{1\le i\ne j\le n} \left[h_{ii}^i - (n-1)(m-1)h_{jj}^i\right]^2 \ge 0. \qquad (34)
\end{aligned}
$$

For contact slant submanifolds, we have $||P||^2 = n\cos^2\theta$; then, we obtain

$$\begin{aligned}||H||^2 \geq & \frac{2(n+2)}{(n-1)(n+1)^2}\tau - \frac{n(n+2)}{(n+1)^2}\frac{c-3}{4} \\ & -3\frac{n^2(n+2)}{(n-1)(n+1)^2}\frac{c+1}{4}\cos^2\theta \\ & +\frac{(n+2)(n-2)}{(n-1)(n+1)^2}\lambda - \frac{n(n+2)}{(n-1)(n+1)^2}s(\xi,\xi) + \frac{(n+2)(n-2)}{n^2-1}\omega(H) \\ & -\frac{2n(n+2)}{(n-1)(n+1)^2}(1+\sin^2\theta)\}. \end{aligned} \tag{35}$$

The equality holds if and only if
$h^k_{ij} = 0$, for all $1 \leq i < j < k \leq n$;
$h^i_{ii} = 3h^i_{jj}$, for all $1 \leq i \neq j \leq n$.

We may choose $e_1, ..., e_n$ in such a way that e_1^* is parallel to H; then, $h^j_{kk} = 0$, for $j > 1$, $k = 1, ..., n$.

It follows that M is a contact H-umbilical submanifold where $\alpha = 3\beta$. □

5. Inequalities for the Casorati Curvatures

Let $\tilde{M}$ be an m-dimensional Lorentzian manifold and M an n-dimensional submanifold. The Casorati curvature $\mathcal{C}$ of M is expressed by

$$\mathcal{C} = \frac{1}{n}\sum_{r=n+1}^{m}\sum_{i,j=1}^{n}\left(h^r_{ij}\right)^2 = \frac{1}{n}||h||^2. \tag{36}$$

For any subspace L in T_pM of dimension $r \geqslant 2$ and $\{e_1, e_2, ..., e_r\}$ an orthonormal basis of L, its scalar curvature $\tau(L)$ is

$$\tau(L) = \sum_{1\leq\alpha<\beta\leq r} K(e_\alpha \wedge e_\beta). \tag{37}$$

The Casorati curvature $\mathcal{C}(L)$ of L is given by

$$\mathcal{C}(L) = \frac{1}{r}\sum_{\gamma=n+1}^{m}\sum_{i,j=1}^{r}\left(h^\gamma_{ij}\right)^2. \tag{38}$$

For n-dimensional submanifolds, S. Decu, S. Haesen, and L. Verstraelen introduced in [13] the notion of normalized Casorati δ-curvatures $\delta_c(n-1)$ and $\delta_c^*(n-1)$ as follows:

$$[\delta_c(n-1)]_p = \frac{1}{2}\mathcal{C}_p + \frac{n+1}{2n}\inf\{\mathcal{C}(L)|L \ \ \texttt{a hyperplane of} \ \ T_pM\},$$

$$[\delta_c^*(n-1)]_p = 2\mathcal{C}_p - \frac{2n-1}{2n}\sup\{\mathcal{C}(L)|L \ \ \texttt{a hyperplane of} \ \ T_pM\}.$$

The same authors [13] proved that, on an n-dimensional submanifold in an m-dimensional real space form with constant sectional curvature c,

$$\rho \leq \delta_c(n-1) + c, \quad (\text{resp. } \rho \leq \delta_c^*(n-1) + c), \tag{39}$$

where ρ denotes the normalized scalar curvature.

Moreover, the equality in (39) holds identically if and only if there exists an orthonormal frame $e_1, e_2, \cdots, e_n, e_{n+1}, \cdots, e_m$ such that, with respect to this frame, the shape operator is $A_{n+1} = \begin{pmatrix} \lambda I_{n-1} & 0 \\ 0 & 2\lambda \end{pmatrix}$, and $A_{n+2} = \cdots = A_m = 0$.

We obtain inequalities for the Casorati curvatures of contact slant submanifolds in a Lorentzian–Sasakian manifold equipped with a semi-symmetric non-metric connection.

The normalized Casorati curvatures $\delta_c(n)$ and $\delta_c^*(n)$ are defined as (see [32])

$$[\delta_c(n)]_p = \frac{1}{2}\mathcal{C}_p + \frac{n+2}{2(n+1)}\inf\{\mathcal{C}(L)|L \quad \texttt{a hyperplane of} \quad T_pM\},$$

$$[\delta_c^*(n)]_p = 2\mathcal{C}_p - \frac{2n+1}{2(n+1)}\sup\{\mathcal{C}(L)|L \quad \texttt{a hyperplane of} \quad T_pM\}.$$

Theorem 4. *Let M be an $(n+1)$-dimensional, $n+1 \geq 3$, contact slant submanifold in a $(2n+1)$-dimensional Lorentzian–Sasakian space form $\tilde{M}(c)$ endowed with a semi-symmetric non-metric connection $\tilde{\nabla}$. Then, the following result:*

(i) The normalized δ-curvature $\delta_c(n)$ satisfies

$$\begin{aligned}\rho \leqslant{}& \delta_c(n) + \frac{n-1}{n+1}\frac{c-3}{4} \\ &+ 3\cos^2\theta\frac{c+1}{4(n+1)} - \frac{n-2}{n(n+1)}\lambda + \frac{1}{n+1}s(\xi,\xi) \\ &- \frac{n-2}{n}\omega(H) + \frac{2}{n+1}(1+\sin^2\theta).\end{aligned} \tag{40}$$

(ii) The normalized δ-curvature $\delta_c^*(n)$ satisfies

$$\begin{aligned}\rho \leqslant{}& \delta_c^*(n) + \frac{n-1}{n+1}\frac{c-3}{4} + 3\cos^2\theta\frac{c+1}{4(n+1)} - \frac{n-2}{n(n+1)}\lambda + \frac{1}{n+1}s(\xi,\xi) \\ &- \frac{n-2}{n}\omega(H) + \frac{2}{n+1}(1+\sin^2\theta).\end{aligned} \tag{41}$$

Moreover, the equality holds in one of the equations, Equation (40) or (41), if and only if M is an invariantly quasi-umbilical submanifold with flat normal connection in $\tilde{M}(c)$.

Proof. Let $\tilde{M}(c)$ be a $(2n+1)$-dimensional Lorentzian–Sasakian space form equipped with a semi-symmetric non-metric connection and M an $(n+1)$-dimensional submanifold of $\tilde{M}(c)$. For any $p \in M$, let $\{e_1, \cdots, e_{n+1}\}$ be an orthonormal basis of T_pM and $\{e_{n+2}, \cdots, e_{2n+1}\}$ an orthonormal basis of $T_p^{\perp}M$.

Equation (30) can be rewritten as

$$\begin{aligned}2\tau(p) ={}&(n+1)^2||H||^2 - (n+1)\mathcal{C} + n(n-1)\frac{c-3}{4} \\ &+3||P||^2\frac{c+1}{4} - (n-2)\lambda + ns(\xi,\xi) - (n+1)(n-2)\omega(H) + 2n(1+\sin^2\theta),\end{aligned} \tag{42}$$

where $||P||^2 = n\cos^2\theta$.

We use the same technique as in the paper [32].

We define the quadratic polynomial $\mathcal{Q}$ in terms of the second fundamental form

$$\begin{aligned}\mathcal{Q} ={}&\frac{1}{2}n(n+1)\mathcal{C} + \frac{1}{2}(n+2)\mathcal{C}(L) - 2\tau(p) + n(n-1)\frac{c-3}{4} \\ &+ 3||P||^2\frac{c+1}{4} - (n-2)\lambda + ns(\xi,\xi) - (n+1)(n-2)\omega(H) + 2n(1+\sin^2\theta),\end{aligned} \tag{43}$$

where L is a hyperplane of T_pM. Without loss of generality, we may assume that L is spanned by $e_1, \cdots, e_n$; then,

$$\mathcal{Q} = \frac{n}{2}\sum_{r=n+2}^{2n+1}\sum_{i,j=1}^{n+1}(h_{ij}^r)^2 + \frac{n+1}{2n}\sum_{r=n+2}^{2n+1}\sum_{i,j=1}^{n}(h_{ij}^r)^2 - 2\tau(p)$$
$$+ n(n-1)\frac{c-3}{4} + 3\|P\|^2\frac{c+1}{4} - (n-2)\lambda + ns(\xi,\xi)$$
$$- (n+1)(n-2)\omega(H) + 2n(1+\sin^2\theta). \tag{44}$$

From (42) and (44), we obtain

$$\mathcal{Q} = \frac{n+2}{2}\sum_{r=n+2}^{2n+1}\sum_{i,j=1}^{n+1}(h_{ij}^r)^2 + \frac{n+1}{2n}\sum_{r=n+2}^{2n+1}\sum_{i,j=1}^{n}(h_{ij}^r)^2 - \sum_{r=n+2}^{2n+1}(\sum_{i,j=1}^{n}h_{ij}^r)^2,$$

from which we derive that

$$\mathcal{Q} = \sum_{n+2}^{2n+1}\sum_{i=1}^{n}\left[\frac{n^2+n+2}{2n}(h_{ii}^r)^2 + (n+2)(h_{in}^r)^2\right]$$
$$+ \sum_{n+2}^{2n+1}\left[\frac{(n+1)(n+2)}{n}\sum_{i<j=1}^{n}(h_{ij}^r)^2 - 2\sum_{i<j=1}^{n}h_{ii}^r h_{jj}^r + \frac{n}{2}(h_{nn}^r)^2\right]. \tag{45}$$

The critical points

$$h^c = (h_{11}^{n+2}, h_{12}^{n+2}, \cdots, h_{n+1n+1}^{n+2}, \cdots, h_{11}^{2n+1}, h_{12}^{2n+1}, \cdots, h_{n+1n+1}^{2n+1}) \tag{46}$$

of $\mathcal{Q}$ are the solutions of the following system of linear homogeneous equations:

$$\begin{cases} \frac{\partial \mathcal{Q}}{\partial h_{ii}^r} = \frac{(n+1)(n+2)}{n}h_{ii}^r - 2\sum_{k=1}^n h_{kk}^r = 0 \\ \frac{\partial \mathcal{Q}}{\partial h_{n+1n+1}^r} = nh_{n+1n+1}^r - 2\sum_{k=1}^n h_{kk}^r = 0 \\ \frac{\partial \mathcal{Q}}{\partial h_{ij}^r} = \frac{2(n+1)(n+2)}{n}h_{ij}^r = 0 \\ \frac{\partial \mathcal{Q}}{\partial h_{in+1}^r} = 2(n+2)h_{in}^r = 0 \end{cases} \tag{47}$$

where $i,j \in \{1,2,\cdots,n\}$, $i \neq j$, and $r \in \{n+2,\cdots,2n+1\}$.

Therefore, every solution h^c has $h_{ij}^r = 0$, for $i \neq j$, and the determinant which corresponds to the first two sets of equations of the above system is zero (there exists a solution for non-totally geodesic submanifolds). Moreover, the Hessian matrix of $H(\mathcal{Q})$ has the following eigenvalues:

$$\lambda_{11} = 0, \qquad \lambda_{22} = \frac{2n^2-n+2}{n},$$
$$\lambda_{33} = \cdots = \lambda_{n+1n+1} = \frac{(n+1)(n+2)}{n},$$
$$\lambda_{ij} = \frac{2(n+1)(n+2)}{n}, \qquad \lambda_{in} = 2(n+2), \quad \forall i,j \in \{1,\cdots,n\}, i \neq j.$$

Thus, we know that $\mathcal{Q}$ is parabolic and reaches the minimum $\mathcal{Q}(h^c) = 0$, for each solution h^c of the system (47). It follows that $\mathcal{Q} \geq 0$, which implies that

$$
\begin{aligned}
2\tau(p) \leqslant & \frac{1}{2}n(n+1)\mathcal{C} + \frac{1}{2}(n+2)\mathcal{C}(L) + n(n-1)\frac{c-3}{4} \\
& + 3\|P\|^2\frac{c+1}{4} - (n-2)\lambda + ns(\xi,\xi) \\
& - (n+1)(n-2)\omega(H) + 2n(1+\sin^2\theta).
\end{aligned} \tag{48}
$$

We divide all terms by $n(n+1)$. Then,

$$
\begin{aligned}
\rho \leqslant & \frac{1}{2}\mathcal{C} + \frac{1}{2}\frac{(n+2)}{n(n+1)}\mathcal{C}(L) + \frac{n-1}{n+1}\frac{c-3}{4} \\
& + 3\cos^2\theta\frac{c+1}{4(n+1)} - \frac{n-2}{n(n+1)}\lambda + \frac{1}{n+1}s(\xi,\xi) \\
& - \frac{n-2}{n}\omega(H) + \frac{2}{n+1}(1+\sin^2\theta),
\end{aligned}
$$

for every hyperplane L tangent to T_pM.

Taking the infimum over all tangent hyperplanes L, we find

$$
\begin{aligned}
\rho \leqslant & \delta_c(n) + \frac{n-1}{n+1}\frac{c-3}{4} \\
& + 3\cos^2\theta\frac{c+1}{4(n+1)} - \frac{n-2}{n(n+1)}\lambda + \frac{1}{n+1}s(\xi,\xi) \\
& - \frac{n-2}{n}\omega(H) + \frac{2}{n+1}(1+\sin^2\theta),
\end{aligned}
$$

which is the result.

We remark that the equality sign holds if and only if

$$
h^r_{ij} = 0, \quad \forall i,j \in \{1,\cdots,n+1\}, i \neq j,\ r \in \{n+2,\cdots,2n+1\}, \tag{49}
$$

$$
h_{n+1n+1} = 2h^r_{11} = \cdots = 2h^r_{nn}, \quad \forall r \in \{n+2,\cdots,2n+1\}. \tag{50}
$$

From Equations (49) and (50), we conclude that the equality sign holds in the inequality (40) if and only if the submanifold M is invariantly quasi-umbilical with a trivial normal connection in $\tilde{M}(c)$.

Similarly to the first part, we can establish the inequality for the second part of the theorem. □

6. An Improved Chen First Inequality for Legendrian Submanifolds in a Lorentzian–Sasakian Space Form Endowed with a Semi-Symmetric Non-Metric Connection

An n-dimensional submanifold M normal to ξ in a $(2n+1)$-dimensional Lorentzian–Sasakian manifold is called a Legendrian submanifold.

For Legendrian submanifolds in a Lorentzian–Sasakian space form $\tilde{M}(c)$ of dimension $2n+1$, admitting a semi-symmetric non-metric connection, we prove a Chen inequality.

Theorem 5. *Let M be an n-dimensional ($n \geq 3$) Legendrian submanifold isometrically immersed in a Lorentzian–Sasakian space form $\tilde{M}(c)$, endowed with a semi-symmetric non-metric connection $\tilde{\nabla}$, and $p \in M$, $\pi \subset T_pM$ a two-plane section. Then, we have*

$$
\begin{aligned}
\tau(p) - K(\pi) \leq & \frac{n^2(2n-3)}{2(2n+3)}\|H\|^2 + \frac{(n-2)(n+1)}{2}\frac{c-3}{4} - \frac{n(n-1)}{2}\omega(H) \\
& - \frac{(n-1)}{2}\lambda + \frac{1}{2}\mathtt{trace}(\mathtt{s}_{|\mathfrak{B}}) + \frac{1}{2}\mathtt{g}(\mathtt{trace}(\mathtt{h}_{|\mathfrak{B}}),\mathtt{P}).
\end{aligned} \tag{51}
$$

Moreover, the equality case of the inequality holds for some two-plane section π at a point $p \in M$ if and only if there exists an orthonormal basis $\{e_1, \cdots, e_n\}$ of the tangent space T_pM at p such that $\pi = \mathtt{span}\{e_1, e_2\}$, and with respect to this basis, the second fundamental form takes the following form:

$$h(e_1,e_1) = a\phi e_1 + 3\lambda\phi e_3, \quad h(e_1,e_3) = 3\lambda\phi e_1, \quad h(e_3,e_j) = 4\lambda\phi e_j.$$
$$h(e_2,e_2) = -a\phi e_1 + 3\lambda\phi e_3, \quad h(e_2,e_3) = 3\lambda\phi e_2, \quad h(e_j,e_k) = 4\lambda\phi e_3\delta_{jk}.$$
$$h(e_1,e_2) = a\phi e_2, \quad h(e_3,e_3) = 12\lambda\phi e_3, \quad h(e_1,e_k) = h(e_2,e_j) = 0.$$

for some numbers a, λ, where $j, k = 4, \cdots, n$.

Proof. Let $p \in M$ and $\pi \subset T_pM$ be a two-plane section and $\{e_1, \cdots, e_n\}$ be an orthonormal basis of the tangent space T_pM at p such that $e_1, e_2 \in \pi$.

Since M is a Legendrian submanifold, we can choose an orthonormal basis $\{e_1^* = e_{n+1} = \phi e_1, e_2^* = e_{n+2} = \phi e_2, \cdots, e_n^* = e_{2n} = \phi e_n, e_{2n+1}^* = \xi\}$.

For the calculation of $\tilde{R}(e_i, e_j, e_j, e_i)$, we use Formula (5), where we put $X = W = e_i$ and $Y = Z = e_j$. $i, j = 1, \cdots n, i \neq j$.

$$\begin{aligned}\tilde{R}(e_i,e_j,e_j,e_i) = {} & \frac{c-3}{4}\{g(e_j,e_j)g(e_i,e_i) - g(e_i,e_j)g(e_j,e_i)\} \\ & + \frac{c+1}{4}[g(e_i,\phi e_j)g(\phi e_j,e_i) - g(e_j,\phi e_j)g(\phi e_i,e_i) \\ & + 2g(e_i,\phi e_j)g(\phi e_j,e_i) \\ & + \eta(e_j)\eta(e_j)g(e_i,e_i) - \eta(e_i)\eta(e_j)g(e_j,e_i) \\ & + g(e_i,e_j)\eta(e_j)\eta(e_i) - g(e_j,e_j)\eta(e_i)\eta(e_i)] \\ & + s(e_i,e_j)g(e_j,e_i) - s(e_j,e_j)g(e_i,e_i),\end{aligned} \tag{52}$$

which implies that

$$\tilde{R}(e_i,e_j,e_j,e_i) = \frac{c-3}{4} - s(e_j,e_j). \tag{53}$$

If we put $X = W = e_i$ and $Y = Z = e_j$ in the Gauss equation for the semi-symmetric non-metric connection, then

$$\begin{aligned}\tilde{R}(e_i,e_j,e_j,e_i) = {} & R(e_i,e_j,e_j,e_i) + g(h(e_i,e_j),h(e_j,e_i)) \\ & - g(h(e_i,e_i),h(e_j,e_j)) + g(P^\perp,h(e_j,e_j))g(e_i,e_i) \\ & - g(P^\perp,h(e_i,e_j))g(e_j,e_i).\end{aligned} \tag{54}$$

Then, by substituting (53) in (54), we obtain

$$\begin{aligned}\frac{c-3}{4} - s(e_j,e_j) = {} & R(e_i,e_j,e_j,e_i) + g(h(e_i,e_j),h(e_j,e_i)) \\ & - g(h(e_i,e_i),h(e_j,e_j)) + g(P^\perp,h(e_j,e_j))g(e_i,e_i) \\ & - g(P^\perp,h(e_i,e_j))g(e_j,e_i),\end{aligned} \tag{55}$$

By summation over $1 \leq i \neq j \leq n$, from the above equation, we have

$$2\tau + \|h\|^2 - n^2\|H\|^2 = n(n-1)\frac{c-3}{4} - n(n-1)\omega(H) - (n-1)\lambda. \tag{56}$$

Since $\overset{\circ}{\nabla}_X \xi = -\phi X$, for any $X \in \Gamma(TM)$, we have $A_\xi X = 0$, which implies that $g(h(X,Y),\xi) = 0$.

We denote by $h_{ij}^r = g(h(e_i,e_j),e_r^*)$, $i, j, r = 1, ..., n$.

It follows that

$$\tau(p) = \sum_{r=1}^{n} \sum_{1\le i<j\le n} \left[h_{ii}^r h_{jj}^r - (h_{ij}^r)^2\right] + n(n-1)\frac{c-3}{8} - \frac{n(n-1)}{2}\omega(H) - \frac{(n-1)}{2}\lambda. \tag{57}$$

Let $\pi = sp\{e_1, e_2\}$ and by the Gauss formula, if we put $X = W = e_1$ and $Y = Z = e_2$, then

$$R(e_1, e_2, e_2, e_1) = \frac{c-3}{4} - s(e_2, e_2) + \sum_{r=1}^{n} [h_{11}^r h_{22}^r - (h_{12}^r)^2] - g(P^{\perp}, h(e_2, e_2)). \tag{58}$$

Similarly,

$$R(e_1, e_2, e_2, e_1) = \frac{c-3}{4} - s(e_1, e_1) - \sum_{r=1}^{n} [h_{11}^r h_{22}^r - (h_{12}^r)^2] + g(P^{\perp}, h(e_1, e_1)). \tag{59}$$

So, from Equations (7), (58), and (59), we have

$$K(\pi) = \frac{c-3}{4} - \frac{1}{2}\mathtt{trace}(s_{|\pi}) - \frac{1}{2}g(\mathtt{trace}(h_{|\pi}), P)] + \sum_{r=1}^{n} [h_{11}^r h_{22}^r - (h_{12}^r)^2]. \tag{60}$$

Equations (57) and (60) imply

$$\begin{aligned}\tau(p) - K(\pi) &= \sum_{r=1}^{n} \sum_{1\le i<j\le n} [h_{ii}^r h_{jj}^r - (h_{ij}^r)^2 - h_{11}^r h_{22}^r + (h_{12}^r)^2] \\ &+ n(n-1)\frac{c-3}{8} - \frac{n(n-1)}{2}\omega(H) - \frac{(n-1)}{2}\lambda \\ &- \frac{c-3}{4} + \frac{1}{2}\mathtt{trace}(s_{|\pi}) + \frac{1}{2}g(\mathtt{trace}(h_{|\pi}), P).\end{aligned}$$

Then, we obtain

$$\begin{aligned}\tau(p) - K(\pi) &= \sum_{r=1}^{n}\left[\sum_{j=3}^{n}(h_{11}^r + h_{22}^r)h_{jj}^r + \sum_{3\le i<j\le n} h_{ii}^r h_{jj}^r - \sum_{j=3}^{n}[(h_{1j}^r)^2 + (h_{2j}^r)^2] - \sum_{3\le i<j\le n}(h_{ij}^r)^2\right] \\ &+ \frac{(n-2)(n+1)}{2}\frac{c-3}{4} - \frac{n(n-1)}{2}\omega(H) - \frac{n-1}{2}\lambda \\ &+ \frac{1}{2}\mathtt{trace}(s_{|\pi}) + \frac{1}{2}g(\mathtt{trace}(h_{|\pi}), P).\end{aligned} \tag{61}$$

Because M is Legendrian, we have $h_{1j}^1 = h_{11}^j$, $h_{1j}^j = h_{jj}^1$, for $3 \le i \le n$, and $h_{ij}^j = h_{jj}^i$, for $2 \le i \ne j \le n$. Then, Equation (61) becomes

$$\tau(p) - K(\pi) = \sum_{r=1}^{n}\left[\sum_{j=3}^{n}(h^r_{11} + h^r_{22})h^r_{jj} + \sum_{3\le i<j\le n} h^r_{ii}h^r_{jj} - \sum_{j=3}^{n}[(h^j_{11})^2 + (h^1_{jj})^2] - \sum_{3\le i<j\le n}(h^i_{jj})^2\right]$$
$$+ \frac{(n-2)(n+1)}{2}\frac{c-3}{4} - \frac{n(n-1)}{2}\omega(H) - \frac{(n-1)}{2}\lambda$$
$$+ \frac{1}{2}\mathtt{trace}(s_{|\pi}) + \frac{1}{2}g(\mathtt{trace}(h_{|\pi}), P). \tag{62}$$

We will use some ideas from [33] to achieve the proof.

We point out the following inequalities,

$$\sum_{j=3}^{n}(h^r_{11} + h^r_{22})h^r_{jj} + \sum_{3\le i<j\le n} h^r_{ii}h^r_{jj} - \sum_{j=3}^{n}(h^r_{jj})^2$$
$$\le \frac{n-2}{2(n+1)}(h^r_{11} + \cdots + h^r_{nn})^2$$
$$\le \frac{2n-3}{2(2n+3)}(h^r_{11} + \cdots + h^r_{nn})^2, \tag{63}$$

for $r = 1, 2$.

The first inequality in (63) is equivalent to

$$\sum_{j=3}^{n}(h^r_{11} + h^r_{22} - 3h^r_{jj})^2 + 3\sum_{3\le i<j\le n}(h^r_{ii} - h^r_{jj})^2 \ge 0.$$

The equality holds if and only if $3h^r_{jj} = h^r_{11} + h^r_{22}, \forall j = 3, \cdots, n$.

The equality holds in the second inequality if and only if $h^r_{11} + h^r_{22} = 0$ and $h^r_{jj} = 0, \forall j = 3, \cdots, n$, and $r = 1, 2$.

For $r = 3, \cdots, n$, we use the inequality

$$\sum_{j=3}^{n}(h^r_{11} + h^r_{22})h^r_{jj} + \sum_{3\le i<j\le n, i} h^r_{ii}h^r_{jj} - \sum_{j=1, j\ne r}^{n}(h^r_{jj})$$
$$\le \frac{2n-3}{2(2n+3)}(h^r_{11} + \cdots + h^r_{nn})^2, \tag{64}$$

which is equivalent to

$$\sum_{3\le j\le n, j\ne r}^{n}[2(h^r_{11} + h^r_{22}) - 3h^r_{jj}]^2 + (2n+3)(h_{11} - h^r_{22})^2$$
$$+ 6\sum_{3\le i<j\le ni, j\ne r}(h^r_{ii} - h^r_{jj})^2 + 2\sum_{j=3}^{n}(h^r_{rr} - h^r_{jj})^2.$$
$$+ 3[h^r_{rr} - 2(h^r_{11} + h^r_{22})]^2 \ge 0.$$

The equality holds if and only if

$$\begin{cases} h^r_{11} = h^r_{22} = 3\lambda^r \\ h^r_{jj} = 4\lambda^r, \forall j = 3, \cdots, n, j \ne r, r = 3, \cdots, n. \\ h^r_{rr} = 12\lambda^r, \lambda^r \in \mathbb{R}, \end{cases} \tag{65}$$

From Equations (62)–(64), we obtain the desired inequality (51). □

7. Conclusions

The study of non-metric connections, particularly semi-symmetric non-metric connections, holds considerable importance in differential geometry. For such submanifolds, the sectional curvature cannot be defined in the standard way, because the tensor field $g(R(X,Y)Z,W)$ is not skew-symmetric in the last two variables.

Recently, in [27], the authors defined a sectional curvature on Riemannian manifolds admitting a semi-symmetric non-metric connection.

In the present paper, we extended this definition in Lorentzian settings, more precisely for Lorentzian-Sasakian manifolds endowed with a semi-symmetric non-metric connection.

Using this sectional curvature, we derived the Chen inequality involving scalar curvature and mean curvature, as well as geometric inequalities for Casorati curvatures, for contact slant submanifolds (which are tangent to the time-like vector ξ) in Lorentzian–Sasakian space forms admitting semi-symmetric non-metric connections. In the last section, we established an improved Chen first inequality for Legendrian submanifolds (which are normal to ξ) in such space forms.

Future research could extend this work by obtaining additional Chen inequalities or refining the current results for special classes of submanifolds in Lorentzian–Sasakian manifolds or, more generally, in pseudo-Riemannian contact metric manifolds admitting semi-symmetric non-metric connections.

Author Contributions: Conceptualization, M.M.; methodology, I.M. and M.M.; software, M.M. and A.O.; validation, I.M. and A.O.; formal analysis, M.M. and I.M.; investigation, A.O. and M.M.; resources, I.M.; writing—original draft preparation, M.M. and A.O.; writing—review and editing, I.M. and A.O.; visualisation, I.M. and A.O.; supervision, I.M.; project administration, I.M. All authors have read and agreed to the published version of the manuscript.

Funding: This research received no external funding.

Data Availability Statement: The original contributions presented in the study are included in the article, further inquiries can be directed to the corresponding author.

Acknowledgments: The first author would like to thank the Ministry of Higher Education and Scientific Research of Sudan and Al Neelain University, Khartoum (Sudan), for their financial support.

Conflicts of Interest: The authors declare no conflicts of interest.

ferences

Friedmann, A.; Schouten, J.A. Über die Geometrie der halbsymmetrischen Übertragungen. *Math. Z.* **1924**, *21*, 211–223. [CrossRef]
Hayden, H.A. Subspaces of a space with torsion. *Proc. Lond. Math. Soc.* **1932**, *34*, 27–50. [CrossRef]
Yano, K. On semi-symmetric metric connection. *Rev. Roum. Math. Pures Appl.* **1970**, *15*, 1579–1586.
Agashe, N.S. A semi-symmetric non-metric connection on a Riemannian manifold. *Indian J. Pure Appl. Math.* **1992**, *23*, 399–409.
Agashe, N.S.; Chafle, M.R. On submanifolds of a Riemannian manifold with a semi-symmetric non-metric connection. *Tensor* **1994**, *55*, 120–130.
Chen, B.Y. Some pinching and classification theorems for minimal submanifolds. *Arch. Math.* **1993**, *60*, 568–578. [CrossRef]
Chen, B.Y. Relations between Ricci curvature and shape operator for submanifolds with arbitrary codimensions. *Glasg. Math. J.* **1999**, *41*, 33–41. [CrossRef]
Alegre, P.; Carriazo, A.; Kim, Y.H.; Yoon, D.W.B.-Y. Chen's inequality for submanifolds of generalized space forms. *Indian J. Pure Appl. Math.* **2007**, *38*, 185–201.
Postavaru, O.; Mihai, I. An optimized Chen first inequality for special slant submanifolds in Lorentz-Sasakian space forms. *Rev. Real Acad. Cienc. Exactas Físicas Nat. Ser. A Matemáticas* **2021**, *115*, 150. [CrossRef]
Defever, F.; Mihai, I.; Verstraelen, L.B.-Y. Chen's inequality for C-totally real submanifolds of Sasakian space forms. *Boll. Unione Mat. Ital.* **1997**, *11*, 365–374.
Carriazo, A.; Fernández, L.M.; Hans-Uber, H.B.B.Y. Chen's inequality for S-space forms: Applications to slant immersions. *Indian J. Pure Appl. Math.* **2003**, *34*, 1287–1298.
Kim, J.S.; Song, Y.M.; Tripathi, M.M. B.-Y. Chen inequalities for submanifolds in generalized complex space forms. *Bull. Korean Math. Soc.* **2003**, *40*, 411–423. [CrossRef]
Decu, S.; Haesen, S.; Verstraelen, L. Optimal inequalities involving Casorati curvatures. *Bull. Transilv. Univ. Brasov Ser. B(NS)* **2007**, *14*, 85–93.

14. Slesar, V.; Şahin, B.; Vîlcu, G.E. Inequalities for the Casorati curvatures of slant submanifolds in quaternionic space forms. *Inequal. Appl.* **2014**, *2014*, 123. [CrossRef]
15. Verstraelen, L. Geometry of submanifolds I, The first Casorati curvature indicatrices. *Kragujev. J. Math.* **2013**, *37*, 5–23.
16. Lee, J.W.; Vîlcu, G.E. Inequalities for generalized normalized δ-Casorati curvatures of slant submanifolds in quaternionic space forms. *Taiwan. J. Math.* **2015**, *19*, 691–702. [CrossRef]
17. Zhang, P.; Zhang, L. Casorati inequalities for submanifolds in a Riemannian manifold of quasi-constant curvature with semi-symmetric metric connection. *Symmetry* **2016**, *8*, 19. [CrossRef]
18. Zhang, P.; Pan, X.; Zhang, L. Inequalities for submanifolds of a Riemannian manifold of nearly quasi-constant curvature with semi-symmetric non-metric connection. *Rev. Un. Mat. Argent.* **2015**, *56*, 1–19.
19. He, G.; Liu, H.; Zhang, L. Optimal inequalities for the Casorati curvatures of submanifolds in generalized space forms endowed with semi-symmetric non-metric connections. *Symmetry* **2016**, *8*, 113. [CrossRef]
20. Dogru, Y. Chen Inequalities for Submanifolds of Some Space Forms Endowed with a Semi-Symmetric Non-Metric Connection. *Jordan J. Math. Stat.* **2013**, *6*, 313–339.
21. Mihai, A.; Özgür, C. Chen inequalities for submanifolds of real space forms with a semi-symmetric metric connection. *Taiwan. J. Math.* **2010**, *14*, 1465–1477. [CrossRef]
22. Mihai, A.; Özgür, C. Chen inequalities for submanifolds of complex space forms and Sasakian space forms endowed with semi-symmetric metric connections. *Rocky Mt. J. Math.* **2011**, *5*, 1653–1673. [CrossRef]
23. Lee, C.W.; Lee, J.W.; Vîlcu, G.E.; Yoon, D.W. Optimal inequalities for the Casorati curvatures of submanifolds of generalized space forms endowed with semi-symmetric metric connections. *Bull. Korean Math. Soc.* **2015**, *52*, 1631–1647. [CrossRef]
24. Lee, J.W.; Lee, C.W.; Yoon, D.W. Inequalities for generalized δ-Casorati curvatures of submanifolds in real space forms endowed with a semi-symmetric metric connection. *Rev. Union Mat. Argent.* **2016**, *57*, 53–62.
25. Mihai, I.; Mohammed, M. Optimal inequalities for submanifolds in trans-Sasakian manifolds endowed with a semi-symmetric metric connection. *Symmetry* **2023**, *15*, 877. [CrossRef]
26. Ikawa, T.; Erdogan, M. Sasakian manifolds with Lorentzian metric. *Kyungpook Math. J.* **1996**, *35*, 517–526.
27. Mihai, A.; Mihai, I. A note on a well-defined sectional curvature of a semi-symmetric non-metric connection. *Int. Electron. J. Math.* **2024**, *17*, 15–23. [CrossRef]
28. Lotta, A. Slant submanifolds in contact geometry. *Bull. Math. Soc. Sci. Math. Roum.* **1996**, *39*, 183–198.
29. Chen, B.Y. *Geometry of Slant Submanifolds*; Katholieke Universiteit: Leuven, Belgium, 1990.
30. Chen, B.Y. Mean curvature and shape operator of isometric immersions in real-space-forms. *Glasg. Math. J.* **1996**, *38*, 87– [CrossRef]
31. Mihai, A.; Rădulescu, I.N. Scalar and Ricci curvatures of special contact slant submanifolds in Sasakian space forms. *Adv. Geom.* **2014**, *14*, 147–159. [CrossRef]
32. Siddiqui, A.N.; Shahid, M.H. A lower bound of normalized scalar curvature for bi-slant submanifolds in generalized Sasakian space forms using Casorati curvatures. *Acta Math. Univ. Comen.* **2018**, *87*, 127–140.
33. Bolton, J.; Dillen, F.; Fastenakels, J.; Vrancken, L. A best possible inequality for curvature-like tensors. *Math. Ineq. Appl.* **2009**, 663–681. [CrossRef]

mathematics

MDPI

Article

Applications of Disaffinity Vectors to Certain Riemannian Manifolds

Hanan Alohali [1,†], Sharief Deshmukh [1,†] and Bang-Yen Chen [2,*,†]

1 Department of Mathematics, College of Science, King Saud University, P.O. Box 2455, Riyadh 11451, Saudi Arabia; halohali@ksu.edu.sa (H.A.); shariefd@ksu.edu.sa (S.D.)
2 Department of Mathematics, Michigan State University, East Lansing, MI 48824, USA
* Correspondence: chenb@msu.edu
† These authors contributed equally to this work.

Abstract: A disaffinity vector on a Riemannian manifold is a vector field whose affinity tensor vanishes. In this paper, we prove that every disaffinity vector on a compact Riemannian manifold is an incompressible vector field. Then, we discover a sufficient condition for an incompressible vector field to be disaffinity. Next, we study trans-Sasakian 3-manifolds whose Reeb vector field is disaffinity and obtain two sufficient conditions for a trans-Sasakian 3-manifold to be homothetic to a Sasakian 3-manifold. Finally, we prove that a complete Riemannian manifold admitting a non-harmonic disaffinity function satisfying the Eikonal equation and a Ricci curvature inequality is isometric to a Euclidean space.

Keywords: disaffinity function; disaffinity vector; incompressible vector field; Eikonal equation; trans-Sasakian manifolds; Sasakian manifold; euclidean space

MSC: 53C21; 53C15

1. Introduction

Throughout this paper, we assume that N^m is a connected Riemannian m-manifold. Let $\mathfrak{X}(N^m)$ be the space of smooth vector fields on N^m.

A disaffinity vector on a Riemannian manifold (N^m) is a vector field (ξ) whose affinity tensor ($\mathcal{L}_\xi \nabla = 0$) vanishes identically (cf., [1]). Hence, a vector field (ζ) on N^m is a *disaffinity vector* if and only if it satisfies (see [1])

$$R(U,\zeta)V = \nabla_U \nabla_V \zeta - \nabla_{\nabla_U V}\zeta, \quad U,V \in \mathfrak{X}(N^m), \tag{1}$$

where ∇ is the Levi-Civita connection and R is the curvature tensor of N^m.

It is shown in [1] that every Killing vector field on a Riemannian manifold is a disaffinity vector; but the converse is not true, since there are disaffinity vectors that are non-Killing. For instance, the following position vector field of a Euclidean m-space ($\mathbb{E}^m$) is a non-Killing disaffinity vector:

$$\xi = \sum_{j=1}^{m} x^j \frac{\partial}{\partial x^j}, \tag{2}$$

where $\{x^1, \ldots, x^m\}$ is a Euclidean coordinate system on $\mathbb{E}^m$. Thus, the class of disaffinity vectors is larger than the class of Killing vector fields.

A function (f) on N^m is called *disaffinity* if its gradient (∇f) is a disaffinity vector. Since every constant function on N^m is a disaffinity function, we call a non-constant disaffinity function a *non-trivial disaffinity function*. Note that the function $f = \frac{1}{2}\|\xi\|^2$, with ξ given by (2), is a non-trivial disaffinity function on $\mathbb{E}^n$; thus, $\xi = \nabla f$ is a disaffinity vector on $\mathbb{E}^n$. Further, it is known that the Reeb vector field (ξ) of a Sasakian structure on the unit sphere

($S^{2m-1}(1)$) is a disaffinity vector (cf., [1]). As we progress, we shall witness that disaffinity vectors on Riemannian manifolds are wide spread and too important to go un-noticed.

In Section 4, we investigate the relations between disaffinity vectors and incompressible vectors on a Riemannian manifold. Note that incompressible vectors play an important role in fluid mechanics (cf., [2–6]). The first result of this section states that every disaffinity vector on a compact Riemannian manifold is incompressible. Then, we show that there exists an incompressible vector that is not a disaffinity vector. This leads to the question of finding conditions under which an incompressible vector field on a compact Riemannian manifold is a disaffinity vector, which is answered in Theorem 3.

In Section 5, we study a trans-Sasakian 3-manifold $(N^3, \varphi, \xi, \eta, g, \alpha, \beta)$ of type (α, β) (cf. [7–9]) and investigate the impact of the Reeb vector field (ξ) being a disaffinity vector. First, for $\alpha \neq 0$, we prove that if ξ is a disaffinity vector, then $\beta = 0$ (Proposition 1). Next, we prove that, as for $\alpha \neq 0$, if ξ is a disaffinity vector and the sectional curvature of plane sections containing ξ is constant, then N^3 is homothetic to a Sasakian 3-manifold (Theorem 4). In addition, we derive another sufficient condition involving the Ricci operator for N^3 to be homothetic to a Sasakian 3-manifold (Theorem 5).

A function (f) on a Riemannian manifold (N^n) is called a *disaffinity function* if its gradient vector (∇f) is a disaffinity vector. A disaffinity function is called non-trivial if it is non-constant. In Section 6, we study the existence of a non-trivial disaffinity function. It is know that the existence of a non-trivial disaffinity function on N^n implies N^n is non-compact. (cf., [1]). A function (f) on N^n is said to satisfy the Eikonal equation if $\|\nabla f\| = 1$ holds (cf., [10]). It is known that Eikonal equations play important roles in geometry and physics. Eikonal equations are also widely used in various scientific and engineering fields, such as seismology, image processing, robotic path planning, and computer graphics (see, e.g., [10–12])). In this section, we prove that if N^m is a complete Riemannian manifold that admits a non-harmonic disaffinity function (h) satisfying the Eikonal equation ($\|\nabla h\| = 1$) and if the Ricci curvature of N^m satisfies

$$Ric(\nabla h, \nabla h) \geq -\frac{1}{m}(\Delta h)^2,$$

then N^m is isometric to a Euclidean space (Theorem 6).

2. Further Examples of Disaffinity Vectors

Now, we provide a non-trivial example of a disaffinity vector on a nearly Kähler 6-sphere $S^6(1)$ (cf., [13]). Let Z be a unit-constant vector in $\mathbb{E}^7$. Consider $S^6(1)$ as the standard unit hypersphere in $\mathbb{E}^7$. If N is the outward unit-normal vector field of $S^6(1) \subset \mathbb{E}^7$, then the shape operator satisfies $A = -I$. Thus, we may express Z, restricted to $S^6(1)$, as

$$Z = \xi + \sigma N, \quad \sigma = \langle Z, N \rangle,$$

where ξ is the tangential part of Z to $S^6(1)$. Differentiating the above equation with respect a vector field (U) on $S^6(1)$, while using fundamental equations of a hypersurface and equating akin parts, we have

$$\nabla_U \xi = -\sigma U, \quad \nabla \sigma = \xi, \tag{3}$$

where ∇_U is the covariant derivative with respect to the Levi-Civita connection on $S^6(1)$ with respect to the induced metric (g). Note that we have an almost complex structure (J) on $S^6(1)$ such that $(S^6(1), J, g)$ is a non-Kähler, nearly Kähler manifold (cf., [13]).

Next, let us consider the vector field ($\mathbf{u} = J\xi$) on $S^6(1)$. Using Equation (3), we find

$$\nabla_U \mathbf{u} = G(U, \xi) - \sigma JU, \tag{4}$$

where $G(U,V) = (\nabla_U J)(V)$. From the above equation and Equation (3), we have

$$\nabla_U \nabla_V \mathbf{u} = (\nabla_U G)(V,\xi) + G(\nabla_U V,\xi) - \sigma G(V,U)$$
$$- U(\sigma) J V - \sigma G(U,V) - \sigma J \nabla_U V,$$

with which, upon using the expression for $(\nabla_U G)(V,\xi)$ (cf., [13]), we arrive at

$$\nabla_U \nabla_V \mathbf{u} - \nabla_{\nabla_U V} \mathbf{u} = g(V,\mathbf{u})U - g(U,V)\mathbf{u}.$$

Then, upon using the expression of curvature tensor for $S^6(1)$, the above equation, in view of Equation (1), confirms that $\mathbf{u}$ is a disaffinity vector on the unit sphere ($S^6(1)$).

To see how disaffinity vectors are so frequent, we provide another example. Consider a non-compact Riemannian manifold ((N^3, g)) constructed in (cf., [14], Examples 2.1, 2.7 and 5.3), where

$$N^3 = \{(x,y,z) \in R^3 : z \neq 0\},$$

and

$$g = \exp(2\sqrt{-k}z)dx^2 + \exp(-2\sqrt{-k}z)dy^2 + dz^2.$$

Then, taking $\xi = \frac{\partial}{\partial z}$, we have the following $(1,1)$-tensor and $\eta = dz$:

$$\varphi = \exp(2\sqrt{-k}z)dx \otimes \frac{\partial}{\partial y} - \exp(-2\sqrt{-k}z)dy \otimes \frac{\partial}{\partial x}$$

It is known that $(N^3, g, \varphi, \xi, \eta)$ is an almost cosymplectic manifold (cf., [14]). Furthermore, using (2.16) in [14] and the expressions of curvature tensor in Example 5.3 in [14], it is easy to see that ξ is a disaffinity vector on $(N^3, g, \varphi, \xi, \eta)$.

3. Some Preliminary Results

On a Riemannian m-manifold N^m with a Levi-Civita connection (∇), the curvature tensor (R) is given by

$$R(U,V)W = [\nabla_U, \nabla_V]W - \nabla_{[U,V]}W, \qquad U,V,W \in \mathfrak{X}(N^m), \tag{5}$$

and the Ricci tensor (Ric) is defined by

$$Ric(U,V) = \sum_l g(R(e_l,U)V,e_l), \tag{6}$$

where $\{e_1, \ldots, e_m\}$ is a local orthonormal frame on N^m. The Ricci operator (S) associated to the Ricci tensor is given by

$$Ric(U,V) = g(SU,V), \tag{7}$$

and the scalar curvature (r) is

$$r = \sum_l Ric(e_l,e_l). \tag{8}$$

The following formula for the gradient (∇r) of r is well known (cf., e.g., [15])

$$\frac{1}{2}\nabla r = \sum_l (\nabla_{e_l} S)(e_l), \tag{9}$$

where

$$(\nabla_U S)(V) = \nabla_U SV - S(\nabla_U V).$$

The Laplace operator (Δ) acting on smooth $f : N^m \to \mathbb{R}$ is given by

$$\Delta f = \operatorname{div}(\nabla f), \tag{10}$$

where ∇f is the gradient of f and the divergence is defined by

$$\operatorname{div} U = \sum_l g(\nabla_{e_l} U, e_l).$$

On a compact Riemannian manifold (N^m), we have Stokes's Theorem.

$$\int_{N^m} (\operatorname{div} U) dV_g = 0, \tag{11}$$

where dV_g is the volume element of N^m, which, taking $U = \nabla f$ for any function (f) on N^m, implies

$$\int_{N^m} \Delta f dV_g = 0. \tag{12}$$

The Hessian operator (H_f) of a function ($f : N^m \to \mathbb{R}$) is defined by

$$H_f U = \nabla_U \nabla f, \tag{13}$$

and the Hessian of f ($Hess(f)$) is related to H_f by

$$Hess(f)(U, V) = g(H_f U, V). \tag{14}$$

Next, consider a contact metric manifold $(N^{2m-1}, \varphi, \xi, \eta, g)$, where φ is a $(1,1)$-tensor field, ξ a unit vector field and η the 1-form dual to ξ with respect to the metric g satisfying

$$\varphi^2 = -I + \eta \otimes \xi, \quad \varphi(\xi) = 0, \quad \eta \circ \varphi = 0, \quad g(\varphi U, \varphi V) = g(U, V) - \eta(U)\eta(V) \tag{15}$$

for $U, V \in \mathfrak{X}(N^{2m-1})$ (cf., [16]). If there exist functions (α, β) on $(N^{2m-1}, \varphi, \xi, \eta, g)$ such that

$$(\nabla_U \varphi)(V) = \alpha(g(U, V)\xi - \eta(V)U) + \beta(g(\varphi U, V)\xi - \eta(V)\varphi U), \tag{16}$$

where $(\nabla\varphi)(U, V) = \nabla_U \varphi V - \varphi(\nabla_U V)$, then N^{2m-1} is called a trans-Sasakian manifold of type (α, β) (cf., [7–9]). From (15) and (16), it follows that

$$\nabla_U \xi = -\alpha\varphi(U) + \beta(U - \eta(U)\xi), \quad U \in \mathfrak{X}(M). \tag{17}$$

A trans-Sasakian manifold is said to be proper if both $\alpha, \beta \neq 0$. In [8], it is shown that there are no proper trans-Sasakian manifolds of dimension ≥ 5. Thus, we are only interested in studying trans-Sasakian 3-manifolds.

We report the following results on trans-Sasakian 3-manifolds from [7].

Lemma 1. *We have $\xi(\alpha) + 2\alpha\beta = 0$ on a trans-Sasakian manifold $(N^3, \varphi, \xi, \eta, g, \alpha, \beta)$ of type (α, β).*

Lemma 2. *The Ricci operator (S) of a trans-Sasakian manifold $(N^3, \varphi, \xi, \eta, g, \alpha, \beta)$ of type (α, β) satisfies*

$$S(\xi) = \varphi(\nabla\alpha) - \nabla\beta + 2(\alpha^2 - \beta^2)\xi - g(\nabla\beta, \xi)\xi.$$

Also, we recall the following result from [17].

Theorem 1. *If a Riemannian manifold (N^{2m-1}) admits a Killing vector field (ξ) of constant length satisfying*

$$k^2(\nabla_U \nabla_V \xi - \nabla_{\nabla_U V} \xi) = g(V, \xi)U - g(U, V)\xi, \quad \forall U, V \in \mathfrak{X}(M)$$

for a non-zero constant (k), then N^{2m-1} is homothetic to a Sasakian manifold.

Let ξ be a vector field on a Riemannian m-manifold $((N^m, g))$. We define a symmetric operator (B) and a skew-symmetric operator (F) related to ξ by

$$(\mathcal{L}_\xi g)(U,V) = 2g(BU,V), \quad d\eta(U,V) = 2g(FU,V), \tag{18}$$

where η is 1-form dual to ξ. Using Equation (20) and

$$2g(\nabla_U \xi, V) = (\mathcal{L}_\xi g)(U,V) + d\eta(U,V),$$

we obtain

$$\nabla_U \xi = BU + FU, \quad U,V \in \mathfrak{X}(N^m). \tag{19}$$

We call the operator (B) the symmetric operator associated to ξ and F the skew-symmetric operator associated to ξ.

4. Incompressible Disaffinity Vectors on a Riemannian Manifold

A vector field (ζ) on a Riemannian manifold is called *incompressible* if its divergence is zero. In fluid mechanics, an incompressible vector field is a vector field that has no sources or sinks. For the importance of incompressible vector fields, we refer to [2–6] and references therein. In this section, we are interested in the relationship between incompressible vector fields and disaffinity vector fields.

Suppose ξ is a disaffinity vector on N^m. Then, by using (1) and (19), we obtain

$$R(U,\xi)V = (\nabla_U B)(V) + (\nabla_U F)(V). \tag{20}$$

Note that the position vector field (ξ) of $\mathbb{E}^m$ appearing in (2) is a disaffinity vector, and its divergence satisfies $\operatorname{div}(\xi) = m$. Thus, the position vector field of $\mathbb{E}^m$ is not incompressible. Consequently, disaffinity vectors are not necessarily incompressible in general. On the other hand, for compact Riemannian manifolds, we have the following.

Theorem 2. *Every disaffinity vector field on a compact Riemannian manifold is incompressible.*

Proof. Let ξ be a disaffinity vector on a Riemannian manifold (N^m) with an associated symmetric operator (B) and skew-symmetric operator (F). By taking the trace in (20), we have

$$Ric(\xi, V) = \sum_j g(V, (\nabla_{e_j} B)(e_j)) - \sum_j g(V, (\nabla_{e_j} F)(e_j)), \tag{21}$$

where we used the symmetry of B, the skew symmetry of F, and a local orthonormal frame $(\{e_1, .., e_m\})$. Observe that (19) implies

$$\|\nabla\xi\|^2 = \|B\|^2 + \|F\|^2, \tag{22}$$

where we used

$$\sum_j g(Be_j, Fe_j) = 0. \tag{23}$$

Equation (18) also implies

$$\frac{1}{4}|\mathcal{L}_\xi g|^2 = \|B\|^2 \tag{24}$$

Note that upon using the symmetry of B, Equations (19) and (23), and a local orthonormal frame ($\{e_1, .., e_m\}$), we find

$$\begin{aligned}\operatorname{div}(B\xi) &= \sum_j g(\nabla_{e_j} B\xi, e_j) = \sum_j g((\nabla_{e_j} B)(\xi) + B(Be_j + Fe_j), e_j) \\ &= \sum_j g(\xi, (\nabla_{e_j} B)(e_j)) + \|B\|^2 + \sum_j g(Fe_j, Be_j) \\ &= \sum_j g(\xi, (\nabla_{e_j} B)(e_j)) + \|B\|^2 \end{aligned} \tag{25}$$

Similarly, by applying (19) and (23), the skew symmetry of F, and a local orthonormal frame ($\{e_1, \ldots, e_m\}$), we have

$$\operatorname{div}(F\xi) = -\sum_j g(\xi, (\nabla_{e_j} F)(e_j)) - \|F\|^2. \tag{26}$$

Now, using Equation (21), we find

$$Ric(\xi, \xi) = \sum_j g(\xi, (\nabla_{e_j} B)(e_j)) - \sum_j g(\xi, (\nabla_{e_j} F)(e_j)).$$

By treating this equation with (25) and (26), we obtain

$$Ric(\xi, \xi) = \operatorname{div}(B\xi) - \|B\|^2 + \operatorname{div}(F\xi) + \|F\|^2.$$

Integrating the above equation and using (11), we conclude

$$\int_{N^m} Ric(\xi, \xi) dV_g = \int_{N^m} \left(\|F\|^2 - \|B\|^2\right) dV_g. \tag{27}$$

Recall the following integral formula (cf., [18]):

$$\int_{N^m} \left(Ric(\xi, \xi) + \frac{1}{2}|\mathcal{L}_\xi g|^2 - \|\nabla\xi\|^2 - (\operatorname{div}\xi)^2\right) dV_g = 0. \tag{28}$$

Then, using Equations (22), (24) and (27), we arrive at

$$\int_{N^m} \left(\|F\|^2 - \|B\|^2 + 2\|B\|^2 - \|B\|^2 - \|F\|^2 - (\operatorname{div}\xi)^2\right) dV_g = 0.$$

Hence, we have

$$\int_{N^m} (\operatorname{div}\xi)^2 dV_g = 0,$$

which implies $\operatorname{div}\xi = 0$. Therefore, ξ is an incompressible vector field. □

Now, we consider an example of an incompressible vector field ($\mathbf{u}$) on $\mathbb{E}^m$ but not a disaffinity vector. We take

$$\mathbf{u} = (x^2)^2 \frac{\partial}{\partial x^1} + (x^1)^2 \frac{\partial}{\partial x^2} + c^3 \frac{\partial}{\partial x^3} + .. + c^m \frac{\partial}{\partial x^m}, \tag{29}$$

where $x^1, \ldots, x^m$ are Euclidean coordinates and $c^3, \ldots, c^m$ are constants. With respect to the Euclidean connection (∇) on $\mathbb{E}^m$, we have

$$\nabla_{\frac{\partial}{\partial x^1}} \mathbf{u} = 2x^1 \frac{\partial}{\partial x^2}, \nabla_{\frac{\partial}{\partial x^2}} \mathbf{u} = 2x^2 \frac{\partial}{\partial x^1}, \nabla_{\frac{\partial}{\partial x^j}} \mathbf{u} = 0, j \geq 3.$$

Thus, we obtain div $(\mathbf{u}) = 0$. Therefore, $\mathbf{u}$ is incompressible. For vector fields (U, V) on $\mathbb{E}^m$, we also have

$$\nabla_V \mathbf{u} = V(x^2)^2 \frac{\partial}{\partial x^1} + V(x^1)^2 \frac{\partial}{\partial x^2},$$

$$\nabla_{\nabla_U V} \mathbf{u} = \nabla_U V(x^2)^2 \frac{\partial}{\partial x^1} + \nabla_U V(x^1)^2 \frac{\partial}{\partial x^2},$$

$$\nabla_U \nabla_V \mathbf{u} = UV(x^2)^2 \frac{\partial}{\partial x^1} + UV(x^1)^2 \frac{\partial}{\partial x^2}.$$

Combining the two above equations and Equation (14) yields

$$\nabla_U \nabla_V \mathbf{u} - \nabla_{\nabla_U V} \mathbf{u} = Hess(x^2)^2(U, V) + Hess(x^1)^2(U, V). \tag{30}$$

Note that

$$Hess(x^1)^2\left(\frac{\partial}{\partial x^1}, \frac{\partial}{\partial x^1}\right) = Hes(x^2)^2\left(\frac{\partial}{\partial x^2}, \frac{\partial}{\partial x^2}\right) = 2.$$

The above equations imply $\nabla_U \nabla_V \mathbf{u} - \nabla_{\nabla_U V} \mathbf{u} \neq 0$, where, as $R(U, \mathbf{u})V = 0$, we see from (1) that $\mathbf{u}$ is not a disaffinity vector.

Next, we seek conditions for an incompressible vector field to be a disaffinity vector.

Theorem 3. *An incompressible vector field (ξ) on a compact Riemannian manifold (N^m) satisfying*

$$\int_{N^m} Ric(\xi, \xi) dV_g \geq \int_{N^m} \|F\|^2 dV_g$$

is a disaffinity vector, where F is the associated skew-symmetric operator.

Proof. Suppose ξ is an incompressible vector field on a compact Riemannian manifold (N^m) with an associated symmetric operator (B) and skew-symmetric operator (F). Then, using (19), we have

$$\text{div}(\xi) = Tr\, B = 0. \tag{31}$$

Using (5) and (19), we find

$$R(U, V)\xi = (\nabla_U B)(V) - (\nabla_V B)(U) + (\nabla_U F)(V) - (\nabla_V F)(U).$$

Taking the trace in this equation and using (31), we arrive at

$$Ric(V, \xi) = \sum_j g(V, (\nabla_{e_j} B)(e_j)) - \sum_j g(V, (\nabla_{e_j} F)(e_j)),$$

that is,

$$Ric(\xi, \xi) = \sum_j g(\xi, (\nabla_{e_j} B)(e_j)) - \sum_j g(\xi, (\nabla_{e_j} F)(e_j)). \tag{32}$$

Note that for the incompressible vector field (ξ), the associated symmetric operator (B), and skew-symmetric operator (F), we have equations similar to Equations (25) and (26). Thus, (32) takes the following form:

$$Ric(\xi, \xi) = \text{div}(B\xi) - \|B\|^2 + \text{div}(B\xi) + \|F\|^2.$$

Integrating the above equation, we arrive at

$$\int_{N^m} \left(Ric(\xi, \xi) - \|F\|^2\right) dV_g = -\int_{N^m} \|B\|^2 dV_g.$$

Using the condition in the hypothesis, we obtain $B = 0$. Therefore, Equation (19) yields

$$\nabla_U \xi = FU. \tag{33}$$

Equations (5) and (33) lend us to

$$R(U,V)\xi = (\nabla_U F)(V) - (\nabla_V F)(U). \tag{34}$$

The second equation in (18) suggests that the 2-form $g(FU,V)$ is closed; thus, we obtain

$$g((\nabla_U F)(V),W) + g((\nabla_V F)(W),U) + g((\nabla_W F)(U),V) = 0,$$

which, upon using the skew symmetry of F and Equation (34) in the first two terms, yields

$$g(R(U,V)\xi,W) + g((\nabla_W F)(U),V) = 0.$$

The above equation implies

$$R(\xi,W)U + (\nabla_W F)(U) = 0,$$

that is,

$$R(W,\xi)U = \nabla_W F(U) - F(\nabla_W U).$$

By applying (33) in the above equation, we find

$$R(W,\xi)U = \nabla_W \nabla_U \xi - \nabla_{\nabla_W U}\xi,$$

which is Equation (1). Consequently, ξ is a disaffinity vector. □

5. Disaffinity Vectors and Trans-Sasakian Manifolds

One of the interesting questions in the geometry of a trans-Sasakian 3-manifold N^3 is to find conditions under which the trans-Sasakian 3-manifold is homothetic to a Sasakian manifold.

First, we prove the following:

Proposition 1. *If the Reeb vector field (ξ) of a trans-Sasakian 3-manifold $(N^3,\varphi,\xi,\eta,g,\alpha,\beta)$ of type (α,β) with $\alpha \neq 0$ is a disaffinity vector, then $\beta = 0$.*

Proof. Let $(N^3,\varphi,\xi,\eta,g,\alpha,\beta)$ be a trans-Sasakian manifold whose Reeb vector field (ξ) is a disaffinity vector. Then, using (17) in Equation (1) and some easy calculations, we arrive at

$$\begin{aligned} R(U,\xi)V = & -U(\alpha)\varphi V - (\alpha^2+\beta^2)g(U,V)\xi + (\alpha^2-\beta^2)\eta(V)U \\ & + 2\alpha\beta\eta(V)\varphi U + U(\beta)V - U(\beta)\eta(V)\xi + 2\beta^2\eta(U)\eta(V)\xi. \end{aligned} \tag{35}$$

Taking $V=\xi$ in the above equation and using (15), we find

$$R(U,\xi)\xi = -(\alpha^2-\beta^2)\varphi^2 U + 2\alpha\beta\varphi U. \tag{36}$$

Taking the inner product with U on both sides of the above equation, we obtain

$$R(U,\xi;\xi,U) = (\alpha^2-\beta^2)g(\varphi U,\varphi V), \tag{37}$$

that is,

$$R(U+V,\xi;\xi,U+V) = (\alpha^2-\beta^2)g(\varphi(U+V),\varphi(U+V)).$$

Using Equation (37), we conclude

$$R(U,\xi;\xi,V) + R(V,\xi;\xi,U) = 2(\alpha^2-\beta^2)g(\varphi U,\varphi V),$$

that is,

$$R(U,\xi;\xi,V) = (\alpha^2-\beta^2)g(\varphi U,\varphi V).$$

The above equation yields

$$R(U,\xi)\xi = -(\alpha^2 - \beta^2)\varphi^2 U, \tag{38}$$

and consequently, Equation (36) yields $\alpha\beta\varphi U = 0$, which yields

$$\alpha\beta\varphi^2 U = 0.$$

Now, by taking the trace of the above equation, we obtain $\alpha\beta = 0$. Since $\alpha \neq 0$, we must have $\beta = 0$. □

Now, recall that an adapted local orthonormal frame on a trans-Sasakian 3-manifold $(N^3, \varphi, \xi, \eta, g, \alpha, \beta)$ is of the form $\{e, \varphi e, \xi\}$, where e is the unit vector field orthogonal to ξ.

Theorem 4. *If the Reeb vector field (ξ) of a trans-Sasakian 3-manifold $(N^3, \varphi, \xi, \eta, g, \alpha, \beta)$ of type (α, β) with $\alpha \neq 0$ is a disaffinity vector and the sectional curvature of plane sections containing ξ is constant, then N^3 is homothetic to a Sasakian manifold.*

Proof. Suppose the Reeb vector field (ξ) of a trans-Sasakian manifold (N^3) with $\alpha \neq 0$ is a disaffinity vector with sectional curvatures of plane sections containing ξ and is constant. Then, Proposition 1 implies $\beta = 0$. Thus, according to Equation (38), we have

$$R(U,\xi)\xi = -\alpha^2\varphi^2 U, \quad U \in \mathfrak{X}(N^3).$$

Setting U in the above equation as a unit vector orthogonal to ξ and taking the inner product with U and using Equation (15), we obtain

$$R(U,\xi;\xi,U) = \alpha^2,$$

and as the sectional curvature ($R(U,\xi;\xi,U)$) is constant, we see that α^2 is constant. Hence, α is a non-zero constant. Then, Equations (17) and (16) are

$$\nabla_U\xi = -\alpha\varphi U, \quad (\nabla_U\varphi)(V) = \alpha(g(U,V)\xi - \eta(V)U). \tag{39}$$

Using the above equation, as α is a constant, we conclude

$$\nabla_U\nabla_V\xi - \nabla_{\nabla_U V}\xi = -\alpha(\nabla_U\varphi)(V) = \alpha^2(\eta(V)U - g(U,V)\xi). \tag{40}$$

Equation (39) also implies

$$\mathcal{L}_\xi g = 0, \tag{41}$$

that is, the unit vector field (ξ) is a Killing vector field, and as $\alpha \neq 0$, (40) becomes

$$\alpha^{-2}(\nabla_U\nabla_V\xi - \nabla_{\nabla_U V}\xi) = g(V,\xi)U - g(U,V)\xi.$$

Therefore, according to Theorem 1, we conclude that N^3 is homothetic to a Sasakian 3-manifold. □

As another application of Proposition 1, we also have the following.

Theorem 5. *If the Reeb vector field (ξ) of a trans-Sasakian 3-manifold $(N^3, \varphi, \xi, \eta, g, \alpha, \beta)$ of type (α, β) with $\alpha \neq 0$ is a disaffinity vector and the Ricci operator (S) satisfies $(\nabla_\xi S)(\xi) = 0$, then N^3 is homothetic to a Sasakian 3-manifold.*

Proof. Suppose the Reeb vector field (ξ) of a trans-Sasakian manifold (N^3) with $\alpha \neq 0$ is a disaffinity vector and the Ricci operator (S) satisfies

$$(\nabla_\xi S)(\xi) = 0. \tag{42}$$

Then, according to Proposition 1, we obtain $\beta = 0$, and according to Lemma 2, we obtain

$$S(\xi) = \varphi(\nabla\alpha) + 2\alpha^2\xi. \tag{43}$$

Note that as $\beta = 0$, according to Lemma 1, we find

$$\xi(\alpha) = 0. \tag{44}$$

As with $\beta = 0$, we have Equation (41), that is, ξ is a Killing vector field and, accordingly, the flow of ξ consists of isometries of the trans-Sasakian 3-manifold (N^3); therefore,

$$\mathcal{L}_\xi(r) = 0, \quad \mathcal{L}_\xi S = 0, \tag{45}$$

where r is the scalar curvature of N^3. Using (45), we have

$$[\xi, S(U)] = S([\xi, U]).$$

Then, upon using (39), we conclude that

$$(\nabla_\xi S)(U) = \nabla_{S(U)}\xi - S(\nabla_U\xi) = \alpha(S(\varphi U) - \varphi(SU)).$$

Now, by taking $U = \xi$ in the above equation and using (42), we arrive at

$$\alpha\varphi(S(\xi)) = 0,$$

which, in view of (15), (43), and (44), implies

$$\alpha\nabla\alpha = 0.$$

Since $\alpha \neq 0$, the above equation implies that α is a non-zero constant. Therefore, as in Theorem 4, N^3 is homothetic to a Sasakian 3-manifold. □

6. Disaffinity Functions and Eikonal Equation

Recall that a function (h) on a Riemannian manifold is said to satisfy the Eikonal equation if it satisfies the following non-linear first-order partial differential equation:

$$\|\nabla h\| = 1. \tag{46}$$

In this section, we prove the following result.

Theorem 6. *Let N^m be a complete Riemannian manifold that admits a non-harmonic disaffinity function (h) satisfying the Eikonal equation ($\|\nabla h\| = 1$). If the Ricci curvature of N^m satisfies*

$$Ric(\nabla h, \nabla h) \geq -\frac{1}{m}(\Delta h)^2, \tag{47}$$

then N^m is isometric to a Euclidean space.

Proof. Suppose N^m is a complete Riemannian m-manifold and h is a non-harmonic, non-trivial disaffinity function on N^m that satisfies the Eikonal Equation (46). Then, it follows from (46) that

$$Ug(\nabla h, \nabla h) = 0,$$

that is, $g(H_h U, \nabla h) = 0$. Thus, we have

$$H_h(\nabla h) = 0. \tag{48}$$

As h is a disaffinity function, ∇h satisfies (1), that is,

$$R(U, \nabla h)V = (\nabla_U H_h)(V). \tag{49}$$

Taking a local orthonormal frame $\{e_1, \ldots, e_n\}$ on N^m and using the symmetry of H_h, Equation (49) implies

$$Ric(\nabla h, V) = \sum_j g(V, (\nabla_{e_j} H_h)(e_j)). \tag{50}$$

Using (5) and (13), we have

$$R(U, V)\nabla h = (\nabla_U H_h)(V) - (\nabla_V H_h)(U). \tag{51}$$

Now, as $\Delta h = TrH_h$, we obtain

$$\begin{aligned} U(\Delta h) &= \sum_j \left(g((\nabla_U H_h)(e_j) + H_h(\nabla_U e_j), e_j) + g(H_h e_j, \nabla_U e_j)\right) \\ &= \sum_j g((\nabla_U H_h)(e_j), e_j) + 2\sum_j g(H_h e_j, \nabla_U e_j). \end{aligned} \tag{52}$$

Since $H_h e_j$ is symmetric, $\nabla_U e_j = \sum_k \omega_j^k(U) e_k$, and connection forms (ω_j^k) are skew-symmetric, we have

$$\sum_j g(H_h e_j, \nabla_U e_j) = 0.$$

Thus, by using Equation (51) and the above equation in (52), we arrive at

$$U(\Delta h) = \sum_j g\Big(\sum_j g(R(U, e_j)\nabla h + (\nabla_{e_j} H_h)(U), e_j)\Big).$$

Using (6) and the symmetry of the Hessian operator (H_h) in the above equation, we obtain

$$U(\Delta h) = -Ric(U, \nabla h) + \sum_j g(U, (\nabla_{e_j} H_h)(e_j)).$$

Applying (50) to the above equation, we find

$$U(\Delta h) = 0. \tag{53}$$

Differentiating Equation (48) yields

$$\nabla_U H_h(\nabla h) = 0,$$

that is,

$$(\nabla_U H_h)(\nabla h) + H_h^2 U = 0.$$

Using Equation (49) in the above equation, we conclude that

$$R(U, \nabla h)\nabla h + H_h^2 U = 0.$$

Now, by taking the trace in the above equation, we obtain

$$Ric(\nabla h, \nabla h) + \|H_h\|^2 = 0.$$

By rearranging the above equation, we have

$$\|H_h\|^2 - \frac{1}{m}(\Delta h)^2 = -\frac{1}{m}(\Delta h)^2 - Ric(\nabla h, \nabla h).$$

Using the bound on the Ricci curvature in the statement, we arrive at

$$\|H_h\|^2 - \frac{1}{m}(\Delta h)^2 \leq 0.$$

Also, according to Schwartz's inequality, we have $\|H_h\|^2 \geq \frac{1}{m}(\Delta h)^2$. Combining it with the above inequality, we obtain the following equality: $\|H_h\|^2 = \frac{1}{m}(\Delta h)^2$, which implies

$$H_h = \frac{1}{m}(\Delta h) I. \quad (54)$$

Note that, according to Equation (53), Δh is a constant. Hence, as h is non-harmonic, Δh is non-zero. Thus, according to (54), we have $Hess(h) = cg$, where c is a non-zero constant. Consequently, N^m is isometric to a Euclidean m-space (cf., [15]). □

7. Epilogue

This work is a modest start in analyzing the combinations of different types of vector fields on Riemannian manifolds with the notion of a disaffinity vector. For instance, in Section 4 we saw the effect of the combination of an incompressible vector field and a disaffinity vector on a Riemannian manifold, and in Section 5, we found that the Reeb vector field (ξ) of a trans-Sasakian manifold (N^3) is also a disaffinity vector and used this combination to obtain conditions under which a N^3 is homothetic to a Sasakian manifold. There is variety of different questions that need to be answered for a Riemannian manifold admitting a disaffinity vector (ξ). For instance, an important type of vector field on a Riemannian manifold is geodesic vector fields whose integral curves are geodesics (cf., [10]). A natural question in this direction could involve finding conditions under which a disaffinity vector (ξ) on a Riemannian manifold is a geodesic vector field, in addition to finding examples of geodesic vector fields that are not disaffinity vectors. Another important question would be to find conditions under which a geodesic vector field is a disaffinity vector.

The next important set of questions pertains to a Riemannian manifold (N^m) that admits a torse-forming vector field (ξ) (cf., [19]). Recall that a torse-forming vector field ($\mathbf{u}$) on N^m satisfies

$$\nabla_U \mathbf{u} = fU + \omega(U)\mathbf{u}, \qquad U \in \mathfrak{X}(N^m),$$

where f is a function defined on N^m that is conformal scalar and ω is 1-form on N^m and the generating form of the torse-forming vector field ($\mathbf{u}$). The following natural questions need to be answered:

(i) Suppose a complete Riemannian manifold (N^m) that admits a torse-forming vector field ($\mathbf{u}$); assuming that the conformal scalar (f) is a non-trivial disaffinity function on N^m, what will be the impact on the geometry of N^m?

(ii) Suppose a complete Riemannian manifold (N^m) that admits a torse-forming vector field ($\mathbf{u}$); assuming that the conformal scalar (f) is a non-trivial disaffinity function that also satisfies the Eikonal equation on N^m, what will be the impact on the geometry of N^m?

Author Contributions: Conceptualization H.A., S.D. and B.-Y.C.; methodology, H.A. and S.D.; software, H.A. and S.D.; validation, S.D. and B.-Y.C.; formal analysis, H.A., S.D. and B.-Y.C.; investigation, H.A. and S.D.; resources, H.A.; data curation, H.A. and S.D.; writing—original draft preparation, H.A., S.D. and B.-Y.C.; writing—review and editing, S.D. and B.-Y.C.; visualization, H.A. and S.D.; supervision, S.D.; project administration, H.A. and S.D.; funding acquisition, H.A. All authors have read and agreed to the published version of the manuscript.

Funding: This research was funded by Project Number RSPD2024R860, King Saud University, Riyadh, Saudi Arabia.

Data Availability Statement: The original contributions presented in this study are included in the article.

Acknowledgments: The authors would like to extend their sincere appreciation for the funding (Supporting Project Number RSPD2024R860) provided by King Saud University, Riyadh, Saudi Arabia.

Conflicts of Interest: The authors declare no conflicts of interest.

ferences

Deshmukh, S.; Ishan, A.; Chen, B.-Y. Disaffinity vectors on a Riemannian manifold and applications. *Mathematics* **2024**, *12*, 3659. [CrossRef]
Bianchini, S.; Gusev, N.A. Steady nearly incompressible vector fields in two-dimension: Chain rule and renormalization. *Arch. Ration. Mech. Anal.* **2016**, *222*, 451–505. [CrossRef]
Ma, T.; Wang, S. Structural classification and stability of divergence-free vector fields. *Phys. D Nonlinear Phenom.* **2002**, *171*, 107–126. [CrossRef]
Ma, T.; Wang, S. *Geometric Theory of Incompressible Flows with Applications to Fluid Dynamics*; Mathematical Surveys and Monographs, 119; American Mathematical Society: Providence, RI, USA, 2005.
Ma, T.; Wang, S. Structural evolution of the Taylor vortices. *M2AN Math. Model. Numer. Anal.* **2000**, *34*, 419–437. [CrossRef]
Ma, T.; Wang, S. Dynamics of Incompressible vector fields. *Appl. Math. Lett.* **1999**, *12*, 39–42. [CrossRef]
Deshmukh, S. Trans-Sasakian manifolds homothetic to Sasakian manifolds. *Mediterr. J. Math.* **2016**, *13*, 2951–2958. [CrossRef]
Marrero, J.C. The local structure of trans-Sasakian manifolds. *Ann. Mat. Pura Appl.* **1992**, *162*, 77–86. [CrossRef]
Oubiña, J.A. New classes of almost contact metric structures. *Publ. Math. Debr.* **1985**, *32*, 187–193. [CrossRef]
Deshmukh, S.; Khan, V.A. Geodesic vector fields and Eikonal equation on a Riemannian manifold. *Indag. Math.* **2019**, *30*, 542–552. [CrossRef]
Mosayebi, P.; Kobzas, D.; Murtha, A.; Jagersand, M. Tumor invasion margin on Riemannian space of brain fibers. *Med. Image Anal.* **2012**, *16*, 361–373. [CrossRef] [PubMed]
Fischer, A.E. Riemannian maps between Riemannian manifolds. *Contemp. Math.* **1992**, *132*, 331–366.
Ejiri, N. Totally real submanifolds in 6-sphere. *Proc. Amer. Math. Soc.* **1981**, *83*, 759–763. [CrossRef]
Maduna, S.H.; Massamba, F. Certain class of almost cosymplectic manifolds with Kählerian leaves. *Mediterr. J. Math.* **2023**, *20*, 21. [CrossRef]
Pigola, S.; Rigoli, M.; Rimoldi, M.; Setti, A.G. Almost Ricci solitons. *Ann. Della Sc. Norm. Super. Pisa* **2011**, *10*, 757–799. [CrossRef]
Blair, D.E. *Contact Manifolds in Riemannian Geometry*; Lecture Notes in Math; Springer: Berlin, Germany; New York, NY, USA, 1976; Volume 509.
Okumura, M. Certain almost contact hypersurfaces in Kählerian manifolds of constant holomorphic sectional curvatures. *Tohoku Math. J.* **1964**, *16*, 270–284. [CrossRef]
Yano, K. *Integral Formulas in Riemannian Geometry*; Marcel Dekker: New York, NY, USA, 1970.
Yano, K. On torse forming direction in a Riemannian space. *Proc. Imp. Acad. Tokyo* **1944**, *20*, 340–345. [CrossRef]

mathematics

MDPI

Review

Geometry of Weak Metric f-Manifolds: A Survey

Vladimir Rovenski

Department of Mathematics, University of Haifa, Haifa 3498838, Israel; vrovenski@univ.haifa.ac.il

Abstract: The interest of geometers in f-structures is motivated by the study of the dynamics of contact foliations, as well as their applications in physics. A weak f-structure on a smooth manifold, introduced by the author and R. Wolak, generalizes K. Yano's f-structure. This generalization allows us to revisit classical theory and discover applications of Killing vector fields, totally geodesic foliations, Ricci-type solitons, and Einstein-type metrics. This article reviews the results regarding weak metric f-manifolds and their distinguished classes.

Keywords: metric f-structure; Killing vector field; totally geodesic foliation; η-Einstein metric; almost $\mathcal{S}$-structure; almost $\mathcal{C}$-structure; η-Ricci soliton; f-K-manifold; β-Kenmotsu; f-manifold.

MSC: 53C15; 53C25; 53D15

1. Introduction

Contact geometry has garnered increasing interest due to its significant role in physics, e.g., ref. [1]. An important class of contact metric manifolds are K-contact manifolds (thus, the structural vector field is Killing) with the following two subclasses: Sasakian and cosymplectic manifolds. Every cosymplectic manifold is locally the product of $\mathbb{R}$ and a Kähler manifold. A Riemannian manifold (M^{2n+1}, g) with contact 1-form η is Sasakian if its Riemannian cone $M \times \mathbb{R}^{>0}$ with metric $t^2 g + dt^2$ is a Kähler manifold. Recent research has been driven by the intriguing question of how Ricci solitons—self-similar solutions of the Ricci flow equation—can be significant for the geometry of contact metric manifolds. Some studies have explored the conditions under which a contact metric manifold equipped with a Ricci-type soliton structure carries a canonical metric, such as an Einstein-type metric, e.g., refs. [2–6].

An f-structure introduced by K. Yano on a smooth manifold M^{2n+s} serves as a higher-dimensional analog of almost complex structures ($s = 0$) and almost contact structures ($s = 1$). This structure is defined by a (1,1)-tensor f of rank $2n$, such that $f^3 + f = 0$, see [7–9]. The tangent bundle splits into two complementary subbundles as follows: $TM = f(TM) \oplus \ker f$. The restriction of f to the $2n$-dimensional distribution $f(TM)$ defines a complex structure. The existence of the f-structure on M^{2n+s} is equivalent to a reduction of the structure group to $U(n) \times O(s)$, see [10]. A submanifold M of an almost complex manifold $(\bar{M}, J)$ that satisfies the condition $\dim(T_x M \cap J(T_x M)) = const > 0$ naturally possesses an f-structure, see [11]. An f-structure is a special case of an almost product structure, defined by two complementary orthogonal distributions of a Riemannian manifold (M, g), with Naveira's 36 distinguished classes, see [12]. Foliations appear when one or both distributions are involutive. An interesting case occurs when the subbundle $\ker f$ is parallelizable, leading to a framed f-structure for which the reduced structure group is $U(n) \times \mathrm{id}_s$. In this scenario, there exist vector fields $\{\xi_i\}_{1\le i\le s}$ spanning $\ker f$

with dual 1-forms $\{\eta^i\}_{1\le i\le s}$, satisfying $f^2 = -\mathrm{id} + \sum_i \eta^i \otimes \xi_i$. Compatible metrics, i.e., $g(fX, fY) = g(X,Y) - \sum_i \eta^i(X)\,\eta^i(Y)$, exist on any framed f-manifold, and we obtain the metric f-structure, see [7–10,13–16].

To generalize concepts and results from almost contact geometry to metric f-manifolds, geometers have introduced and studied various broad classes of metric f-structures. A notable class is Kenmotsu f-manifolds, see [17] (Kenmotsu manifolds when $s = 1$, see [18]), characterized in terms of warped products of $\mathbb{R}^s$ and a Kähler manifold. A metric f-manifold is termed a $\mathcal{K}$-manifold if it is normal and $d\Phi = 0$, where $\Phi(X,Y) := g(X, fY)$. Two important subclasses of $\mathcal{K}$-manifolds are $\mathcal{C}$-manifolds if $d\eta^i = 0$ and $\mathcal{S}$-manifolds if $d\eta^i = \Phi$ for any i, see [10]. Omitting the normality condition, we obtain almost $\mathcal{K}$-manifolds, almost $\mathcal{S}$-manifolds and almost $\mathcal{C}$-manifolds, e.g., refs. [19–21]. The distribution $\ker f$ of a $\mathcal{K}$-manifold is tangent to a $\mathfrak{g}$-foliation with flat totally geodesic leaves. An f-K-contact manifold is an almost $\mathcal{S}$-manifold, whose characteristic vector fields are Killing vector fields; the structure is an intermediate between an almost $\mathcal{S}$-structure and S-structure, see [15,22]. Note that there are no Einstein metrics on f-K-contact manifolds. The interest of geometers in f-structures is also motivated by the study of the dynamics of contact foliations. Contact foliations generalize, to higher dimensions, the flow of the Reeb vector field on contact manifolds, and $\mathcal{K}$-structures are a particular case of uniform contact structures, see [23,24].

Extrinsic geometry is concerned with the properties of submanifolds (as being totally geodesic) that depend on the second fundamental form, which, roughly speaking, describes how the submanifolds are located in the ambient Riemannian manifold. The extrinsic geometry of foliations (i.e., involutive distributions) is a field of Riemannian geometry which studies the properties expressed by the second fundamental tensor of the leaves. Although the Riemann tensor belongs to the field of intrinsic geometry, a special component called mixed sectional curvature is a part of the extrinsic geometry of foliations. Totally geodesic foliations have simple extrinsic geometry and appear on manifolds with degenerate tensor fields, see [25]. A key problem posed in this context in [7] is identifying suitable structures on manifolds that lead to totally geodesic foliations.

In [26], we initiated the study of weak f-structures on a smooth $(2n+s)$-dimensional manifold, that is, the linear complex structure on the subbundle $\mathcal{D} = f(TM)$ of a metric f-structure is replaced with a nonsingular skew-symmetric tensor. These generalize the metric f-structure (the weak almost contact metric structure for $s = 1$, see [27]) and its satellites allow us to look at classical theory in a new way and find new applications of Killing, totally geodesic foliations, Einstein-type metrics, and Ricci-type solitons.

The article reviews the results of our works [25,26,28–32] regarding the geometry of weak metric f-manifolds and their distinguished classes. It is organized as follows. Section 2 (following the Introduction) presents the basics of weak metric f-manifolds and introduces their important subclasses. It also investigates the normality condition and derives the covariant derivative of f using a new tensor $\mathcal{N}^{(5)}$, showing that the distribution $\ker f$ of a weak almost $\mathcal{S}$-manifold and a weak almost $\mathcal{C}$-manifold is tangent to a $\mathfrak{g}$-foliation with an abelian Lie algebra. Section 3 presents the basic characteristics of weak almost $\mathcal{S}$-manifolds and shows that these manifolds are endowed with totally geodesic foliations. Section 4 discusses weak $\mathcal{C}$-structures and shows that the weak metric f-structure is a weak $\mathcal{S}$-structure if and only if it is an $\mathcal{S}$-structure. Section 5 characterizes weak f-K-contact manifolds among all weak almost $\mathcal{S}$-manifolds by the property known for f-K-contact manifolds, and presents the sufficient conditions under which a Riemannian manifold endowed with a set of unit Killing vector fields is a weak f-K-contact manifold. We also find the Ricci curvature of a weak f-K-contact manifold in the $\ker f$-directions, showing that there are no Einstein weak f-K-contact manifolds for $s > 1$ and that the mixed sectional curvature is positive. Using this, the weak f-K-contact structure can be deformed to the f-K-contact structure, and we obtain a topological obstruction (including the Adams number) to the existence of weak f-K-contact manifolds. Section 6 presents the sufficient conditions for a weak f-K-contact manifold with a generalized gradient Ricci soliton to be a quasi-Einstein or Ricci flat manifold. In Section 7, we find that the sufficient conditions for a compact weak f-K-contact manifold with the η-Ricci structure of constant scalar curvature is η-Einstein. Sections 8 and 9 show that a weak β-Kenmotsu f-manifold is locally a twisted product of $\mathbb{R}^s$ and a weak Kähler manifold, and in the case of an additional η-Ricci soliton structure, we explore their potential to be η-Einstein manifolds of a constant scalar curvature. The proofs (of some results) given in the article for the convenience of the reader use the properties of the new tensors, as well as the constructions needed in the classical case.

2. Preliminaries

In this section, we review the basics of the weak metric f-structure, see [26,28]. First, let us generalize the notion of a framed f-structure [8,9,13,14,16,33], called an f-structure with complemented frames in [10] or an f-structure with parallelizable kernel in [7].

Definition 1. *A framed weak f-structure on a smooth manifold* M^{2n+s} $(n, s > 0)$ *is a set* (f, Q, ξ_i, η^i)*, where* f *is a* $(1,1)$*-tensor of rank* $2n$*,* Q *is a nonsingular* $(1,1)$*-tensor,* ξ_i $(1 \le i \le s)$ *are structure vector fields, and* η^i $(1 \le i \le s)$ *are 1-forms, satisfying*

$$f^2 = -Q + \sum\nolimits_i \eta^i \otimes \xi_i, \quad \eta^i(\xi_j) = \delta^i_j, \quad Q\,\xi_i = \xi_i. \tag{1}$$

Then, theequality $f^3 + fQ = 0$ *holds. If there exists a Riemannian metric* g *on* M^{2n+s} *such that*

$$g(fX, fY) = g(X, QY) - \sum\nolimits_i \eta^i(X)\, \eta^i(Y), \quad X, Y \in \mathfrak{X}_M, \tag{2}$$

then, (f, Q, ξ_i, η^i, g) *is a weak metric f-structure, and* g *is called a compatible metric.*

Assume that a $2n$-dimensional contact distribution $\mathcal{D} = \bigcap_i \ker \eta^i$ is f-invariant. Note that for the framed weak f-structure, $\mathcal{D} = f(TM)$ is true, and

$$f\,\xi_i = 0, \quad \eta^i \circ f = 0, \quad \eta^i \circ Q = \eta^i, \quad [Q, f] = 0. \tag{3}$$

Using the above, the distribution $\mathcal{D}^\perp = \ker f$ is spanned by $\{\xi_1, \dots, \xi_s\}$ and is invariant for Q.

Remark 1. *The concept of an almost paracontact structure is analogous to the concept of an almost contact structure and is closely related to an almost product structure. Similarly to* (1)*, we define a framed weak para-f-structure, see details in [31], by*

$$f^2 = Q - \sum\nolimits_i \eta^i \otimes \xi_i, \quad \eta^i(\xi_j) = \delta^i_j, \quad Q\,\xi_i = \xi_i,$$

and we assume that a $2n$*-dimensional contact distribution* $\mathcal{D} = \bigcap_i \ker \eta^i$ *is* f*-invariant.*

The framed weak f-structure is called normal if the following tensor is zero:

$$\mathcal{N}^{(1)} = [f, f] + 2\sum\nolimits_i d\eta^i \otimes \xi_i. \tag{4}$$

The Nijenhuis torsion of a (1,1)-tensor S and the exterior derivative of a 1-form ω are given by

$$[S,S](X,Y) = S^2[X,Y] + [SX,SY] - S[SX,Y] - S[X,SY], \quad X,Y \in \mathfrak{X}_M,$$
$$d\omega(X,Y) = \frac{1}{2}\{X(\omega(Y)) - Y(\omega(X)) - \omega([X,Y])\}, \quad X,Y \in \mathfrak{X}_M.$$

Using the Levi–Civita connection ∇ of g, one can rewrite $[S,S]$ as

$$[S,S](X,Y) = (S\nabla_Y S - \nabla_{SY}S)X - (S\nabla_X S - \nabla_{SX}S)Y. \tag{5}$$

The following tensors $\mathcal{N}_i^{(2)}, \mathcal{N}_i^{(3)}$, and $\mathcal{N}_{ij}^{(4)}$ on framed weak f-manifolds, see [28,30], are well known in the classical theory, see [10], and are expressed as follows:

$$\mathcal{N}_i^{(2)}(X,Y) := (\pounds_{fX}\,\eta^i)(Y) - (\pounds_{fY}\,\eta^i)(X) = 2\,d\eta^i(fX,Y) - 2\,d\eta^i(fY,X),$$
$$\mathcal{N}_i^{(3)}(X) := (\pounds_{\xi_i} f)X = [\xi_i, fX] - f[\xi_i, X],$$
$$\mathcal{N}_{ij}^{(4)}(X) := (\pounds_{\xi_i}\,\eta^j)(X) = \xi_i(\eta^j(X)) - \eta^j([\xi_i, X]) = 2\,d\eta^j(\xi_i, X).$$

Remark 2. *Let $M^{2n+s}(f,Q,\xi_i,\eta^i)$ be a weak framed f-manifold. Consider the product manifold $\bar M = M^{2n+s} \times \mathbb{R}^s$, where $\mathbb{R}^s$ is a Euclidean space with a basis $\partial_1,\ldots,\partial_s$, and define tensors J and $\bar Q$ on $\bar M$ assuming $J(X, \sum_i a^i\partial_i) = (fX - \sum_i a^i\xi_i, \sum_j \eta^j(X)\partial_j)$ and $\bar Q(X, \sum_i a^i\partial_i) = (QX, \sum_i a^i\partial_i)$ for $a_i \in C^\infty(M)$. It can be shown that $J^2 = -\bar Q$. The tensors $\mathcal{N}_i^{(2)}, \mathcal{N}_i^{(3)}, \mathcal{N}_{ij}^{(4)}$ appear when we derive the integrability condition $[J,J] = 0$ and express the normality condition $\mathcal{N}^{(1)} = 0$ for (f,Q,ξ_i,η^i).*

A framed weak f-manifold admits a compatible metric if f has a skew-symmetric representation, i.e., for any $x \in M$, there exist a frame $\{e_i\}$ on a neighborhood $U_x \subset M$, for which f has a skew-symmetric matrix, see [26]. For a weak metric f-manifold, the tensor f is skew-symmetric and Q is self-adjoint and positive definite. Putting $Y = \xi_i$ in (2), and using $Q\,\xi_i = \xi_i$, we obtain $\eta^i(X) = g(X,\xi_i)$. Hence, ξ_i is orthogonal to $\mathcal{D}$ for any compatible metric. Thus, $TM = \mathcal{D} \oplus \mathcal{D}^\perp$—the sum of two complementary orthogonal subbundles.

A distribution $\widetilde{\mathcal{D}} \subset TM$ is called totally geodesic if and only if its second fundamental form vanishes, i.e., $\nabla_X Y + \nabla_Y X \in \widetilde{\mathcal{D}}$ for any vector fields $X,Y \in \widetilde{\mathcal{D}}$—this is the case when any geodesic of M that is tangent to $\widetilde{\mathcal{D}}$ at one point is tangent to $\widetilde{\mathcal{D}}$ at all its points, e.g., Section 1.3.1 [25]. According to the Frobenius theorem, any involutive distribution is tangent to (the leaves of) a foliation. Any involutive and totally geodesic distribution is tangent to a totally geodesic foliation. A foliation whose orthogonal distribution is totally geodesic is called a Riemannian foliation.

A "small" (1,1)-tensor $\widetilde{Q} = Q - \mathrm{id}_{TM}$ measures the difference between weak and classical f-structures. By (3), we obtain

$$[\widetilde{Q}, f] = 0, \quad \widetilde{Q}\,\xi_i = 0, \quad \eta^i \circ \widetilde{Q} = 0.$$

Proposition 1 (see [28])**.** *The normality condition for a weak metric f-structure implies*

$$\mathcal{N}_i^{(3)} = \mathcal{N}_{ij}^{(4)} = 0, \quad \mathcal{N}_i^{(2)}(X,Y) = \eta^i([\widetilde{Q}X, fY]), \tag{6}$$
$$\nabla_{\xi_i}\xi_j \in \mathcal{D}, \quad [X,\xi_i] \in \mathcal{D} \quad (X \in \mathcal{D}). \tag{7}$$

Moreover, $\mathcal{D}^\perp$ is a totally geodesic distribution.

The coboundary formula for exterior derivative of a 2-form Φ is

$$d\Phi(X,Y,Z) = \frac{1}{3}\{X\,\Phi(Y,Z) + Y\,\Phi(Z,X) + Z\,\Phi(X,Y) \\ - \Phi([X,Y],Z) - \Phi([Z,X],Y) - \Phi([Y,Z],X)\}. \tag{8}$$

Note that $d\Phi(X,Y,Z) = -\Phi([X,Y],Z)$ for $X,Y \in \mathcal{D}^\perp$. Therefore, for a weak metric f-structure, the distribution $\mathcal{D}^\perp$ is involutive if and only if $d\Phi = 0$.

Only one new tensor $\mathcal{N}^{(5)}$ (vanishing at $\widetilde{Q} = 0$), which supplements the sequence of well-known tensors $\mathcal{N}^{(1)}, \mathcal{N}_i^{(2)}, \mathcal{N}_i^{(3)}, \mathcal{N}_{ij}^{(4)}$, is needed to study the weak metric f-structure.

Proposition 2 (see [28]). *For a weak metric f-structure we obtain*

$$2\,g((\nabla_X f)Y,Z) = 3\,d\Phi(X,fY,fZ) - 3\,d\Phi(X,Y,Z) + g(\mathcal{N}^{(1)}(Y,Z),fX) \\ + \sum\nolimits_i (\mathcal{N}_i^{(2)}(Y,Z)\,\eta^i(X) + 2\,d\eta^i(fY,X)\,\eta^i(Z) - 2\,d\eta^i(fZ,X)\,\eta^i(Y)) \\ + \mathcal{N}^{(5)}(X,Y,Z), \tag{9}$$

where a skew-symmetric with respect to Y and Z tensor $\mathcal{N}^{(5)}(X,Y,Z)$ is defined by

$$\mathcal{N}^{(5)}(X,Y,Z) = fZ\,(g(X,\widetilde{Q}Y)) - fY\,(g(X,\widetilde{Q}Z)) \\ + g([X,fZ],\widetilde{Q}Y) - g([X,fY],\widetilde{Q}Z) + g([Y,fZ] - [Z,fY] - f[Y,Z],\ \widetilde{Q}X).$$

For particular values of the tensor $\mathcal{N}^{(5)}$ we obtain

$$\mathcal{N}^{(5)}(X,\xi_i,Z) = -\mathcal{N}^{(5)}(X,Z,\xi_i) = g(\mathcal{N}_i^{(3)}(Z),\ \widetilde{Q}X), \\ \mathcal{N}^{(5)}(\xi_i,Y,Z) = g([\xi_i,fZ],\widetilde{Q}Y) - g([\xi_i,fY],\widetilde{Q}Z), \\ \mathcal{N}^{(5)}(\xi_i,\xi_j,Y) = \mathcal{N}^{(5)}(\xi_i,Y,\xi_j) = 0. \tag{10}$$

Similar to the classical case, we introduce broad classes of weak metric f-structures.

Definition 2. *A weak metric f-structure (f,Q,ξ_i,η^i,g) is called*
(i) A weak $\mathcal{K}$-structure if it is normal and $d\Phi = 0$.
We define two subclasses of weak $\mathcal{K}$-manifolds as follows:
(ii) Weak $\mathcal{C}$-manifolds if $d\eta^i = 0$ for any i.
(iii) Weak $\mathcal{S}$-manifolds if the following is valid:

$$\Phi = d\eta^1 = \ldots = d\eta^s \quad (\text{hence, } d\Phi = 0). \tag{11}$$

Omitting the normality condition, we obtain the following: a weak metric f-structure is called
(i) A weak almost $\mathcal{K}$-structure if $d\Phi = 0$.
(ii) A weak almost $\mathcal{C}$-structure if Φ and η^i $(1 \le i \le s)$ are closed forms.
(iii) A weak almost $\mathcal{S}$-structure (or, a weak f-contact structure), if (11) *is valid.*
A weak almost $\mathcal{S}$-structure, whose structure vector fields ξ_i are Killing, i.e., the tensor

$$(\pounds_{\xi_i} g)(X,Y) := \xi_i(g(X,Y)) - g([\xi_i,X],Y) - g(X,[\xi_i,Y]) = g(\nabla_Y\,\xi_i,X) + g(\nabla_X\,\xi_i,Y),$$

vanishes, is called a weak f-K-contact structure.

For a weak almost $\mathcal{K}$-structure (and its special cases, a weak almost $\mathcal{S}$-structure and a weak almost $\mathcal{C}$-structure), the distribution $\mathcal{D}^\perp$ is involutive. Moreover, for a weak almost $\mathcal{S}$-structure and a weak almost $\mathcal{C}$-structure, we obtain $[\xi_i, \xi_j] = 0$; in other words, the distribution $\mathcal{D}^\perp$ of these manifolds is tangent to a $\mathfrak{g}$-foliation with an abelian Lie algebra.

Remark 3. *Let $\mathfrak{g}$ be a Lie algebra of dimension s. We can say that a foliation $\mathcal{F}$ of dimension s on a smooth connected manifold M is a $\mathfrak{g}$-foliation if there exist complete vector fields $\xi_1, \ldots, \xi_s$ on M, which, when restricted to each leaf of $\mathcal{F}$, form a parallelism of this submanifold with a Lie algebra isomorphic to $\mathfrak{g}$, see, for example, refs. [26,34].*

The following diagram (well known for classical structures) summarizes the relationships between some classes of weak metric f-manifolds considered in this article:

weak metric f-manifold	$\overset{\Phi=d\eta^i}{\supset}$	weak almost $\mathcal{S}$-manifold	$\overset{\xi_i\text{-Killing}}{\supset}$	weak f-K-contact	$\overset{\mathcal{N}^{(1)}=0}{\supset}$	weak $\mathcal{S}$-manifold

.

For $s = 1$, we obtain the following diagram:

weak almost contact metric manifold	$\overset{\Phi=d\eta}{\supset}$	weak contact metric manifold	$\overset{\xi\text{-Killing}}{\supset}$	weak K-contact manifold	$\overset{\mathcal{N}^{(1)}=0}{\supset}$	weak Sasakian manifold

.

3. Geometry of Weak Almost $\mathcal{S}$-Manifolds

For a weak almost $\mathcal{S}$-structure, the distribution $\mathcal{D}$ is not involutive, since we have

$$g([X, fX], \xi_i) = 2\, d\eta^i(fX, X) = g(fX, fX) > 0 \quad (X \in \mathcal{D} \setminus \{0\}).$$

Proposition 3 (see Theorem 2.2 in [28]). *For a weak almost $\mathcal{S}$-structure, the tensors $\mathcal{N}_i^{(2)}$ and $\mathcal{N}_{ij}^{(4)}$ vanish; moreover, $\mathcal{N}_i^{(3)}$ vanishes if and only if ξ_i is a Killing vector field.*

By $\mathcal{N}_{ij}^{(4)} = 0$, we have $g(X, \nabla_{\xi_i} \xi_j) + g(\nabla_X \xi_i, \xi_j) = 0$ for all $X \in \mathfrak{X}_M$. Symmetrizing the above equality (with $i \leftrightarrow j$) and using $g(\xi_i, \xi_j) = \delta_{ij}$ yields $\nabla_{\xi_i} \xi_j + \nabla_{\xi_j} \xi_i = 0$. From this and the equality $[\xi_i, \xi_j] = 0$, it follows that weak almost $\mathcal{S}$-manifolds satisfy

$$\nabla_{\xi_i} \xi_j = 0, \qquad 1 \le i, j \le s.$$

Corollary 1. *For a weak almost $\mathcal{S}$-structure, the distribution $\mathcal{D}^\perp$ is tangent to a $\mathfrak{g}$-foliation with totally geodesic flat (that is $R_{\xi_i, \xi_j} \xi_k = 0$) leaves.*

The following corollary of Propositions 2 and 3 generalizes well-known results with $Q = \mathrm{id}_{TM}$, e.g., Proposition 1.4 [10] and Proposition 2.1 [35].

Proposition 4. *For a weak almost $\mathcal{S}$-structure we obtain*

$$2\, g((\nabla_X f)Y, Z) = g(\mathcal{N}^{(1)}(Y, Z), fX) + 2\, g(fX, fY)\, \bar\eta(Z) - 2\, g(fX, fZ)\, \bar\eta(Y) + \mathcal{N}^{(5)}(X, Y, Z), \tag{12}$$

where $\bar\eta = \sum_i \eta^i$. Taking $X = \xi_i$ in (12), we obtain

$$2\, g((\nabla_{\xi_i} f)Y, Z) = \mathcal{N}^{(5)}(\xi_i, Y, Z), \quad 1 \le i \le s. \tag{13}$$

The tensor $\mathcal{N}_i^{(3)}$ is important for weak almost $\mathcal{S}$-manifolds, see Proposition 3. Therefore, we define the tensor field $\mathbf{h} = (h_1, \ldots, h_s)$, where

$$h_i = \frac{1}{2}\mathcal{N}_i^{(3)} = \frac{1}{2}\pounds_{\xi_i} f.$$

Using $[\xi_i, \xi_j] = 0$ and $f\xi_j = 0$, we obtain $(\pounds_{\xi_i} f)\xi_j = [\xi_i, f\xi_j] - f[\xi_i, \xi_j] = 0$; therefore, $h_i\,\xi_j = 0$ is true. For $X \in \mathcal{D}$, using the equality $[\xi_i, \xi_j] = 0$, we derive the following:

$$0 = 2\Phi(\xi_i, X) = 2\eta^j(\xi_i, X) = g(\nabla_X\,\xi_j, \xi_i);$$

therefore, $g(\nabla_X\,\xi_j, \xi_i) = 0$ for all $X \in \mathfrak{X}_M$. Next, we calculate

$$(\pounds_{\xi_i} f)Y = (\nabla_{\xi_i} f)Y - \nabla_{fY}\,\xi_i + f\nabla_Y\,\xi_i.$$

Thus, using $g((\nabla_{\xi_i} f)Y, \xi_j) = 0$, see (13) with $Z = \xi_j$, we obtain $\eta^j \circ h_i = 0$ for all $1 \le i, j \le s$:

$$(\eta^j \circ h_i)Y = g((\pounds_{\xi_i} f)Y, \xi_j) = g((\nabla_{\xi_i} f)Y, \xi_j) - g(\nabla_{fY}\,\xi_i, \xi_j) + g(f\nabla_Y\,\xi_i, \xi_j) = 0.$$

For an almost $\mathcal{S}$-structure, the tensor h_i is self-adjoint, trace-free, and anti-commutes with f, i.e., $h_i f + f\,h_i = 0$, see [35]. We generalize this result for a weak almost $\mathcal{S}$-structure.

Proposition 5 (see [28,29])**.** *For a weak almost $\mathcal{S}$-structure (f, Q, ξ_i, η^i, g), the tensor h_i and its conjugate tensor h_i^* satisfy*

$$\begin{aligned} g((h_i - h_i^*)X, Y) &= \frac{1}{2}\mathcal{N}^{(5)}(\xi_i, X, Y), \quad X, Y \in TM, \\ h_i f + f\,h_i &= -\frac{1}{2}\pounds_{\xi_i} Q, \\ h_i Q - Q\,h_i &= \frac{1}{2}\,[f, \pounds_{\xi_i} Q]. \end{aligned}$$

Let us consider the following condition (which is trivially satisfied by metric f-manifolds):

$$\pounds_{\xi_i} Q = 0. \tag{14}$$

The following corollary generalizes the known property of almost $\mathcal{S}$-manifolds.

Corollary 2. *Let a weak almost $\mathcal{S}$-manifold satisfy* (14)*, then, $h_i = h_i^*$ and* $\operatorname{trace} h_i = 0$ *for all i.*

Proof. Under the conditions and Proposition 5, h_i commutes with Q. Since the self-adjoint tensor Q is positive definite, then, h_i is also self-adjoint, that is, $h_i = h_i^*$. If $h_i X = \lambda X$, then using $h_i f = -f h_i$ (by assumptions and Proposition 5), we obtain $h_i f X = -\lambda f X$. Thus, if λ is an eigenvalue of h_i, then $-\lambda$ is also an eigenvalue of h_i; hence, $\operatorname{trace} h_i = 0$. □

Definition 3 (see [28])**.** *Framed weak f-structures (f, Q, ξ_i, η^i) and (f', Q', ξ_i, η^i) on a smooth manifold are said to be homothetic if the following conditions:*

$$f = \sqrt{\lambda}\, f', \tag{15a}$$

$$Q\,|_{\mathcal{D}} = \lambda\, Q'\,|_{\mathcal{D}}, \tag{15b}$$

are valid for some real $\lambda > 0$. Weak metric f-structures (f, Q, ξ_i, η^i, g) and $(f', Q', \xi_i, \eta^i, g')$ are said to be homothetic if they satisfy conditions (15a,b) and

$$g\,|_{\mathcal{D}} = \lambda^{-\frac{1}{2}}\, g'\,|_{\mathcal{D}}, \quad g(\xi_i, \cdot) = g'(\xi_i, \cdot). \tag{16}$$

Proposition 6 (see [28]). *Let a framed weak f-structure (f, Q, ξ_i, η^i) satisfy $Q\,|_{\mathcal{D}} = \lambda\, \mathrm{id}_{\mathcal{D}}$ for some real $\lambda > 0$. Then, (f', ξ_i, η^i) is a framed f-structure, where f' is given by* (15)*. Moreover, if (f, Q, ξ_i, η^i, g) is a weak almost $\mathcal{S}$-structure and* (15) *and* (16) *are valid, then (f', ξ_i, η^i, g') is an almost $\mathcal{S}$-structure.*

Denote by $\mathrm{Ric}(X,Y) = \mathrm{trace}(Z \to R_{Z,X}Y)$ the Ricci tensor, where $R_{X,Y} = (\nabla_X\nabla_Y - \nabla_Y\nabla_X - \nabla_{[X,Y]})Z$ is the curvature tensor. The Ricci operator is given by $g(\mathrm{Ric}^\sharp X, Y) = \mathrm{Ric}(X,Y)$. The scalar curvature of g is given by $r = \mathrm{trace}_g \mathrm{Ric}$.

Remark 4. *For almost $\mathcal{S}$-manifolds, we have, see Proposition 2.6 [35],*

$$\mathrm{Ric}^\sharp(\xi_i) = \sum\nolimits_{i=1}^{2n} (\nabla_{e_i} f)\, e_i = 2\,n\,\bar{\xi}. \tag{17}$$

Can one generalize (17) *for weak almost $\mathcal{S}$-manifolds?*

For a weak almost $\mathcal{S}$-manifold, the *splitting tensor* $C : \mathcal{D}^\perp \times \mathcal{D} \to \mathcal{D}$ is defined by

$$C_\xi(X) = -(\nabla_X\, \xi)^\top \quad (X \in \mathcal{D},\ \xi \in \mathcal{D}^\perp,\ \|\xi\| = 1), \tag{18}$$

where $^\top : TM \to \mathcal{D}$ is the orthoprojector, see [29]. The splitting tensor is decomposed as $C_\xi = A_\xi + T^\sharp_\xi$, where the skew-symmetric operator $T^\sharp_\xi$ and the self-adjoint operator A_ξ are defined using the integrability tensor $T(X,Y) = \frac{1}{2}(\nabla_X Y - \nabla_Y X)^\perp$ and the second fundamental form $b(X,Y) = \frac{1}{2}(\nabla_X Y + \nabla_Y X)^\perp$ of $\mathcal{D}$ by

$$g(T^\sharp_\xi X, Y) = g(T(X,Y), \xi), \quad g(A_\xi X, Y) = g(b(X,Y), \xi), \quad X, Y \in \mathcal{D}. \tag{19}$$

Since $\mathcal{D}^\perp$ defines a totally geodesic foliation, see Corollary 1, then the distribution $\mathcal{D}$ is totally geodesic if and only if C_ξ is skew-symmetric, and $\mathcal{D}$ is integrable if and only if the tensor C_ξ is self-adjoint. Thus, $C_\xi \equiv 0$ if and only if $\mathcal{D}$ is integrable and defines a totally geodesic foliation; in this case, by de Rham Decomposition Theorem, the manifold splits (is locally the product of Riemannian manifolds, defined by distributions $\mathcal{D}$ and $\mathcal{D}^\perp$), e.g., ref. [25].

Theorem 1. *The splitting tensor of a weak almost $\mathcal{S}$-manifold has the following view:*

$$C_{\xi_i} = f + Q^{-1} f h_i^* \quad (i = 1, \ldots, s).$$

The mixed scalar curvature of an almost product manifold $M(\mathcal{D}_1, \mathcal{D}_2, g)$ is the function

$$r_{\rm mix} = \sum\nolimits_{a,i} g(R_{E_a, \mathcal{E}_i}\, E_a, \mathcal{E}_i),$$

where $\{\mathcal{E}_i,\, E_a\}$ is an adapted orthonormal frame, i.e., $\{E_a\} \subset \mathcal{D}_1$ and $\{\mathcal{E}_i\} \subset \mathcal{D}_2$. Let b_i and H_i be the second fundamental form and the mean curvature vector, and let T_i be the integrability tensor of the distribution $\mathcal{D}_i$. The following formula:

$$r_{\rm mix} = \mathrm{div}(H_1 + H_2) - \|b_1\|^2 - \|b_2\|^2 + \|H_1\|^2 + \|H_2\|^2 + \|T_1\|^2 + \|T_2\|^2, \tag{20}$$

has many applications in Riemannian, Kähler and Sasakian geometries, see [25].

Theorem 2. *For the weak almost $\mathcal{S}$-structure on a closed manifold M^{2n+s} satisfying condition* (14)*, the following integral formula is true:*

$$\int_M \Big\{ \sum\nolimits_i \big(\mathrm{Ric}(\xi_i, \xi_i) + \|Q^{-1} f h_i\|^2 - (\,\mathrm{trace}\, Q^{-1} f h_i)^2 \big) - s\,\|f\|^2 \Big\}\, d\,\mathrm{vol} = 0. \tag{21}$$

Proof. According to (20), set $\mathcal{D}_1 = \mathcal{D}$ and $\mathcal{D}_2 = \mathcal{D}^\perp$. By the assumptions, Proposition 5 and Corollary 2, $h_i^* = h_i$, $h_i f + f h_i = 0$ and $[h_i, Q] = 0$. Then, $b_2 = H_2 = T_2 = 0$ and

$$b_1(X,Y) = \sum\nolimits_i g(Q^{-1} f h_i X, Y)\, \xi_i, \quad H_1 = \sum\nolimits_i \mathrm{trace}\,(Q^{-1} f h_i)\, \xi_i, \quad T_1(X,Y) = g(fX, Y)\, \bar{\xi},$$

where $\bar{\xi} = \sum_i \xi_i$. For a weak almost $\mathcal{S}$-manifold, we have $r_{\rm mix} = \sum_i \mathrm{Ric}(\xi_i, \xi_i)$. Thus, (21) is the counterpart of (20) integrated on a closed manifold using the Divergence Theorem. □

Definition 4 (see [36])**.** *An even-dimensional Riemannian manifold $(\tilde{M}, \tilde{g})$ equipped with a skew-symmetric $(1,1)$-tensor J such that J^2 is negative-definite is called a weak Hermitian manifold. This manifold is called weak Kählerian if $\tilde{\nabla} J = 0$, where $\tilde{\nabla}$ is the Levi–Civita connection of $\tilde{g}$.*

An involutive distribution is *regular* if every point of the manifold has a neighborhood such that any integral submanifold passing through the neighborhood passes through only once, see, for example, ref. [21]. The next theorem states that a compact manifold with a regular weak almost $\mathcal{S}$-structure is a principal torus bundle over a weak Hermitian manifold, and we believe that its proof using Proposition 3 is similar to the proof of Theorem 4.2 [21].

Theorem 3. *Let M^{2n+s} be a compact manifold equipped with a regular weak almost $\mathcal{S}$-structure $(f, Q, \tilde{\xi}_i, \tilde{\eta}^i, \tilde{g})$. Then, there exists a weak almost $\mathcal{S}$-structure (f, Q, ξ_i, η^i, g) on M for which the structure vector fields $\xi_1, \ldots, \xi_s$ are the infinitesimal generators of a free and effective $\mathbb{T}^s$-action on M. Moreover, the quotient $N = M/\mathbb{T}^s$ is a smooth weak Hermitian manifold of dimension $2n$.*

4. Geometry of Weak $\mathcal{K}$-Manifolds and Their Two Subclasses

The following result generalizes Theorem 1.1 [10].

Theorem 4. *On a weak $\mathcal{K}$-manifold the structure vector fields $\xi_1, \ldots, \xi_s$ are Killing and*

$$\nabla_{\xi_i} \xi_j = 0, \quad 1 \le i, j \le s; \tag{22}$$

thus, the distribution $\mathcal{D}^\perp$ is tangent to a totally geodesic Riemannian foliation with flat leaves.

Proof. By Proposition 1, $\mathcal{D}^\perp$ is totally geodesic and $\mathcal{N}_i^{(3)} = \pounds_{\xi_i} f = 0$. Using $\iota_{\xi_i} \Phi = 0$ and condition $d\Phi = 0$ in the identity $\pounds_{\xi_i} = \iota_{\xi_i}\, d + d\, \iota_{\xi_i}$, we obtain $\pounds_{\xi_i} \Phi = 0$. By direct calculation we obtain the following:

$$(\pounds_{\xi_i} \Phi)(X,Y) = (\pounds_{\xi_i} g)(X, fY) + g(X, (\pounds_{\xi_i} f) Y). \tag{23}$$

Thus, from (23) we obtain $(\pounds_{\xi_i} g)(X, fY) = 0$. To show $\pounds_{\xi_i} g = 0$, we will examine $(\pounds_{\xi_i} g)(fX, \xi_j)$ and $(\pounds_{\xi_i} g)(\xi_k, \xi_j)$. Using $\pounds_{\xi_i} \eta^j = 0$, we obtain $(\pounds_{\xi_i} g)(fX, \xi_j) = (\pounds_{\xi_i} \eta^j) fX - g(fX, [\xi_i, \xi_j]) = -g(fX, [\xi_i, \xi_j]) = 0$. Next, using Proposition 1, we obtain

$$(\pounds_{\xi_i} g)(\xi_k, \xi_j) = -g(\xi_i, \nabla_{\xi_k} \xi_j + \nabla_{\xi_j} \xi_k) = 0.$$

Thus, ξ_i is a Killing vector field, i.e., $\pounds_{\xi_i} g = 0$. From $d\Phi(X, \xi_i, \xi_j) = 0$ and (8) we obtain $g([\xi_i, \xi_j], fX) = 0$, i.e., $\ker f$ is integrable. Finally, from this and Proposition 1 we obtain (22); thus, the sectional curvature $K(\xi_i, \xi_j)$ vanishes. □

By Proposition 4 with $\mathcal{N}^{(1)} = 0$, we obtain the following.

Corollary 3. *For a weak $\mathcal{S}$-structure (f, Q, ξ_i, η^i, g) we obtain*

$$g((\nabla_X f)Y, Z) = g(fX, fY)\,\bar{\eta}(Z) - g(fX, fZ)\,\bar{\eta}(Y) + \frac{1}{2}\mathcal{N}^{(5)}(X, Y, Z). \tag{24}$$

Moreover, ξ_i are Killing vector fields and $\mathcal{D}^\perp$ is tangent to a Riemannian totally geodesic foliation.

Using Corollary 3, we obtain a rigidity theorem for an $\mathcal{S}$-structure.

Theorem 5. *A weak metric f-structure is a weak $\mathcal{S}$-structure if and only if it is an $\mathcal{S}$-structure.*

Proof. Let (f, Q, ξ_i, η^i, g) be a weak $\mathcal{S}$-structure. Since $\mathcal{N}^{(1)} = 0$, by Proposition 1, we obtain $\mathcal{N}^{(3)}_i = 0$. By (10), we obtain $\mathcal{N}^{(5)}(\cdot, \xi_i, \cdot) = \mathcal{N}^{(5)}(\cdot, \cdot, \xi_i) = 0$. Since f is skew-symmetric, applying (24) with $Z = \xi_i$ in (5) yields

$$g([f, f](X, Y), \xi_i) = g((\nabla_{fX} f)Y, \xi_i) - g((\nabla_{fY} f)X, \xi_i) = -2\,g(QX, fY).$$

From this and $\mathcal{N}^{(1)} = 0$, we obtain $g(\widetilde{Q}X, fY) = 0$ for all $X, Y \in \mathfrak{X}_M$; therefore, $\widetilde{Q} = 0$. □

For $s = 1$, from Theorem 5, we have the following

Corollary 4 (see [37]). *A weak almost contact metric structure on M^{2n+1} is weak Sasakian if and only if it is a Sasakian structure (i.e., a normal contact metric structure) on M^{2n+1}.*

Next, we study a weak almost $\mathcal{C}$-structure.

Proposition 7. *For a weak $\mathcal{C}$-structure (f, Q, ξ_i, η^i, g), we obtain*

$$2\,g((\nabla_X f)Y, Z) = \mathcal{N}^{(5)}(X, Y, Z). \tag{25}$$

A $\mathcal{K}$-structure is a $\mathcal{C}$-structure if and only if f is parallel, e.g., Theorem 1.5 [10]. The following theorem extends this result and characterizes weak $\mathcal{C}$-manifolds using the condition $\nabla f = 0$.

Theorem 6. *A weak metric f-structure (f, Q, ξ_i, η^i, g) with conditions $\nabla f = 0$ and*

$$[\xi_i, \xi_j]^\perp = 0, \quad 1 \le i, j \le s,$$

is a weak $\mathcal{C}$-structure with the property $\mathcal{N}^{(5)} = 0$.

Proof. Using condition $\nabla f = 0$, from (5), we obtain $[f, f] = 0$. Hence, from (4), we obtain $\mathcal{N}^{(1)}(X, Y) = 2\sum_i d\eta^i(X, Y)\,\xi_i$, and from (5), with $S = f$ and $Y = \xi_i$, we obtain

$$\nabla_{fX}\,\xi_i - f\,\nabla_X\,\xi_i = 0, \quad X \in \mathfrak{X}_M. \tag{26}$$

From (8), we calculate

$$3\,d\Phi(X,Y,Z) = g((\nabla_X f)Z,Y) + g((\nabla_Y f)X,Z) + g((\nabla_Z f)Y,X);$$

hence, using condition $\nabla f = 0$ again, we obtain $d\Phi = 0$. Next,

$$\mathcal{N}_i^{(2)}(Y,\xi_j) = -\eta^i([fY,\xi_j]) = g(\xi_j, f\nabla_{\xi_i}Y) = 0.$$

Thus, setting $Z = \xi_j$ in Proposition 4 and using the condition $\nabla f = 0$ and the properties $d\Phi = 0$, $\mathcal{N}_i^{(2)}(Y,\xi_j) = 0$, and $\mathcal{N}^{(1)}(X,Y) = 2\sum_i d\eta^i(X,Y)\,\xi_i$, we find $0 = 2\,d\eta^i(fY,X) - \mathcal{N}^{(5)}(X,\xi_i,Y)$. By (10) and (26), we obtain

$$\mathcal{N}^{(5)}(X,\xi_i,Y) = g([\xi_i,fY] - f[\xi_i,Y],\, \widetilde{Q}X) = g(\nabla_{fY}\,\xi_i - f\,\nabla_Y\,\xi_i,\, \widetilde{Q}X) = 0;$$

hence, $d\eta^i(fY,X) = 0$. From this and condition $g([\xi_i,\xi_j],\xi_k) = 2\,d\eta^k(\xi_j,\xi_i) = 0$, we obtain $d\eta^i = 0$. By the above, $\mathcal{N}^{(1)} = 0$. Thus, (f,Q,ξ_i,η^i,g) is a weak $\mathcal{C}$-structure. Finally, from (25) and condition $\nabla f = 0$, we obtain $\mathcal{N}^{(5)} = 0$. □

Example 1. *Let M be a $2n$-dimensional smooth manifold and $J : TM \to TM$ an endomorphism of rank $2n$ such that $\nabla J = 0$. To construct a weak $\mathcal{C}$-structure on $M \times \mathbb{R}^s$ or $M \times T^s$, where T^s is an s-dimensional torus, take any point $(x,t_1,\dots,t_s)$ of either space and set $\xi_i = (0,\partial/\partial t_i)$, $\eta^i = (0,dt_i)$, and*

$$f(X,Y) = (JX,0), \quad Q(X,Y) = (-J^2X,\, Y).$$

where $X \in M_x$ and $Y \in \{\mathbb{R}^s_t, T^s_t\}$. Then, (1) *holds and Theorem 6 can be used.*

5. Geometry of Weak f-K-Contact Manifolds

Here, we characterize weak f-K-contact manifolds among all weak almost $\mathcal{S}$-manifolds and find conditions under which a Riemannian manifold endowed with a set of unit Killing vector fields becomes a weak f-K-contact manifold.

Lemma 1. *For a weak f-K-contact manifold we obtain*

$$\begin{aligned}
\mathcal{N}^{(1)}(\xi_i,\cdot) &= 0,\\
\mathcal{N}^{(5)}(\xi_i,\cdot,\cdot) &= \mathcal{N}^{(5)}(\cdot,\xi_i,\cdot) = 0,\\
\pounds_{\xi_i}Q &= \nabla_{\xi_i}Q = 0,\\
\nabla_{\xi_i}f &= 0.
\end{aligned}$$

Recall the following property of f-K-contact manifolds, see [10,22]:

$$\nabla\,\xi_i = -f, \quad 1 \le i \le s. \tag{27}$$

Using Proposition 3 and Theorem 1, we obtain the following.

Theorem 7. *A weak almost $\mathcal{S}$-structure is weak f-K-contact if and only if* (27) *is true.*

From Proposition 5 and Theorem 1, using Lemma 1, we obtain $h_i = h_i^* = 0\ (1 \le i \le s)$.

The mapping $R_{\xi_i} : X \mapsto R_{X,\xi_i}\,\xi_i$ $(\xi \in \mathcal{D}^\perp,\ \|\xi\| = 1)$ is called the *Jacobi operator* in the ξ_i-direction, e.g., ref. [25]. For a weak almost $\mathcal{S}$-manifold, by Proposition 4, we obtain $R_{\xi_i}(X) \in \mathcal{D}$.

Theorem 8. *Let (M^{2n+s}, g) be a Riemannian manifold with orthonormal Killing vector fields $\xi_i, \ldots, \xi_s$ such that $d\,\eta^1 = \ldots = d\,\eta^s$ (where η^i is the 1-form dual to ξ_i) and the Jacobi operators R_{ξ_i} $(i \le s)$ are positive definite on the distribution $\mathcal{D} = \bigcap_i \ker \eta^i$. Then, the manifold is weak f-K-contact, and its structural tensors are as follows:*

$$f = -\nabla\,\xi_i, \quad QX = R_{\xi_i}(X) \quad (X \in \mathcal{D},\ 1 \le i \le s).$$

Proof. Since ξ_i are Killing vector fields, we obtain the following property (11):

$$d\eta^i(X,Y) = (1/2)\,(g(\nabla_X\,\xi_i, Y) - g(\nabla_Y\,\xi_i, X)) = -g(\nabla_Y\,\xi_i, X) = g(X, fY).$$

Set $QX = R_{\xi_i}(X)$ for some i and all $X \in \mathcal{D}$. Since ξ_i is a unit Killing vector field, we obtain $\nabla_{\xi_i}\xi_i = 0$ and $\nabla_X\nabla_Y\,\xi_i - \nabla_{\nabla_X Y}\,\xi_i = R_{X,\xi_i}Y$, see [9]. Thus, $f\,\xi_i = 0$ is true, and

$$f^2 Y = \nabla_{\nabla_Y\,\xi_i}\,\xi_i = R_{\xi_i,Y}\,\xi_i = -R_{\xi_i}(Y) = -QY, \quad Y \in \mathcal{D}.$$

By the conditions, the tensor Q is positive definite on the subbundle $\mathcal{D}$. Therefore, the rank of f restricted to $\mathcal{D}$ is $2\,n$. Set $Q\,\xi_i = \xi_i$ $(1 \le i \le s)$. Thus, (1) and (2) are true. □

The sectional curvature of a plane containing unit vectors $\xi \in \mathcal{D}^\perp$ and $X \in \mathcal{D}$ is called mixed sectional curvature. The mixed sectional curvature of an almost $\mathcal{S}$-manifold is a spacial case of mixed sectional curvature of almost product manifolds, for example, ref. [25]. Note that the mixed sectional curvature of an f-K-contact manifold is constant and equal to 1.

Proposition 8. *A weak f-K-contact structure (f, Q, ξ_i, η^i, g) of constant mixed sectional curvature, satisfying $K(\xi_i, X) = \lambda > 0$ for all $X \in \mathcal{D}$ and some $\lambda = const \in \mathbb{R}$, is homothetic to an f-K-contact structure (f', ξ_i, η^i, g') after the transformation (15a,b)–(16).*

Example 2. *According to Theorem 8, we can search for examples of weak f-K-contact (not f-K-contact) manifolds that can be found among Riemannian manifolds of positive sectional curvature admitting $s \ge 1$ mutually orthogonal unit Killing vector fields. Set $s = 1$, and let M^{n+1} be an ellipsoid with induced metric g of $\mathbb{R}^{2n+2}$ $(n \ge 1)$,*

$$M = \Big\{(u_1, \ldots, u_{2n+2}) \in \mathbb{R}^{2n+2} : \sum\nolimits_{i=1}^{n+1} u_i^2 + a \sum\nolimits_{i=n+2}^{2n+1} u_i^2 = 1\Big\},$$

where $0 < a = const \neq 1$. The sectional curvature of (M, g) is positive. It follows that

$$\xi = (-u_2, u_1, \ldots, -u_{n+1}, u_n, -\sqrt{a}\,u_{n+3}, \sqrt{a}\,u_{n+2}, \ldots, -\sqrt{a}\,u_{2n+2}, \sqrt{a}\,u_{2n+1})$$

is a Killing vector field on $\mathbb{R}^{2n+2}$, whose restriction to M has unit length. Since M is invariant under the flow of ξ, then ξ is a unit Killing vector field on (M, g).

Since $K(\xi_i, \xi_j) = 0$ for a weak almost $\mathcal{S}$-manifold, the Ricci curvature in the ξ_j-direction is given by

$$\mathrm{Ric}(\xi_j, \xi_j) = \sum\nolimits_{i=1}^{2n} g(R_{e_i,\xi_j}\,\xi_j, e_i),$$

where (e_i) is a local orthonormal basis of $\mathcal{D}$. The next proposition generalizes some particular properties of f-K-contact manifolds for the case of weak f-K-contact manifolds.

Proposition 9. *Let $M^{2n+s}(f,Q,\xi_i,\eta^i,g)$ be a weak f-K-contact manifold, then, for all i,j we obtain*

$$R_{\xi_i,Y} = \nabla_Y f \quad (Y \in \mathfrak{X}_M), \tag{28}$$
$$R_{\xi_i,Y}\,\xi_j = f^2 Y \quad (Y \in \mathfrak{X}_M), \tag{29}$$
$$\mathrm{Ric}^\sharp\,\xi_i = \mathrm{div}\,f, \tag{30}$$
$$\mathrm{Ric}(\xi_i,\xi_j) = \mathrm{trace}\,Q > 0. \tag{31}$$

Proposition 10. *There are no Einstein weak f-K-contact manifolds with $s > 1$.*

Proof. A weak f-K-contact manifold with $s > 1$ and $\xi' = \frac{\xi_1+\xi_2}{\sqrt{2}}$, satisfies the following:

$$\mathrm{Ric}(\xi',\xi') = \frac{1}{2}\sum\nolimits_{i,j=1}^{2}\mathrm{Ric}(\xi_i,\xi_j) \overset{(31)}{=} 2\,\mathrm{trace}\,Q. \tag{32}$$

If the manifold is an Einstein manifold, then for the unit vector field ξ', we obtain $\mathrm{Ric}(\xi',\xi') = \mathrm{Ric}(\xi_i,\xi_i) = \mathrm{trace}\,Q$. Comparing this with (32) yields $\mathrm{trace}\,Q = 0$—a contradiction. □

For a f-K-contact manifold, Equations (30) and (17) give $\mathrm{Ric}^\sharp(\xi_i) = 2\,n\,\bar{\xi}$ $(1 \le i \le s)$ and $\mathrm{Ric}(\xi_i,\xi_i) = 2\,n$.

Proposition 11. *For a weak f-K-contact manifold, the mixed sectional curvature is positive, as follows:*

$$K(\xi_i, X) = g(QX,X) > 0 \quad (X \in \mathcal{D},\ \|X\| = 1),$$

and the Ricci curvature satisfies the following: $\mathrm{Ric}(\xi_i,\xi_j) > 0$ *for all* $1 \le i,j \le s$.

Proof. From (29), we obtain $K(\xi, X) = g(fX, fX) > 0$ for any unit vectors $\xi \in \mathcal{D}^\perp$, $X \in \mathcal{D}$. Using (1) and non-singularity of f on $\mathcal{D}$, from (31) we obtain

$$\mathrm{Ric}(\xi_j,\xi_i) = \mathrm{trace}\,Q = -\sum\nolimits_{p=1}^{2n} g(f^2 e_p, e_p) = \sum\nolimits_{p=1}^{2n} g(f e_p, f e_p) > 0,$$

where (e_p) is a local orthonormal frame of $\mathcal{D}$, thus the second statement is valid. □

Theorem 9. *A weak f-K-contact manifold with conditions $(\nabla\,\mathrm{Ric})(\xi_i,\,\cdot) = 0$ $(1 \le i \le s)$ and $\mathrm{trace}\,Q = const$ is an Einstein manifold and $s = 1$.*

Proof. Differentiating (31) and using (27) and the conditions, we have

$$0 = \nabla_Y(\mathrm{Ric}(\xi_i,\xi_i)) = (\nabla_Y\,\mathrm{Ric})(\xi_i,\xi_i) + 2\,\mathrm{Ric}(\nabla_Y\,\xi_i,\xi_i) = -2\,\mathrm{Ric}(fY,\xi_i),$$

hence $\mathrm{Ric}(Y,\xi_i) = (\,\mathrm{trace}\,Q)\,\eta^i(Y)$. Differentiating this, then using

$$X(\eta^i(Y)) = g(\nabla_X\xi_i, Y) = -g(fX,Y) + g(\nabla_X Y, \xi_i)$$

and assuming $\nabla_X Y = 0$ at a point $x \in M$, gives

$$\begin{aligned}(\,\mathrm{trace}\,Q)\,g(fX,Y) &= \nabla_Y(\mathrm{Ric}(X,\xi_i)) \\ &= (\nabla_Y\,\mathrm{Ric})(X,\xi_i) + 2\,\mathrm{Ric}(X,\nabla_Y\,\xi_i) = -2\,\mathrm{Ric}(X,fY).\end{aligned}$$

Thus, $\mathrm{Ric}(X,fY) = (\,\mathrm{trace}\,Q)\,g(X,fY)$. Hence, $\mathrm{Ric}(X,Y) = (\,\mathrm{trace}\,Q)\,g(X,Y)$ for any vectors $X,Y \in T_xM$; therefore, (M,g) is an Einstein manifold. By Proposition 10, $s = 1$. □

The partial Ricci curvature tensor, see [26], is self-adjoint and is given by

$$\mathrm{Ric}^\top(X) = \sum\nolimits_{i=1}^{s}(R_{X^\top,\xi_i}\,\xi_i)^\top.$$

Note that $\mathrm{Ric}^\top = s\,\mathrm{id}_{\mathcal{D}}$ holds for f-K-contact manifolds. For weak f-K-contact manifolds, the tensor $\mathrm{Ric}^\top$ is positive definite, see Proposition 11. In [26], applying the flow of metrics for a $\mathfrak{g}$-foliation, a deformation of a weak almost $\mathcal{S}$-structure with positive partial Ricci curvature onto the the classical structures of the same kind was constructed.

The next theorem, using Proposition 11 and the method of [26], shows that a weak f-K-contact manifold can be deformed into an f-K-contact manifold.

Theorem 10. *Let $M^{2n+s}(f_0, Q_0, \xi_i, \eta^i, g_0)$ be a weak f-K-contact manifold. Then, there exist metrics g_τ ($\tau \in \mathbb{R}$), such that each $(f_t, Q_\tau, \xi_i, \eta^i, g_\tau)$ is a weak f-K-contact structure satisfying*

$$Q_\tau = (1/s)\,\mathrm{Ric}^\top_\tau, \quad f_\tau\,|_{\mathcal{D}} = T^\sharp_{\xi_i}(\tau). \tag{33}$$

Moreover, g_τ converges exponentially fast, as $\tau \to -\infty$, to a metric $\hat{g}$ with $\mathrm{Ric}^\top_{\hat{g}} = s\,\mathrm{id}_{\mathcal{D}}$ that gives an f-K-contact structure on M.

Proof. For a weak f-K-contact manifold, the tensor $\mathrm{Ric}^\top$ is positive definite. Thus, we can apply the method of Theorem 1 [26]. Consider the partial Ricci flow, see [25],

$$\partial_\tau g_\tau = -2\,(\mathrm{Ric}^\top)^\flat_{g_\tau} + 2\,s\,g^\top_\tau, \tag{34}$$

where the tensor $g^\top$ is given by $g^\top(X,Y) = g(X^\top, Y^\top)$. We obtain the following, based on $\mathcal{D}$ for $\tau = 0$:

$$\mathrm{Ric}^\top = -\sum\nolimits_i (T^\sharp_{\xi_i})^2 = -s\,f^2 = s\,Q,$$

and find $T^\sharp_{\xi_i}\,\mathrm{Ric}^\top = \mathrm{Ric}^\top\,T^\sharp_{\xi_i}$, see also [26]. Thus, $\sum_i T^\sharp_{\xi_i}\,\mathrm{Ric}^\top\,T^\sharp_{\xi_i} = -(\mathrm{Ric}^\top)^2$. By the above, we obtain the following ordinary differential equation:

$$\partial_\tau\,\mathrm{Ric}^\top = 4\,\mathrm{Ric}^\top(\mathrm{Ric}^\top - s\,\mathrm{id}_{\mathcal{D}}).$$

According to ODE theory, there exists a unique solution $\mathrm{Ric}^\top_\tau$ for $\tau \in \mathbb{R}$; thus, a solution g_τ of (34) exists for $\tau \in \mathbb{R}$ and it is unique. Observe that $(f(\tau), \xi_i, \eta^i, Q_\tau)$ with $f(\tau), Q_\tau$, given in (33), is a weak f-K-contact structure on (M, g_τ). By the uniqueness of the solution, the flow (34) preserves the directions of the eigenvectors of $\mathrm{Ric}^\top$, and each eigenvalue $\mu_i > 0$ satisfies the ODE $\dot{\mu}_i = 4\mu_i\,(\mu_i - s)$ with $\mu_i(0) > 0$. This ODE has the following solution:

$$\mu_i(\tau) = \frac{\mu_i(0)\,s}{\mu_i(0) + \exp(4\,s\,\tau)(s - \mu_i(0))}$$

(a function $\mu_i(\tau)$ on M for $\tau \in \mathbb{R}$) with $\lim\limits_{\tau\to-\infty}\mu_i(\tau) = s$. Thus, $\lim\limits_{\tau\to-\infty}\mathrm{Ric}^\top(\tau) = s\,\mathrm{id}_{\mathcal{D}}$. Let $\{e_i(\tau)\}$ be a g_τ-orthonormal frame of $\mathcal{D}$ of eigenvectors associated with $\mu_i(\tau)$, we then obtain $\partial_\tau e_i = (\mu_i - s)e_i$. Since $e_i(\tau) = z_i(\tau)\,e_i(0)$ with $z_i(0) = 1$, then $\partial_\tau \log z_i(\tau) = \mu_i(\tau) - s$. By the above, we obtain $z_i(\tau) = (\mu_i(\tau)/\mu_i(0))^{1/4}$. Hence,

$$g_t(e_i(0), e_j(0)) = z_i^{-1}(\tau) z_j^{-1}(\tau)\,g_t(e_i(\tau), e_j(\tau)) = \delta_{ij}(\mu_i(0)\mu_j(0)/(\mu_i(\tau)\mu_j(\tau)))^{1/4}.$$

$$\hat{g}(e_i(0), e_j(0)) = \delta_{ij}\sqrt{\mu_i(0)/s}. \quad \square$$

Denote by $\rho(n)-1$ the maximal number of point-wise linearly independent vector fields on a sphere S^{n-1}. The topological invariant $\rho(n)$, called the Adams number, is

$$\rho((\text{odd})\,2^{4b+c}) = 8b + 2^c \ \text{ for any integers } \ b \geq 0,\ 0 \leq c \leq 3,$$

see Table 1, and the inequality $\rho(n) \leq 2\,\log_2 n + 2$ is valid, for example, Section 1.4.4 [25].

Table 1. The number of vector fields on the $(2n-1)$ sphere.

$2n-1$	1	3	5	7	9	11	13	15	17	19	21	23	25	27	29
$\rho(2n)-1$	1	3	1	7	1	3	1	8	1	3	1	7	1	3	1

There are not many theorems in differential geometry that use $\rho(n)$. Applying the Adams number, we obtain a topological obstruction to the existence of weak f-K-contact manifolds.

Theorem 11. *For a weak f-K-contact manifold $M^{2n+s}(f, Q, \xi_i, \eta^i, g)$ we have $s < \rho(2n)$.*

Proof. For a weak almost $\mathcal{S}$-structure, the following Riccati equation is true, e.g., ref. [25]:

$$\nabla_\xi\, C_\xi + (C_\xi)^2 + R_\xi = 0 \quad (\xi \in \mathcal{D}^\perp). \tag{35}$$

Since the splitting tensor C_ξ is skew-symmetric for a weak f-K-contact manifold, i.e., $C_\xi = T^\sharp_\xi$ and $A_\xi = 0$, see (19), and the Jacobi operator R_ξ is self-adjoined, (35) reduces to two equations on $\mathcal{D}$, as follows:

$$\nabla_\xi\, T^\sharp_\xi = 0 \text{ (the skew-symmetric part)}, \quad (T^\sharp_\xi)^2 = -R_\xi \text{ (the self-adjoint part)}.$$

By this and Proposition 11, we obtain $C_\xi(Y) \neq 0$ for any $\xi \neq 0$ and $Y \neq 0$. Note that a skew-symmetric linear operator $T^\sharp_\xi$ can only have zero real eigenvalues. Thus, for any point $x \in M$, the continuous vector fields $C_{\xi_i}(Y)$ ($\|Y\| = 1,\ 1 \leq i \leq s$), are tangent to the unit sphere $\mathbb{S}^{2n-1}_x \subset \mathcal{D}^\perp_x$. If $s \geq \rho(2n)$. Then, these vector fields are linearly dependent at point $\tilde{Y} \in \mathbb{S}^{2n-1}_x$ with weights λ_i, i.e., $\sum_i \lambda_i C_{\xi_i}(\tilde{Y}) = 0$. Then, the splitting tensor has a real eigenvector as follows: $C_\xi(\tilde{Y}) = \lambda\,\tilde{Y}$, where $\xi = \sum_i \lambda_i\,\xi_i \neq 0$ and $\lambda = g(C_\xi(\tilde{Y}), \tilde{Y}) = 0$, a contradiction. Thus, the inequality $s < \rho(2n)$ holds. □

6. Weak f-K-Contact Structure Equipped with a Generalized Ricci Soliton

The following three lemmas are used in the proof of Theorem 12 given below.

Lemma 2 (see Lemma 3.1 in [4]). *For a weak f-K-contact manifold the following holds:*

$$(\pounds_{\xi_i}(\pounds_X\, g))(Z, \xi_i) = g(X, Z) + g(\nabla_{\xi_i}\nabla_{\xi_i} X, Z) + Zg(\nabla_{\xi_i} X, \xi_i)$$

for any $1 \leq i \leq s$ and all vector fields X, Z such that Z is orthogonal to $\mathcal{D}^\perp$.

Proof. This uses $\nabla_{\xi_i}\xi_i = 0$ and (29). □

Lemma 3 (e.g., ref. [4]). *Let σ be a smooth function on a Riemannian manifold $(M; g)$. Then for any vector fields ξ and Z on M we have the following:*

$$\pounds_\xi(d\sigma \otimes d\sigma)(Z, \xi) = Z(\xi(\sigma))\,\xi(\sigma) + Z(\sigma)\,\xi(\xi(\sigma)).$$

Recall the following property of f-K-contact manifolds:

$$\mathrm{Ric}(Y,\xi)=0\quad(Y\in\mathcal{D},\ \xi\in\mathcal{D}^{\perp}).\tag{36}$$

Lemma 4 (see [4]). *Let a weak f-K-contact manifold satisfy* (36) *and admit the generalized gradient Ricci soliton structure. Then,*

$$\nabla_{\xi_i}\nabla\sigma=(\lambda+c_2\,\mathrm{trace}\,Q)\,\xi_i-c_1\xi_i(\sigma)\,\nabla\sigma.$$

The following theorem generalizes Theorem 3.1 [4].

Theorem 12. *Let a weak f-K-contact manifold with the properties* $\mathrm{trace}\,Q=const$ *and* (36) *satisfy the following generalized gradient Ricci soliton equation with $c_1(\lambda+c_2\,\mathrm{trace}\,Q)\neq-1$:*

$$\mathrm{Hess}_{\sigma}-c_2\,\mathrm{Ric}=\lambda\,g-c_1d\sigma\otimes d\sigma\tag{37}$$

for some $\sigma\in C^{\infty}(M)$ and $c_1,c_2,\lambda\in\mathbb{R}$. Then, $\sigma=const$, and if $c_2\neq0$, then, our manifold is an Einstein manifold and $s=1$.

Proof. Set $Z\in\mathcal{D}$. Using Lemma 2 with $X=\nabla f$, we obtain

$$2\,(\pounds_{\xi_i}(\mathrm{Hess}_{\sigma}))(Z,\xi_i)=Z(\sigma)+g(\nabla_{\xi_i}\nabla_{\xi_i}\nabla\sigma,Z)+Zg(\nabla_{\xi_i}\nabla\sigma,\xi_i).\tag{38}$$

Using Lemma 4 in (38) and the properties $\nabla_{\xi_i}\xi_i=0$ and $g(\xi_i,\xi_i)=1$, yields

$$\begin{aligned}2\,(\pounds_{\xi_i}(\mathrm{Hess}_{\sigma}))(Z,\xi_i)&=Z(\sigma)+a\,g(\nabla_{\xi_i}\xi_i,Z)\\&-c_1g(\nabla_{\xi_i}(\xi_i(\sigma)\nabla\sigma),Z)+a\,Z(g(\xi_i,\xi_i))-c_1Z(\xi_i(\sigma)^2)\\&=Z(\sigma)-c_1g(\nabla_{\xi_i}(\xi_i(\sigma)\nabla\sigma),Z)-c_1Z(\xi_i(\sigma)^2),\end{aligned}\tag{39}$$

where $a=\lambda+c_2\,\mathrm{trace}\,Q$. Using Lemma 4 with $Z\in\mathcal{D}$, from (39) we deduce

$$2\,(\pounds_{\xi_i}(\mathrm{Hess}_{\sigma}))(Z,\xi_i)=Z(\sigma)-c_1\xi_i(\xi_i(\sigma))\,Z(\sigma)+c_1^2\xi_i(\sigma)^2Z(\sigma)-c_1Z(\xi_i(\sigma)^2).\tag{40}$$

Since ξ_i is a Killing vector field, we obtain $\pounds_{\xi_i}g=0$. This implies $\pounds_{\xi_i}\mathrm{Ric}=0$. Using the above fact and applying the Lie derivative to Equation (37), gives

$$2\,(\pounds_{\xi_i}(\mathrm{Hess}_{\sigma}))(Z,\xi_i)=-2\,c_1(\pounds_{\xi_i}(d\sigma\otimes d\sigma))(Z,\xi_i).\tag{41}$$

Using (40), (41), and Lemma 3, we obtain

$$Z(\sigma)\big(1+c_1\xi_i(\xi_i(\sigma))+c_1^2\,\xi_i(\sigma)^2\big)=0.\tag{42}$$

Using Lemma 4, we obtain

$$c_1\xi_i(\xi_i(\sigma))=c_1\,\xi_i(g(\xi_i,\nabla\sigma))=c_1g(\xi_i,\nabla_{\xi_i}\nabla\sigma)=c_1a-c_1^2\,\xi_i(\sigma)^2.\tag{43}$$

Applying (42) in (43), we obtain $Z(\sigma)(c_1a+1)=0$. This implies $Z(\sigma)=0$ provided by $c_1a+1\neq0$. Hence, $\nabla\sigma\in\mathcal{D}^{\perp}$. Taking the covariant derivative of $\nabla\sigma=\sum_i\xi_i(\sigma)\,\xi_i$ and using (27) and $\bar\xi=\sum_i\xi_i$, yields

$$g(\nabla_X\nabla\sigma,Z)=\textstyle\sum_i X(\xi_i(\sigma))\,\eta^i(Z)-\bar\xi(\sigma)\,g(\sigma X,Z),\quad X,Z\in\mathfrak{X}_M.$$

From this, by symmetry of Hess_{σ}, i.e., $g(\nabla_X \nabla\sigma, Z) = g(\nabla_Z \nabla\sigma, X)$, we obtain the equality $\bar{\xi}(\sigma)\, g(\sigma X, Z) = 0$. For $Z = \sigma X$ with some $X \neq 0$, since $g(\sigma X, \sigma X) > 0$, we obtain $\bar{\xi}(\sigma) = 0$. Replacing (ξ_i) with another orthonormal frame from $\mathcal{D}^\perp$ preserves the weak f-K-contact structure and allows reaching any direction $\bar{\xi}$ in $\mathcal{D}^\perp$. So $\nabla\sigma = 0$, i.e., $\sigma = const$. Thus, from (37) and $c_2 \neq 0$, the claim holds. □

Motivated by Proposition 10, we consider quasi-Einstein manifolds, defined by

$$\mathrm{Ric}(X,Y) = a\, g(X,Y) + b\, \mu(X)\, \mu(Y),$$

where a and b are nonzero real scalars, and μ is a 1-form of unit norm. If μ is the differential of a function, then we obtain a gradient quasi-Einstein manifold.

The following theorem generalizes (and uses) Theorem 12.

Theorem 13. *Let a weak f-K-contact manifold with the properties* $\mathrm{trace}\, Q = const$ *and* (36) *satisfy the following generalized gradient Ricci soliton equation:*

$$\mathrm{Hess}_{\sigma_1} - c_2 \,\mathrm{Ric} = \lambda\, g - c_1 d\sigma_2 \otimes d\sigma_2 \tag{44}$$

with $c_1 a \neq -1$, where $a = \lambda + c_2\, \mathrm{trace}\, Q$. Then, $\tilde{\sigma} = \sigma_1 + c_1 a\, \sigma_2$ is constant and

$$-c_1 a\, \mathrm{Hess}_{\sigma_2} = -c_1\, d\sigma_2 \otimes d\sigma_2 + c_2\, \mathrm{Ric} + \lambda\, g. \tag{45}$$

Furthermore,

1. *If $c_1 a \neq 0$, then* (45) *reduces to* $\mathrm{Hess}_{\sigma_2} = \frac{1}{a}\, d\sigma_2 \otimes d\sigma_2 - \frac{c_2}{c_1 a}\, \mathrm{Ric} - \frac{\lambda}{c_1 a}\, g$. *By Theorem 12, if $c_1 a \neq -1$, then $\sigma_2 = const$; moreover, if $c_2 \neq 0$, then (M, g) is an Einstein manifold and $s = 1$.*
2. *If $a = 0$ and $c_1 \neq 0$, then* (45) *reduces to* $0 = c_2\, \mathrm{Ric} - c_1\, d\sigma_2 \otimes d\sigma_2 + \lambda\, g$. *If $c_2 \neq 0$ and $\sigma_2 \neq const$, then we obtain a gradient quasi-Einstein manifold.*
3. *If $c_1 = 0$, then* (45) *reduces to* $0 = c_2\, \mathrm{Ric} + \lambda\, g$, *and for $c_2 \neq 0$ we obtain an Einstein manifold and $s = 1$.*

Proof. Similarly to Lemma 4, we obtain

$$\nabla_{\xi_i} \nabla\sigma_1 = a\, \xi_i - c_1 \xi_i(\sigma_2)\, \nabla\sigma_2. \tag{46}$$

Using (46) and Lemmas 2 and 3, and slightly modifying the proof of Theorem 12, we find that the vector field $\nabla\tilde{\sigma}$ belongs to $\mathcal{D}^\perp$, where $\tilde{\sigma} = \sigma_1 + c_1 a \sigma_2$. As in the proof of Theorem 12, we obtain $d\tilde{\sigma} = 0$, i.e., $d\sigma_1 = -c_1 a\, d\sigma_2$. Applying this in (44), we obtain (45). Finally, from (45), we obtain the required three cases specified in the theorem. □

7. Compact Weak f-K-Contact Manifold Equipped with an η-Ricci Soliton

The following concepts were introduced in [30,32].

Definition 5. *An η-Ricci soliton is a weak metric f-manifold $M(f, Q, \xi_i, \eta^i, g)$ satisfying*

$$\frac{1}{2} \pounds_V\, g + \mathrm{Ric} = \lambda\, g + \mu \sum\nolimits_i \eta^i \otimes \eta^i + (\lambda + \mu) \sum\nolimits_{i \neq j} \eta^i \otimes \eta^j \tag{47}$$

for some smooth vector field V on M and functions $\lambda, \mu \in C^\infty(M)$. A weak metric f-manifold $M(f, Q, \xi_i, \eta^i, g)$ is said to be η-Einstein, if

$$\mathrm{Ric} = a\, g + b \sum\nolimits_i \eta^i \otimes \eta^i + (a + b) \sum\nolimits_{i \neq j} \eta^i \otimes \eta^j \quad \textit{for some } a, b \in C^\infty(M). \tag{48}$$

Remark 5. *For a Killing vector field V, e.g., $V = \xi_i$ or $V = \bar\xi$, Equation (47) reduces to (48). Taking the trace of (48), gives the scalar curvature $r = (2n+s)\,a + s\,b$. For $s = 1$, (48) and (47) give the following well-known definitions: from (48) we obtain an η-Einstein structure $\mathrm{Ric} = a\,g + b\,\eta\otimes\eta$, and (47) gives an η-Ricci soliton $\frac{1}{2}\pounds_V\,g + \mathrm{Ric} = \lambda\,g + \mu\,\eta\otimes\eta$ on an almost contact metric manifold.*

We will use the generalized Pohozaev–Schoen identity.

Lemma 5 (e.g., ref. [38]). *Let E be a divergence free symmetric (0,2)-tensor, $E^0 = E - \frac{1}{n}(\,\mathrm{trace}_g\,E)\,g$, and V a vector field on a compact Riemannian manifold (M^n, g) without a boundary. Then,*

$$\int_M V(\,\mathrm{trace}_g\,E)\,\mathrm{d\,vol} = \frac{n}{2}\int_M g(E^0, \pounds_V\,g)\,\mathrm{d\,vol}. \tag{49}$$

The constancy of the scalar curvature r of a Riemannian manifold is important for proving the triviality of compact generalized η-Ricci solitons, see [6]. The next theorem extends this result for weak f-K-contact manifolds.

Theorem 14 (see [30]). *Let $M^{2n+s}(f,\xi_i,\eta^i,Q,g)$ be a compact weak f-K-contact manifold with $r = const$ and $\mathrm{trace}\,Q = const$. Suppose that (g,V,λ,ν) represents an η-Ricci soliton. Then, M is an η-Einstein manifold.*

Proof. Taking the trace of (48), yields $r = (2n+s)\,\alpha + s\,\beta$. Using (31) in (48), we obtain $\alpha+\beta = \mathrm{trace}\,Q$. The solution of the linear system is $\alpha = \frac{1}{2n}(r - s\,\mathrm{trace}\,Q)$ and $\beta = \frac{1}{2n}((2n+s)\,\mathrm{trace}\,Q - r)$. Define a traceless (0,2)-tensor

$$E := \mathrm{Ric} - \frac{r - s\,\mathrm{trace}\,Q}{2n}\,g + \frac{r-(2n+s)\,\mathrm{trace}\,Q}{2n}\sum\nolimits_i \eta^i\otimes\eta^i - \mathrm{trace}\,Q\sum\nolimits_{i\ne j}\eta^i\otimes\eta^j.$$

Using this, we express η-Ricci soliton (47) as

$$\begin{aligned}\frac{1}{2}\pounds_V\,g + E = (\lambda - \frac{r - s\,\mathrm{trace}\,Q}{2n})g + (\nu + \frac{r-(2n+s)\,\mathrm{trace}\,Q}{2n})\sum\nolimits_i \eta^i\otimes\eta^i\\ + (\lambda+\nu-\mathrm{trace}\,Q)\sum\nolimits_{i\ne j}\eta^i\otimes\eta^j.\end{aligned} \tag{50}$$

We obtain the equality $\mathrm{div}_g\,E = 0$. Since E is traceless, we obtain $E^0 = E$ and $V(\,\mathrm{trace}_g E) = 0$. Then, applying (49), gives $\int_M g(E, \pounds_V\,g)\,\mathrm{d\,vol} = 0$. Using (50) and (31) in the above integral formula, and using the following equalities: $g(E,g) = \mathrm{trace}_g E = 0$ and $g(E,\eta^i\otimes\eta^j) = E(\xi_i,\xi_j) = 0$, we obtain $\int_M g(E,E)\,\mathrm{d\,vol} = 0$. By this, $E = 0$ is true. Therefore, our manifold is an η-Einstein manifold. □

8. Geometry of Weak β-Kenmotsu f-Manifolds

The next definition generalizes the notions of β-Kenmotsu manifolds and Kenmotsu f-manifolds ($\beta = 1$, $s > 1$), see [5,17], and weak β-Kenmotsu manifolds ($s = 1$), see [36].

Definition 6 ([32]). *A normal (i.e., $\mathcal{N}^{(1)} = 0$) weak metric f-manifold $M^{2n+s}(f,Q,\xi_i,\eta^i,g)$ is called a weak β-Kenmotsu f-manifold (a weak Kenmotsu f-manifold when $\beta\equiv 1$) if*

$$(\nabla_X f)Y = \beta\{g(fX,Y)\,\bar\xi - \bar\eta(Y)fX\}\quad (X,Y\in\mathfrak{X}_M), \tag{51}$$

where $\bar\xi = \sum_i \xi_i$, $\bar\eta = \sum_i \eta^i$, and β is a nonzero smooth function on M.

Remark 6. *Many of the results below on weak β-Kenmotsu f-manifolds hold if we replace* (51) *with a weaker condition involving non-zero smooth functions β_i $(1 \le i \le s)$ on M, as follows:*

$$(\nabla_X f)Y = \sum\nolimits_i \beta_i\{g(fX,Y)\,\xi_i - \eta^i(Y)fX\} \quad (X,Y \in \mathfrak{X}_M).$$

Note that $\bar\eta(\xi_i) = \eta^i(\bar\xi) = 1$ and $\bar\eta(\bar\xi) = s$. Taking $X = \xi_j$ in (51) and using $f\,\xi_j = 0$, we obtain $\nabla_{\xi_j} f = 0$, which implies $f(\nabla_{\xi_i}\xi_j) = 0$, and so $\nabla_{\xi_i}\xi_j \in \mathcal{D}^\perp$. This and the 1st equality in (7) give

$$\nabla_{\xi_i}\xi_j = 0 \quad (1 \le i,j \le s), \tag{52}$$

thus, $\mathcal{D}^\perp$ (of weak β-Kenmotsu f-manifolds) is tangent to a totally geodesic $\mathfrak{g}$-foliation with an abelian Lie algebra.

Proposition 12. *A weak metric f-manifold with condition* (51) *is a weak β-Kenmotsu f-manifold if and only if the following formula holds:*

$$\nabla_X\,\xi_i = \beta\{X - \sum\nolimits_j \eta^j(X)\,\xi_j\} \quad (1 \le i \le s, \quad X \in \mathfrak{X}_M). \tag{53}$$

Proof. ($\Rightarrow$) Consider a weak β-Kenmotsu f-manifold. Taking $Y = \xi_i$ in (51) and using $g(fX,\xi_i) = 0$ and $\bar\eta(\xi_i) = 1$, we obtain $f(\nabla_X\,\xi_i - \beta X) = 0$. Since f is non-degenerate on $\mathcal{D}$ and has rank $2\,n$, we obtain $\nabla_X\,\xi_i - \beta X = \sum_p c^p\,\xi_p$. The inner product with ξ_j gives $g(\nabla_X\,\xi_i,\xi_j) = \beta\,g(X,\xi_j) - c^j$. Using (7) and (52), we find $g(\nabla_X\,\xi_i,\xi_j) = g(\nabla_{\xi_i}X,\xi_j) = 0$; hence, $c^j = \beta\,\eta^j(X)$. This proves (53).

($\Leftarrow$) Consider a weak metric f-manifold with conditions (51) and (53). It follows from (5) with $S = f$ and (51) that

$$[f,f](X,Y) = 2\beta\,g(X,fY)\,f\,\bar\xi = 0.$$

Using (53) in the formula for $d\eta^j$ gives

$$d\eta^j(X,Y) = \sum\nolimits_j\{g(\nabla_X\,\xi_j),Y) - g(\nabla_Y\,\xi_j),X)\} = 0.$$

Therefore, the manifold is normal and hence a weak β-Kenmotsu f-manifold. □

It follows from (53) and $\beta \ne 0$ that every structure vector field ξ_i (of a weak β-Kenmotsu f-manifold) is not a Killing vector field.

Theorem 15. *A weak metric f-manifold is a weak β-Kenmotsu f-manifold if and only if the following conditions are valid:*

$$\mathcal{N}^{(1)} = 0, \quad d\eta^i = 0, \quad d\Phi = 2\,\beta\,\bar\eta\wedge\Phi, \quad \mathcal{N}^{(5)}(X,Y,Z) = 2\,\beta\,\bar\eta(X)g(fY,\widetilde{Q}Z). \tag{54}$$

Proof. ($\Rightarrow$) For a weak β-Kenmotsu f-manifold, using (53), we obtain

$$(\nabla_X\,\eta^i)Y = Xg(\xi_i,Y) - g(\xi_i,\nabla_X\,Y) = g(\nabla_X\,\xi_i,Y) = \beta\{g(X,Y) - \sum\nolimits_j \eta^j(X)\,\eta^j(Y)\} \tag{55}$$

for all $X,Y \in \mathfrak{X}_M$. By (55), $(\nabla_X\,\eta^i)Y = (\nabla_Y\,\eta^i)X$ is true. Thus, for $X,Y \in \mathcal{D}$ we obtain

$$0 = (\nabla_X\,\eta^i)Y - (\nabla_Y\,\eta^i)X = -\beta\,g([X,Y],\xi_i)$$

that means $d\eta^i(X,Y) = 0$ for all $i = 1,\ldots,s$ and $X, Y \in \mathcal{D}$, that is, the distribution $\mathcal{D}$ is involutive. By this and $\mathcal{N}^{(4)}_{ij} = 0$, see (6), we find $d\eta^i = 0$. Using (51) and (8), we obtain

$$3\, d\Phi(X,Y,Z) = 2\,\beta\{\bar\eta(X) g(fZ,Y) + \bar\eta(Y) g(fX,Z) + \bar\eta(Z) g(fY,X)\}.$$

We also have the following:

$$3(\bar\eta \wedge \Phi)(X,Y,Z) = \bar\eta(X) g(fZ,Y) + \bar\eta(Y) g(fX,Z) + \bar\eta(Z) g(fY,X).$$

Thus, $d\Phi = 2\,\beta\,\bar\eta \wedge \Phi$ is valid. By (5) with $S = f$, and (51), we obtain $[f,f] = 0$; thus, $\mathcal{N}^{(1)} = 0$. Finally, from (9), using (1) and (2), we obtain

$$\begin{aligned}
&g((\nabla_X f)Y,Z) - \frac{1}{2}\mathcal{N}^{(5)}(X,Y,Z) = 3\,\beta\{(\bar\eta\wedge\Phi)(X,fY,fZ) - (\bar\eta\wedge\Phi)(X,Y,Z)\}\\
&= \beta\{-\bar\eta(X) g(QZ,fY) + \bar\eta(X) g(Z,fY) - \bar\eta(Y) g(fX,Z) - \bar\eta(Z) g(X,fY)\}\\
&= \beta\{\bar\eta(Z) g(fX,Y) - \bar\eta(Y) g(fX,Z) - \bar\eta(X) g(fY,\widetilde{Q}Z)\}.
\end{aligned}$$

From this, using (51), we obtain $\mathcal{N}^{(5)}(X,Y,Z) = 2\,\beta\,\bar\eta(X) g(fY,\widetilde{Q}Z)$.

($\Leftarrow$) Conversely, for a weak metric f-manifold, using (1) and (54) in (9), we obtain

$$\begin{aligned}
&2\,g((\nabla_X f)Y,Z) = 6\,\beta\,(\bar\eta\wedge\Phi)(X,fY,fZ) - 6\,\beta\,(\bar\eta\wedge\Phi)(X,Y,Z) + 2\,\beta\,\bar\eta(X) g(\widetilde{Q}fY,Z)\\
&= 2\beta\{-\bar\eta(X) g(fY,QZ) - \bar\eta(X) g(fZ,Y) - \bar\eta(Y) g(fX,Z) - \bar\eta(Z) g(fY,X) + \bar\eta(X) g(fY,\widetilde{Q}Z)\}\\
&= 2\,\beta\{g(fX,Y)\, g(\bar\xi,Z) - \bar\eta(Y) g(fX,Z)\},
\end{aligned}$$

thus (51) is true. □

Theorem 16. *A weak β-Kenmotsu f-manifold is locally a twisted product* $\mathbb{R}^s \times_\sigma \bar M$ *(a warped product when* $X(\beta) = 0$ *for* $X \in \mathcal{D}$*), where* $\bar M(\bar g, J)$ *is a weak Kähler manifold.*

Proof. By (52), the distribution $\mathcal{D}^\perp$ is tangent to a totally geodesic foliation. By $d\eta^j = 0$ (see the proof of Theorem 15), the distribution $\mathcal{D}$ is tangent to a foliation. By (53), the splitting tensor (18) is conformal as follows: $C_{\xi_i} X = -\beta X$ $(X \in \mathcal{D})$. Hence, $\mathcal{D}$ is tangent to a totally umbilical foliation with the mean curvature vector $H = -\beta\,\bar\xi$. By Theorem 1 in [39], our manifold is locally a twisted product. If $X(\beta) = 0$ $(X \in \mathcal{D})$, then we locally obtain a warped product, see Proposition 3 in [39]. By (2), the (1,1)-tensor $J = f|_{\mathcal{D}}$ is skew-symmetric and J^2 is negative definite. Using (51), we find $(\bar\nabla_X J)Y = \pi_{2*}((\nabla_X f)Y) = 0$ for $X, Y \in \mathcal{D}$, thus $\bar\nabla J = 0$. □

Example 3. *Let* $\bar M(\bar g, J)$ *be a weak Kähler manifold and* $\sigma = c\,e^{\beta\sum t_i}$ *a function on Euclidean space* $\mathbb{R}^s(t_1,\ldots,t_s)$*, where* β, c *are nonzero constants. Then, the warped product manifold* $M = \mathbb{R}^s \times_\sigma \bar M$ *has a weak metric f-structure which satisfies* (51)*. Using* (5) *with* $S = J$*, for a weak Kähler manifold, we obtain* $[J,J] = 0$*; hence,* $\mathcal{N}^{(1)} = 0$ *is true.*

Corollary 5. *A weak Kenmotsu f-manifold* $M^{2n+s}(f,Q,\xi_i,\eta^i,g)$ *is locally a warped product* $\mathbb{R}^s \times_\sigma \bar M$*, where* $\sigma = c\,e^{\sum t_i}$ $(c = const \neq 0)$ *and* $\bar M(\bar g, J)$ *is a weak Kähler manifold.*

To simplify the calculations in the rest of the paper, we assume that $\beta = const$.

Proposition 13. *The following formulas hold for weak β-Kenmotsu f-manifolds with $\beta = const$:*

$$
\begin{aligned}
&R_{X,Y}\,\xi_i = \beta^2\{\bar\eta(X)Y - \bar\eta(Y)X + \textstyle\sum_j (\bar\eta(Y)\eta^j(X) - \bar\eta(X)\eta^j(Y))\xi_j\} \quad (X,Y \in \mathfrak{X}_M),\\
&\mathrm{Ric}^\sharp\,\xi_i = -2\,n\,\beta^2\bar\xi,\\
&(\nabla_{\xi_i}\mathrm{Ric}^\sharp)X = -2\,\beta\,\mathrm{Ric}^\sharp\,X - 4\,n\,\beta^3\{s(X - \textstyle\sum_j \eta^j(X)\,\xi_j) + \bar\eta(X)\bar\xi\} \quad (X \in \mathfrak{X}_M),\\
&\xi_i(r) = -2\,\beta\{r + 2\,s\,n(2\,n+1)\,\beta^2\}.
\end{aligned}
$$

The following theorem generalizes Theorem 1 [5] with $\beta \equiv 1$ and $Q = \mathrm{id}$.

Theorem 17. *Let $M^{2n+s}(f, Q, \xi_i, \eta^i, g)$ be a weak β-Kenmotsu f-manifold satisfying $\beta = const$. If $\nabla_{\xi_i}\mathrm{Ric}^\sharp = 0$, then (M, g) is an η-Einstein manifold* (48) *of constant scalar curvature $r = -2\,s\,n(2\,n+1)\,\beta^2$.*

Proof. By conditions and Proposition 13, $\mathrm{Ric}^\sharp\,Y = 2\,n\,\beta^2\{s(Y - \sum_j \eta^j(Y)\,\xi_j) + \bar\eta(Y)\bar\xi\}$; thus, $r = -2\,s\,n(2\,n+1)\,\beta^2$. Since (48) with $a = 2\,s\,n\,\beta^2$ and $b = 2\,(1-s)\,n\,\beta^2$ holds, (M, g) is η-Einstein. □

9. η-Ricci Solitons on Weak β-Kenmotsu f-Manifolds

The following lemmas are used in the proof of Theorem 18 given below.

Lemma 6. *Let $M^{2n+s}(f, Q, \xi_i, \eta^i, g)$ be a weak β-Kenmotsu f-manifold with $\beta = const$. If g represents an η-Ricci soliton* (47)*, then $\lambda + \mu = -2\,n\,\beta^2$.*

Proof. For a weak β-Kenmotsu f-manifold equipped with an η-Ricci soliton (47), using (7), we obtain

$$
(\pounds_V\, g)(\xi_i, \xi_j) = g(\xi_i, [V, \xi_j]) = 0.
$$

Thus, using (47) in the Lie derivative of $g(\xi_i, \xi_j) = \delta_{ij}$, we obtain $\mathrm{Ric}(\xi_i, \xi_j) = \lambda + \mu$. Finally, using the equality $\mathrm{Ric}(\xi_i, \xi_j) = -2\,n\,\beta^2$, see Proposition 13, we achieve the result. □

Lemma 7 (see [32])**.** *Let $M^{2n+s}(f, Q, \xi_i, \eta^i, g)$ be a weak β-Kenmotsu f-manifold with $\beta = const$. If g represents an η-Ricci soliton* (47)*, then $(\pounds_V R)_{X,\xi_j}\xi_i = 0$ for all i, j.*

Lemma 8. *On an η-Einstein* (48) *weak β-Kenmotsu f-manifold with $\beta = const$, we obtain*

$$
\mathrm{Ric}^\sharp X = \Big(s\,\beta^2 + \frac{r}{2\,n}\Big)X - \Big((2\,n+s)\,\beta^2 + \frac{r}{2\,n}\Big)\textstyle\sum_i \eta^j(X)\,\xi_j - 2\,n\beta^2 \sum_{i\neq j}\eta^i(X)\,\xi_j. \quad (56)
$$

Proof. Tracing (48) gives $r = (2\,n+s)\,a + sb$. Putting $X = Y = \xi_i$ in (48) and using Proposition 13, yields $a + b = -2\,n\,\beta^2$. Thus, $a = s\,\beta^2 + \frac{r}{2n}$ and $b = -(2\,n+s)\,\beta^2 - \frac{r}{2n}$, and (48) gives (56). □

Next, we consider an η-Einstein weak β-Kenmotsu f-manifold as an η-Ricci soliton.

Theorem 18. *Let g represent an η-Ricci soliton* (47) *on a weak β-Kenmotsu f-manifold with* $\dim M > 3$ *and $\beta = const$. If the manifold is also η-Einstein* (48)*, then $a = -2\,s\,n\beta^2$, $b = 2(s-1)n\beta^2$, and the scalar curvature is $r = -2\,s\,n(2\,n+1)\,\beta^2$.*

Definition 7. *A vector field V on a weak metric f-manifold is called a contact vector field, if there exists a function $\rho \in C^\infty(M)$ such that*

$$\pounds_X \eta^i = \rho\, \eta^i,$$

and if $\rho = 0$, i.e., the flow of X preserves the forms η^i, and then V is a strict contact vector field.

We consider the interaction of a weak β-Kenmotsu f-structure with an η-Ricci soliton whose potential vector field V is a contact vector field, or V is collinear to $\bar\xi$.

Theorem 19. *Let $M^{2n+s}(f, Q, \xi_i, \eta^i, g)$ be a weak β-Kenmotsu f-manifold with $\dim M > 3$ and $\beta = const$. If g represents an η-Ricci soliton (47) with a contact potential vector field V, then V is strict contact and the manifold is η-Einstein (48) with $a = -2\,s\,n\beta^2$, $b = 2(s-1)n\beta^2$ of constant scalar curvature $r = -2\,s\,n(2\,n+1)\beta^2$.*

Theorem 20. *Let $M^{2n+s}(f, Q, \xi_i, \eta^i, g)$ be a weak β-Kenmotsu f-manifold with $\dim M > 3$ and $\beta = const$. If g represents an η-Ricci soliton (47) with a potential vector field V collinear to $\bar\xi$, as follows: $V = \delta\,\bar\xi$ for a smooth function $\delta \neq 0$ on M, then, $\delta = const$ and the manifold is η-Einstein (48) with $a = -2\,s\,n\beta^2$ and $b = 2(s-1)n\beta^2$ of constant scalar curvature $r = -2\,s\,n(2\,n+1)\beta^2$.*

Proof. Using (51) in the derivative of $V = \delta\,\bar\xi$, yields

$$\nabla_X V = X(\delta)\,\bar\xi + \delta\,\beta(X - \textstyle\sum_j \eta^j(X)\,\xi_j), \quad X \in \mathfrak{X}_M.$$

Using this and calculations

$$(\pounds_{\delta\bar\xi}\, g)(X,Y) = \delta(\pounds_{\bar\xi}\, g)(X,Y) + X(\delta)\,\bar\eta(Y) + Y(\delta)\,\bar\eta(X),$$
$$(\pounds_{\bar\xi}\, g)(X,Y) = 2\,s\,\beta\{g(X,Y) - \textstyle\sum_j \eta^j(X)\,\eta^j(Y)\},$$

we transform the η-Ricci soliton Equation (47) into

$$\begin{aligned} 2\,\mathrm{Ric}(X,Y) &= -X(\delta)\,\bar\eta(Y) - Y(\delta)\,\bar\eta(X) + 2(\lambda - \delta\beta)\,g(X,Y) \\ &+ 2(\delta\beta + \mu)\textstyle\sum_j \eta^j(X)\,\eta^j(Y) - 4\,n\beta^2 \textstyle\sum_{i\neq j} \eta^i(X)\,\eta^j(Y), \quad X,Y \in \mathfrak{X}_M. \end{aligned} \tag{57}$$

Inserting $X = Y = \xi_i$ in (57) and using Proposition 13 and $\lambda + \mu = -2\,n\,\beta^2$, see Lemma 6, we obtain $\xi_i(\delta) = 0$. It follows from (57) and Proposition 13 that $X(\delta) = 0$ $(X \in \mathcal{D})$. Thus, δ is constant on M, and (57) reads as

$$\mathrm{Ric} = (\lambda - \delta\beta)\,g + (\delta\beta + \mu)\textstyle\sum_j \eta^j \otimes \eta^j - 2\,n\beta^2 \textstyle\sum_{i\neq j} \eta^i \otimes \eta^j.$$

This shows that (M, g) is an η-Einstein manifold with $a = \lambda - \delta\beta$ and $b = \mu + \delta\beta$ in (48). Therefore, from Theorem 18, we conclude that $\lambda = \delta\beta - 2\,s\,n\beta^2$, $\mu = -\delta\beta + 2(s-1)n\beta^2$, and the scalar curvature of (M, g) is $r = -2\,s\,n(2\,n+1)\beta^2$. □

Remark 7. *For the case* $\dim M = 3$ *(thus, $s = 1$) in Theorems 18–20, see [36].*

10. Conclusions and Future Directions

This review paper demonstrates that the weak metric f-structure is a valuable tool for exploring various geometric properties on manifolds, including Killing vector fields, totally geodesic foliations, twisted products, Ricci-type solitons, and Einstein-type metrics. Several results for metric f-manifolds have been extended to manifolds with weak structures, providing new insights and applications.

In conclusion, we pose the following open questions: 1. Is the condition "the mixed sectional curvature is positive" sufficient for a weak metric f-manifold to be weak f-K-contact? 2. Does a weak metric f-manifold of a dimension greater than 3 have some positive mixed sectional curvature? 3. Is a compact weak f-K-contact Einstein manifold an $\mathcal{S}$-manifold? 4. When is a given weak f-K-contact manifold a mapping torus (see [22]) of a manifold of a lower dimension? 5. When does a weak metric f-manifold equipped with a Ricci-type soliton structure, carry a canonical (e.g., of constant sectional curvature or Einstein-type) metric? 6. Can Theorems 13–16 be extended to the case where β is not constant? These questions highlight the potential for further exploration and development in the field of weak metric f-manifolds, encouraging continued research and discovery.

We delegate the following to the future:

- The study of integral formulas, variational problems, and extrinsic geometric flows for weak metric f-manifolds and their distinguished classes using the methodology of [25].
- The study of weak nearly $\mathcal{S}$- and weak nearly $\mathcal{C}$-manifolds as well as weak nearly Kenmotsu f-manifolds, and the generalization to the case $s > 1$ of our results on weak nearly Sasakian/cosymplectic manifolds, see the survey in [27].
- The study of geometric inequalities with the mutual curvature invariants and with Chen-type invariants (and the case of equality in them) for submanifolds in weak metric f-manifolds and in their distinguished classes using the methodology of [40].

Funding: This research received no external funding.

Data Availability Statement: Data are contained within the article.

Conflicts of Interest: The author declares no conflicts of interest.

References

1. Gaset, J.; Gràcia, X.; Muñoz-Lecanda, M.C.; Rivas, X.; Román-Roy, N. A contact geometry framework for field theories wi dissipation. *Ann. Phys.* **2020**, *414*, 168092. [CrossRef]
2. Baishya, K.K.; Chowdhury, P.R.; Mikeš, J.; Peška, P. On almost generalized weakly symmetric Kenmotsu manifolds. *Acta Ur Palacki. Olomuc. Fac. Rerum Nat. Math.* **2016**, *55*, 5–15.
3. Hui, S.K.; Mikes, J.; Mandal, P. Submanifolds of Kenmotsu manifolds and Ricci solitons. *J. Tensor Soc.* **2016**, *10*, 79–89. [CrossR
4. Ghosh, G.; De, U.C. Generalized Ricci soliton on K-contact manifolds. *Math. Sci. Appl. E-Notes* **2020**, *8*, 165–169. [CrossRef]
5. Ghosh, A. Ricci soliton and Ricci almost soliton within the framework of Kenmotsu manifold. *Carpathian Math. Publ.* **2019**, 59–69. [CrossRef]
6. Ghosh, A. K-contact and (k, μ)-contact metric as a generalized η-Ricci soliton. *Math. Slovaca* **2023**, *73*, 185–194.
7. Falcitelli, M.; Ianus, S.; Pastore, A. *Riemannian Submersions and Related Topics*; World Scientific: London, UK, 2004.
8. Yano, K. *On a Structure f Satisfying $f^3 + f = 0$, Technical Report No.* 12; University of Washington: Washington, DC, US 1961.
9. Yano, K.; Kon, M. *Structures on Manifolds*; Vol. 3 of Series in Pure Math; World Scientific Publishing Co.: Singapore, 1985.
10. Blair, D.E. Geometry of manifolds with structural group $U(n) \times O(s)$. *J. Diff. Geom.* **1970**, *4*, 155–167.
11. Ludden, G.D. Submanifolds of manifolds with an f-structure. *Kodai Math. Semin. Rep.* **1969**, *21*, 160–166. [CrossRef]
12. Naveira, A. A classification of Riemannian almost product manifolds. *Rend. Math.* **1983**, *3*, 577–592.
13. Brunetti, L.; Pastore, A.M. S-manifolds versus indefinite S-manifolds and local decomposition theorems. *Int. Electron. J. Gec* **2016**, *9*, 1–8. [CrossRef]
14. Cabrerizo, J.L.; Fernández, L.M.; Fernández, M. The curvature tensor fields on f-manifolds with complemented frames. *Stiint. Univ. Al. I. Cuza Iasi* **1990**, *36*, 151–161.
15. Di Terlizzi, L. On the curvature of a generalization of contact metric manifolds. *Acta Math. Hung.* **2006**, *110*, 225–239. [CrossR
16. Di Terlizzi, L.; Pastore, A.M.; Wolak, R. Harmonic and holomorphic vector fields on an f-manifold with parallelizable kernel. *Stiint. Univ. Al. I. Cuza Iausi Ser. Noua, Mat.* **2014**, *60*, 125–144. [CrossRef]
17. Sari, R.; Turgut Vanli, A. Generalized Kenmotsu manifolds. *Commun. Math. Appl.* **2016**, *7*, 311–328.
18. Kenmotsu, K. A class of almost contact Riemannian manifolds. *Tôhoku Math. J.* **1972**, *24*, 93–103.

Cappelletti Montano, B.; Di Terlizzi, L. D-homothetic transformations for a generalization of contact metric manifolds. *Bull. Belg. Math. Soc.–Simon Stevin* **2007**, *14*, 277–289. [CrossRef]
Carriazo, A.; Fernández, L.M.; Loiudice, E. Metric f-contact manifolds satisfying the (k,μ)-nullity condition. *Mathematics* **2020**, *8*, 891. [CrossRef]
Fitzpatrick, S. On the geometry of almost $\mathcal{S}$-manifolds. *ISRN Geom.* **2011**, *2011*, 879042. [CrossRef]
Goertsches, O.; Loiudice, E. On the topology of metric f-K-contact manifolds. *Monatshefte für Math.* **2020**, *192*, 355–370. [CrossRef]
de Almeida, U.N.M. Generalized k-contact structures. *J. Lie Theory* **2024**, *34*, 113–136.
Finamore, D. Contact foliations and generalised Weinstein conjectures. *Ann. Global Anal. Geom.* **2024**, *65*, 27. [CrossRef]
Rovenski, V.; Walczak, P.G. *Extrinsic Geometry of Foliations*; Progress in Mathematics; Birkhäuser: Cham, Switzerland, 2021; Volume 339.
Rovenski, V.; Wolak, R. New metric structures on $\mathfrak{g}$-foliations. *Indag. Math.* **2022**, *33*, 518–532. [CrossRef]
Rovenski, V. Weak almost contact structures: A survey. *arXiv* **2024**, arXiv:2408.13827. [CrossRef]
Rovenski, V. Metric structures that admit totally geodesic foliations. *J. Geom.* **2023**, *114*, 32. [CrossRef]
Rovenski, V. On the splitting tensor of the weak f-contact structure. *Symmetry* **2023**, *15*, 1215. [CrossRef]
Rovenski, V. Einstein-type metrics and Ricci-type solitons on weak f-K-contact manifolds. In *Differential Geometric Structures and Applications*; Rovenski, V., Walczak, P., Wolak, R., Eds.; Springer Proceedings in Mathematics and Statistics, 440; Springer: Cham, Switzerland, 2023; pp. 29–51.
Rovenski, V. Geometry of a weak para-f-structure. *U.P.B. Sci. Bull. Ser. A Appl. Math. Phys.* **2023**, *85*, 11–20.
Rovenski, V. η-Ricci solitons and η-Einstein metrics on weak β-Kenmotsu f-manifolds. *arXiv* **2024**, arXiv:2412.14125.
Goldberg, S.I.; Yano, K. On normal globally framed f-manifolds. *Tohoku Math. J.* **1970**, *22*, 362–370. [CrossRef]
Alekseevsky, D.; Michor, P. Differential geometry of $\mathfrak{g}$-manifolds. *Differ. Geom. Appl.* **1995**, *5*, 371–403. [CrossRef]
Duggal, K.L.; Ianus, S.; Pastore, A.M. Maps interchanging f-structures and their harmonicity. *Acta Appl. Math.* **2001**, *67*, 91–115. [CrossRef]
Patra, D.S.; Rovenski, V. Weak β-Kenmotsu manifolds and η-Ricci solitons. In *Differential Geometric Structures and Applications*; Rovenski, V., Walczak, P., Wolak, R., Eds.; Springer Proceedings in Mathematics and Statistics, 440; Springer: Cham, Switzerland, 2023; pp. 53–72.
Rovenski, V.; Patra, D.S. On the rigidity of the Sasakian structure and characterization of cosymplectic manifolds. *Differ. Geom. Its Appl.* **2023**, *90*, 102043.
Gover, A.R.; Orsted, B. Universal principles for Kazdan-Warner and Pohozaev-Schoen type identities. *Commun. Contemp. Math.* **2013**, *15*, 1350002. [CrossRef]
Ponge, R.; Reckziegel, H. Twisted products in pseudo-Riemannian geometry. *Geom. Dedicata* **1993**, *48*, 15–25. [CrossRef]
Rovenski, V. Geometric inequalities for a submanifold equipped with distributions. *Mathematics* **2022**, *10*, 4741. [CrossRef]

mathematics

MDPI

Article

Optimal Inequalities Characterizing Totally Real Submanifolds in Quaternionic Space Form

Fatimah Alghamdi [1,*,†] and Akram Ali [2,†]

1 Department of Mathematics and Statistics, College of Science, University of Jeddah, Jeddah 21589, Saudi Arabia
2 Department of Mathematics, College of Science, King Khalid University, Abha 61413, Saudi Arabia; akali@kku.edu.sa
* Correspondence: fmalghamdi@uj.edu.sa
† These authors contributed equally to this work.

Abstract: In the present paper, we investigate some pinching inequalities on the scalar curvature of a totally real submanifold in quaternionic space form that leads to a topological conclusion of the submanifold. In addition, we construct another inequality which includes the mean curvature and the length of the second fundamental form.

Keywords: totally real submanifold; second fundamental form; rigidity theorems; scalar curvature; space form; mean curvature

MSC: 53C40; 57R19; 53C20; 53C23; 53C42; 57R60

1. Introduction

Exploring the topology of closed submanifolds in space forms presents an intriguing area of study. In contrast to standard techniques in Riemannian geometry, our findings on the topology of these submanifolds are derived by applying specific constraints on the scalar curvature and mean curvatures. In this context, the squared norm of the second fundamental form emerges with significant importance in the study of minimal submanifolds. Therefore, the study of rigidity theorems was essential in the groundbreaking work of Simons [1], Lawson [2], and Chern et al. [3] on minimal submanifolds in spheres. For example, let the square norm of the second fundamental form be presented by $\mathcal{B}$ and a unit sphere by $\mathbb{S}^{n+m}$ with codimension m. If there is a compact minimal submanifold $\mathcal{N}^n$ in $\mathbb{S}^{n+m}$ with pinching condition $0 \leq \mathcal{B} \leq \left(\frac{n}{2-\frac{1}{m}}\right)$, then either $\mathcal{B} = 0$ or $\mathcal{B} = \left(\frac{n}{2-\frac{1}{m}}\right)$, and $\mathcal{N}$ is the Clifford hypersurface or the Veronese surface in $\mathbb{S}^4$. Later, Li [4] and Chen-Xu [5] improved the pinching number $\frac{n}{(2-1/m)}$ to $\frac{2n}{3}$. They showed that if $0 \leq \mathcal{B} \leq \frac{2n}{3}$, then either $\mathcal{B} = 0$ or $\mathcal{B} = \frac{2n}{3}$, and $\mathcal{N}$ is the Veronese surface in $\mathbb{S}^4$. This topic has garnered significant attention following the early work by Simons [1] and subsequent developments (e.g., [2,4–6]). This study was extended to the rigidity theorems for minimal submanifolds and submanifolds with parallel mean curvature in space forms; see [7–13], among others.

Space forms are fundamental objects in differential geometry, providing rich settings for studying the relationship between curvature and topology. Over the past two decades, many researchers have investigated the geometry of submanifolds in various types of space forms, particularly through eigenvalue estimates and curvature inequalities. For example, Carriazo et al. [14] established early results on totally real submanifolds in quaternionic space forms. Later, Mutlu and Şentürk [15], as well as Lee and Vîlcu [16], contributed

https://doi.org/10.3390/math131016

further to the development of curvature inequalities in the quaternionic setting. More recent contributions have appeared in this direction [17–20].

In complex and contact geometries, Ali et al. [21] studied eigenvalue inequalities for the p-Laplacian on C-totally real submanifolds in Sasakian space forms, and Ali et al. [22] also focused on Lagrangian submanifolds in complex space forms. Li et al. [23] derived Reilly-type inequalities for semi-slant submanifolds, while Alluhaibi and Ali [24] considered totally real submanifolds in generalized complex space forms involving the p-Laplacian operator. More recent work by Mehraj Ahmad [25] and Siddiqi et al. [26] focused on quarter-symmetric and statistical connection settings.

However, rigidity theorems for totally real submanifolds in quaternionic space forms remain relatively underexplored. Motivated by these developments, this paper establishes new rigidity theorems for totally real submanifolds in quaternionic space forms. In particular, we derive optimal inequalities involving scalar curvature, the squared norm of the second fundamental form, and the mean curvature. These results enrich rigidity theory and offer new insights into the interplay between intrinsic and extrinsic geometry in quaternionic settings.

2. Preliminaries

An almost quaternionic structure on $(\mathcal{M},\sigma)$ is defined as having a local canonical basis $\{J_1,J_2,J_3\}$ on a rank 3-subbundle σ of $\mathrm{End}(T\mathcal{M})$ for a differentiable manifold $\mathcal{M}$ such that where $s, s+1, s+2$ are taken modulo 3, and the following conditions hold:

$$J_s^2 = -Id, \quad J_sJ_{s+1} = -J_{s+1}J_s = J_{s+2}. \tag{1}$$

and

$$\tilde{g}(J_sX_1,J_sY_1) = \tilde{g}(X_1,Y_1), \tag{2}$$

for all vector fields X_1, Y_1 on $\mathcal{M}$, where $\mathcal{M}$ stands for a Riemannian manifold with real dimension $4m$. Moreover, $(\mathcal{M},\sigma,\tilde{g})$ is said to be an almost quaternionic Hermitian manifold (see [16] for more details). Let there exist 1-form $\theta_1,\theta_2,\theta_3$, and let σ be parallel with respect to the Levi–Civita connection $\tilde{\nabla}$ satisfying the following:

$$\nabla_{X_1}J_s = \theta_{s+2}(X_1)J_{s+1} - \theta_{s+1}(X_1)J_{s+2}, \tag{3}$$

For any vector field X_1 on $\mathcal{M}$, and where s, $s+1$, and $s+2$ are taken modulo 3, the triple $(\mathcal{M},\sigma,\tilde{g})$ denotes a quaternionic Kähler manifold [15]. The curvature tensor $\widetilde{R}$ of a quaternionic space form $\mathcal{M}(c)$, which has constant quaternionic sectional curvature c [14], is given by:

$$\begin{aligned}\widetilde{R}(X_1,Y_1)Z_1 =& \frac{c}{4}\Big\{\tilde{g}(Z_1,Y_1)X_1 - \tilde{g}(X_1,Z_1)Y_1 + \sum_{s=1}^{3}\Big(\tilde{g}(Z_1,J_sY_1)J_sX_1 \\ &- \tilde{g}(Z_1,J_sX_1)J_sY_1 + 2\tilde{g}(X_1,J_sY_1)J_sZ_1\Big)\Big\}\end{aligned} \tag{4}$$

for all vector fields X_1, Y_1, Z_1 on $\mathcal{M}$ and any local basis $\{J_1,J_2,J_3\}$ of σ.

Assume that $\mathcal{N}$ is a submanifold of an almost Hermitian quaternionic manifold $\mathcal{M}(c)$ with an induced metric g. Thus, the Weingarten and Gauss formulas are provided by

$$(i)\ \tilde{\nabla}_{X_2}Y_2 = \nabla_{X_2}Y_2 + \mathcal{B}(X_2,Y_2), \quad (ii)\ \tilde{\nabla}_{X_2}Z_2^{\perp} = -A_{Z_2^{\perp}}X_2 + \nabla^{\perp}_{X_2}Z_2^{\perp}, \tag{5}$$

for each $X_2, Y_2 \in \Gamma(T\mathcal{N})$ and $Z_2^{\perp} \in \Gamma(T^{\perp}\mathcal{N})$, where ∇ and $\nabla^{\perp}$ denote the induced connections on the tangent bundle $T\mathcal{N}$ and the normal bundle $T^{\perp}\mathcal{N}$ of $\mathcal{N}$, respectively. The second fundamental form $\mathcal{B}$ and the shape operator $A_{Z_2^{\perp}}$ are associated with the normal vector field $Z_2^{\perp}$ for the immersion of $\mathcal{N}$ into $\mathcal{M}(c)$. They are related as follows:

$$g(\mathcal{B}(X_2, Y_2), Z_2^{\perp}) = g(A_{Z_2^{\perp}} X_2, Y_2), \tag{6}$$

for any $X_2 \in \Gamma(T\mathcal{N})$. If the almost complex structure J_s carries each tangent space of the submanifold $\mathcal{N}$ of the quaternion-Kähler manifold $\mathcal{M}$ into its normal space, then the submanifold $\mathcal{N}$ is called a totally real submanifold of $\mathcal{M}$. In this case, $g(J_s X_2, Y_2) = 0$ for each X_2, Y_2 tangent to $\mathcal{N}$, (see [14,15]). Then, (4) is reduced to the following

$$g(R(X_2, Y_2)Z_2, W_2) = \frac{c}{4}\Big\{g(Z_2, Y_2)g(X_2, W_2) - g(X_2, Z_2)g(Y_2, W_2)\Big\}. \tag{7}$$

for any $X_2, Y_2, Z_2, W_2 \in \Gamma(T\mathcal{N})$. The Gauss equation is defined for the curvature tensors $\tilde{R}$ and R of $\mathcal{N}^m$ and $\mathcal{M}$, respectively.

$$\begin{aligned} \tilde{R}(X_2, Y_2, Z_2, W_2) =& R(X_2, Y_2, Z_2, W_2) + \tilde{g}(\mathcal{B}(X_2, W_2), \mathcal{B}(Y_2, Z_2)) \\ &- \tilde{g}(\mathcal{B}(X_2, Z_2), \mathcal{B}(Y_2, W_2)). \end{aligned} \tag{8}$$

Let us consider an orthonormal frame $\{e_1, \ldots, e_m, e_{m+1}, \ldots, e_{4m+h}, e_{1^*} = J_s e_1, \ldots, e_{m^*} = J_s e_m, e_{(m+1)^*} = J_s e_{m+1}, \ldots, e_{(4m+h)^*} = J_s e_{4m+h}\}$ on the ambient space $\mathcal{M}^{4m+h}(c)$, where the codimension is h. On the submanifold $\mathcal{N}^m$, we assume that the vectors $e_1, \ldots, e_m$ span the tangent bundle $T\mathcal{N}^m$. The indices are assigned accordingly.

$$\begin{aligned} \mathcal{A}, \mathcal{B}, \mathcal{C} \cdots =& 1, \cdots, 4m+h, 1^* \cdots, 4m+h^*, \\ i^*, j^*, k^* =& 4m+1, \cdots, 4m+h, 1^*, \cdots, 4m+h^*. \\ & 1 \leq a, b, c, \ldots \leq m, \end{aligned}$$

Equation (7) is expressed in terms of local coordinates:

$$\widetilde{K}_{ABCD} = \frac{c}{4}\Big(\delta_{AC}\delta_{BD} - \delta_{AD}\delta_{BC}\Big). \tag{9}$$

Then, the curvature tensor for the submanifold is given by the following:

$$R_{abcl} = \widetilde{K}_{abcl} + \sum_i \Big(\mathcal{B}^i_{ac}\mathcal{B}^i_{bl} - \mathcal{B}^i_{al}\mathcal{B}^i_{bc}\Big), \tag{10}$$

where $\mathcal{B}$ denotes the second fundamental form of the submanifold.

The Ricci curvature for a totally real submanifold is expressed as follows:

$$R_{ab} = \frac{c}{4}(m-1)\delta_{ab} + \sum_k \left(\mathcal{B}^k_{ab}\sum_c \mathcal{B}^k_{cc} - \sum_c \mathcal{B}^k_{ac}\mathcal{B}^k_{cb}\right). \tag{11}$$

Let Π represent the squared length of the second fundamental form $\mathcal{B}$ of the submanifold $\mathcal{N}^m$, and it is defined by

$$\Pi = \sum_{abk} (\mathcal{B}^k_{ab})^2. \tag{12}$$

Correspondingly, the mean curvature of $\mathcal{N}^m$ is given by

$$\xi = \frac{1}{m} \sum_{ak} \mathcal{B}^k_{aa} e_k \tag{13}$$

From the above, we fix some notation

$$\Pi = \|\mathcal{B}\|^2, \quad \zeta = |\xi|, \quad \zeta_i = (\mathcal{B}^i_{ab})_{m\times m}. \tag{14}$$

where ξ is the mean curvature vector.

Let us assume that e_{m+1} is parallel to ξ; then, we have

$$tr\zeta_{m+1} = m\zeta, \quad tr\zeta_i = 0, \quad i \neq m+1 \tag{15}$$

where tr stands for the trace of the matrix $\zeta_i = (\mathcal{B}^i_{ab})$. Taking (9), (10), and (15) into account, we have scalar curvature as follows:

$$\mathrm{R} = \frac{1}{4} m(m-1)c + m^2\zeta^2 - \Pi \tag{16}$$

where ζ is the mean curvature of $\mathcal{N}^m$. Since ζ is constant, it can be concluded that the scalar curvature R is constant if and only if Π is constant according to (16). Let $\mathcal{B}^k_{abc}$ denote the second covariant derivative of $\mathcal{B}^k_{ab}$; we have

$$\sum_c \mathcal{B}^i_{abc}\theta_c = \mathrm{d}\mathcal{B}^i_{ab} - \sum_c \mathcal{B}^i_{cb}\theta_{ca} - \sum_c \mathcal{B}^i_{ac}\theta_{cb} - \sum_j \mathcal{B}^j_{ab}\theta_{ji} \tag{17}$$

where $\{\theta_a\}$ is the dual frame of $\mathcal{N}^m$. Taking the exterior derivative of the above equation, we obtain

$$\begin{aligned}\sum_l \mathcal{B}^i_{abcl}\theta_l =& \mathrm{d}\mathcal{B}^i_{abc} - \sum_l \mathcal{B}^i_{abc}\theta_{la} + \sum_j \mathcal{B}^j_{abc}\theta_{ji} \\ &- \sum_l \mathcal{B}^i_{alc}\theta_{lb} - \sum_l \mathcal{B}^i_{abl}\theta_{lc}.\end{aligned} \tag{18}$$

Moreover, the Laplacian of $\mathcal{B}^i_{ab}$ is

$$\begin{aligned}\Delta\mathcal{B}^i_{ab} =& \sum_c \mathcal{B}^i_{abcc} = \sum_c \mathcal{B}^i_{ccab} + \sum_{cd}\left(\mathcal{B}^i_{cd}\mathrm{R}_{dabc} + \mathcal{B}^i_{da}\mathrm{R}_{dcbc}\right) \\ &- \sum_{jc} \mathcal{B}^j_{ca}\mathrm{R}_{ijbc}.\end{aligned} \tag{19}$$

The DDVV conjecture, also known as the normal scalar curvature conjecture, establishes a pointwise inequality that relates the scalar curvature, the normal scalar curvature, and the mean curvature of a submanifold in a real space form. This conjecture has been confirmed and is now known as the DDVV inequality, stated as follows:

Lemma 1 ([27,28]). *If we let* $\mathrm{T}_1, \cdots, \mathrm{T}_n$ *be* $(m \times m)$*-symmetric matrices, then*

$$\sum_{r,s=1}^{n} \|[\mathrm{T}_r, \mathrm{T}_s]\|^2 \leq \left(\sum_{r=1}^{n} \|\mathrm{T}_r\|^2\right)^2$$

such that equality holds if and only if the following matrices are satisfied:

$$\mathrm{T}_r = P\begin{pmatrix} 0 & \mu & 0\cdots 0 \\ \mu & 0 & 0\cdots 0 \\ 0 & 0 & 0\cdots 0 \\ \vdots & \vdots & \vdots\ddots\vdots \\ 0 & 0 & 0\cdots 0 \end{pmatrix} P^t, \quad \mathrm{T}_s = P\begin{pmatrix} \mu & 0 & 0\cdots 0 \\ 0 & -\mu & 0\cdots 0 \\ 0 & 0 & 0\cdots 0 \\ \vdots & \vdots & \vdots\ddots\vdots \\ 0 & 0 & 0\cdots 0 \end{pmatrix} P^t$$

where P is an orthogonal $(m \times m)$-matrix, and $[\mathrm{T}_r, \mathrm{T}_s] = \mathrm{T}_r\mathrm{T}_s - \mathrm{T}_s\mathrm{T}_r$ is the commutator of the matrices $\mathrm{T}_r, \mathrm{T}_s$.

Lemma 2 ([4])**.** *Let $\mathcal{T}_1, \mathcal{T}_2, \cdots \mathcal{T}_n$ be $(m \times m)$-symmetric matrices. Then,*

$$-2\sum_{i,j=1}^{n}\left(tr(\mathcal{T}_i^2\mathcal{T}_j^2) - tr(\mathcal{T}_i\mathcal{T}_j)^2\right) \geq \sum_{i,j=1}^{n}[tr(\mathcal{T}_i\mathcal{T}_j)]^2 - \frac{3}{2}\left(\sum_{i=1}^{n} tr(\mathcal{T}_i^2)\right)^2. \tag{20}$$

3. Main Results

In this section, we present our main results concerning totally real submanifolds in quaternionic space forms.

Theorem 1. *Let $\mathcal{N}^m$ be an m-dimensional, compact, totally real submanifold $\mathcal{N}^m$ of a quaternionic space forms $\mathcal{M}^{4m+h}(c)$. If the mean curvature vector ξ of $\mathcal{N}^m$ is parallel and the the mean curvature ζ of $\mathcal{N}^m$ satisfies the inequality*

$$\mathrm{R}_{\mathcal{N}} \geq \left(\frac{m+2h-1}{2(m+2h)}\right)\left(\frac{c}{4} + \zeta^2\right), \tag{21}$$

then, under these conditions, $\mathcal{N}^m$ is a totally umbilical sphere $\mathbb{S}^m\left(\frac{1}{\sqrt{\frac{c}{4}+\zeta^2}}\right)$.

Proof. Assume that $\mathcal{N}^m$ is a totally real submanifold of quaternionic space form $\mathcal{M}(c)$ with parallel mean curvature vector ξ. Consider e_{m+1} such that it is parallel to ξ . Then, from (15), we have

$$\nabla^{\perp}\xi = \mathrm{d}\zeta e_{m+1} + \zeta\nabla^{\perp}e_{m+1} = \mathrm{d}\zeta e_{m+1} + \zeta\sum_j \theta_{m+1j}e_j = 0 \tag{22}$$

From the structure equation and (22), we have

$$\begin{aligned} \mathrm{d}\theta_{m+1j} &= -\sum_k \theta_{m+1k}\wedge\theta_{kj} + \frac{1}{2}\sum_{cl}\mathrm{R}_{m+1jcl}\theta_c\wedge\theta_l \\ &= \frac{1}{2}\sum_{cl}\mathrm{R}_{m+1jcl}\theta_c\wedge\theta_l = 0 \end{aligned} \tag{23}$$

From Equation (19), and considering that $\mathcal{N}^m$ has a parallel mean curvature vector and satisfies $\sum_c \zeta^i_{ccab} = 0$, we obtain the following:

$$\begin{aligned} \frac{1}{2}\Delta\Pi_\zeta &= \sum_{abc}(\mathcal{B}^{m+1}_{abc})^2 + \sum_{ab}\mathcal{B}^{m+1}_{ab}\Delta\mathcal{B}^{m+1}_{ab} \\ &= \sum_{abc}(\mathcal{B}^{m+1}_{abc})^2 + \sum_{abcl}\mathcal{B}^{m+1}_{ab}\left(\mathcal{B}^{m+1}_{cl}R_{labc} + \mathcal{B}^{m+1}_{la}\mathrm{R}_{lcbc}\right). \end{aligned} \tag{24}$$

If 2-plane $\pi \subset T_p\mathcal{N}$ at the point $p \in \mathcal{N}^m$, the sectional curvature is denoted by $\mathrm{R}_{\mathcal{N}}(p,\pi)$. Then, the following is fixed:

$$\mathrm{R}_{\min}(p) = \min_{\pi \subset T_p\mathcal{N}} \mathrm{R}_{\mathcal{N}}(p,\pi).$$

Therefore, we consider $\{e_i\}$ the orthonormal fields such that $\mathcal{B}^{m+1}_{ab} = \lambda_a \delta_{ab}$ for non-zero eigenvalues λ_i. Then, we obtain

$$\begin{aligned}\sum_{abcl} \mathcal{B}^{m+1}_{ab}\left(\mathcal{B}^{m+1}_{cl}\mathrm{R}_{labc} + \mathcal{B}^{m+1}_{la}\mathrm{R}_{lcbc}\right) =& \frac{1}{2}\sum_{ab}(\lambda_a - \lambda_b)^2 \mathrm{R}_{abab} \\ \geq& \frac{1}{2}\sum_{ab}\left(\lambda_a - \lambda_b\right)^2 \mathrm{R}_{\min}.\end{aligned} \tag{25}$$

Taking (24) and (25) into account, we have

$$\frac{1}{2}\Delta\Pi_{\zeta} \geq \sum_{abc}(\mathcal{B}^{m+1}_{abc})^2 + \frac{1}{2}\sum_{ab}(\lambda_a - \lambda_b)^2 \mathrm{R}_{\min}. \tag{26}$$

It follows from $\mathrm{R}_{\mathcal{N}} \geq \frac{m+2h-1}{2(m+2h)}\left(\frac{c}{4} + \zeta^2\right)$ with Hopf's lemma that Π_{ζ} is a constant, and we derive

$$\frac{1}{2}\sum_{ab}(\lambda_a - \lambda_b)^2 \mathrm{R}_{\min} = 0. \tag{27}$$

It is implied that $\lambda_a = \lambda_b$. Then, $\mathcal{N}^m$ is pseudo-umbilical.

Again from (19), $\sum_c \zeta^i_{ccab} = 0$, and the mean curvature vector of $\mathcal{N}^m$ is parallel; thus, one constructs

$$\begin{aligned}\frac{1}{2}\Delta\tau =& \sum_{i\neq m+1}\sum_{abc}(\mathcal{B}^i_{abc})^2 + \sum_{i\neq m+1}\sum_{abcl}\mathcal{B}^i_{ab}\left(\mathcal{B}^i_{cl}\mathrm{R}_{labc} + \mathcal{B}^i_{la}\mathrm{R}_{lcbc}\right) \\ &- \sum_{i\neq m+1}\sum_{jabc}\mathcal{B}^i_{ab}\mathcal{B}^j_{ca}\mathrm{R}_{ijbc}\end{aligned} \tag{28}$$

where τ is a scalar curvature in $\mathcal{N}^m$. From (10) and (15), we have

$$\begin{aligned}&\sum_{i\neq m+1}\sum_{abcl}\mathcal{B}^i_{ab}\left(\mathcal{B}^i_{cl}\mathrm{R}_{labc} + \mathcal{B}^i_{la}\mathrm{R}_{lcbc}\right) \\ &\qquad = m\left(\frac{c}{4} + \zeta^2\right)\tau + \sum_{ij\neq m+1}\left(\mathrm{tr}(\zeta_i\zeta_j)^2 - \mathrm{tr}(\zeta_i^2\zeta_j^2)\right) \\ &\qquad\quad - \sum_{ij\neq m+1}\left(\mathrm{tr}(\zeta_i\zeta_j)\right)^2\end{aligned} \tag{29}$$

From (10), we derive

$$\sum_{i\neq m+1}\sum_{jabc}\mathcal{B}^i_{ab}\mathcal{B}^j_{ca}\mathrm{R}_{ijbc} = -\sum_i tr\zeta^2_{i*} - \sum_{i,j\neq m+1}\left(tr(\zeta_i\zeta_j)^2 - tr(\zeta_i^2\zeta_j^2)\right) \tag{30}$$

Inserting (30) and (29) into (28), for any real number b', we obtain

$$\begin{aligned}\frac{1}{2}\Delta\tau =& \sum_{i\neq m+1}\sum_{abc}(\mathcal{B}^i_{abc})^2 + \sum_i tr\zeta^2_{i*} - b'm\left(\frac{c}{4} + \zeta^2\right)\tau \\ &+ (1+b')\sum_{i\neq m+1}\sum_{abcl}\mathcal{B}^i_{ab}(\mathcal{B}^i_{cl}R_{labc} + \mathcal{B}^i_{la}R_{lcbc}) + b'\sum_{ij\neq m+1}\left(\mathrm{tr}(\zeta_i\zeta_j)\right)^2\end{aligned}$$

$$+(1-b')\sum_{ij\neq m+1}\left(\mathrm{tr}(\zeta_i\zeta_j)^2-\mathrm{tr}(\zeta_i^2\zeta_j^2)\right) \tag{31}$$

For a fixed index i, we consider an orthonormal frame field $\{e_i\}$ satisfying $\mathcal{B}^i_{ab}=\lambda^i_a\delta_{ab}$. From the above assumption, together with Equation (15), it follows that

$$\begin{aligned}\sum_{abcl}\mathcal{B}^i_{ab}\left(\mathcal{B}^i_{cl}\mathrm{R}_{labc}+\mathcal{B}^i_{la}\mathrm{R}_{lcbc}\right)=&\frac{1}{2}\sum_{ab}(\lambda^i_a-\lambda^i_b)^2\mathrm{R}_{abab}\\ \geq&\frac{1}{2}\sum_{ab}(\lambda^i_a-\lambda^i_b)^2\mathrm{R}_{\min}\\ =&m\mathrm{tr}\zeta_i^2\mathrm{R}_{\min}\end{aligned}$$

which implies that

$$\sum_{i\neq m+1}\sum_{abcl}\mathcal{B}^i_{ab}\left(\mathcal{B}^i_{cl}\mathrm{R}_{labc}+\mathcal{B}^i_{la}\mathrm{R}_{lcbc}\right)\geq m\tau\mathrm{R}_{min}. \tag{32}$$

By applying the DDVV inequality stated in Lemma 1, we derive the following:

$$\begin{aligned}\sum_{ij\neq m+1}\left\{\mathrm{tr}(\zeta_i^2\zeta_j^2)-\mathrm{tr}(\zeta_i\zeta_j)^2\right\}=&\frac{1}{2}\sum_{ij\neq m+1}\mathrm{tr}(\zeta_i\zeta_j-\zeta_j\zeta_i)^2\\ \leq&\frac{1}{2}(\sum_{i\neq m+1}\mathrm{tr}\zeta_i^2)^2\\ =&\frac{1}{2}\tau^2.\end{aligned} \tag{33}$$

Also, we have

$$\sum_{ij\neq m+1}\left(\mathrm{tr}(\zeta_i\zeta_j)\right)^2\geq\frac{\tau^2}{m+2h-1}. \tag{34}$$

Choosing $b'=\frac{m+2h-1}{m+2h+1}$ in (31) and making use of (32)–(34) leads to the following result:

$$\frac{1}{2}\Delta\tau\geq\left\{-\left(\frac{m+2h-1}{m+2h+1}\right)\left(\frac{c}{4}+\zeta^2\right)+\left(\frac{2m+4h}{m+2h+1}\right)\mathrm{R}_{min}\right\}m\tau \tag{35}$$

Under the assumption given by (21), the following result is obtained:

$$\frac{1}{2}\Delta\tau\geq 0$$

According to Hopf's lemma. This leads to the conclusion that $\Delta\tau=0$. Hence, we obtain the following:

$$\tau=0\quad or\quad \mathrm{R}_{\mathcal{N}}=\left(\frac{m+2h-1}{2(m+2h)}\right)\left(\frac{c}{4}+\zeta^2\right).$$

In the case where $\tau=0$, the submanifold $\mathcal{N}^m$ becomes totally umbilical. On the other hand, when $\tau\neq 0$, from (10), we derive

$$\mathrm{R}_{abab}=\frac{c}{4}+\zeta^2,$$

which leads to the conclusion that $\mathcal{N}^m$ is a totally umbilical sphere $\mathbb{S}^m\left(\frac{1}{\sqrt{\frac{c}{4}+\zeta^2}}\right)$.

Moreover, the inequalities (32)–(35) all become equalities when the sectional curvature satisfies

$$tr\zeta_i^2 = tr\zeta_j^2 \quad (i,j \neq m+1), \quad \text{and} \quad \sum_a tr\zeta_{a^*}^2 = 0.$$

Next, we demonstrate that the second case cannot occur. For this, we must examine the case when equality holds in (33), which indicates either that all components ζ_i vanish or that exactly two of them are nonzero, with $i \neq m+1$.

Now, assume that the inequalities (34) and (35) also hold as equalities. Then, we must have

$$\operatorname{tr}\zeta_i^2 = \operatorname{tr}\zeta_j^2 \quad (i,j \neq m+1), \qquad \text{and} \qquad \sum_a \operatorname{tr}\zeta_{a^*}^2 = 0. \tag{36}$$

These conditions imply that $\mathcal{N}^m$ must be totally umbilical, satisfying

$$\mathrm{R}_{abab} = \frac{c}{4} + \zeta^2, \tag{37}$$

since all ζ_i with $i \neq m+1$ are zero. However, this leads to a contradiction. This completes the proof of the theorem. □

In the subsequent results, we have the following.

Theorem 2. *Let $\mathcal{N}^m$ be an m-dimensional, totally real submanifold of a quaternionic space form $\mathcal{M}^{4m+h}(c)$. If $\mathrm{J}\xi$ has a normal relation to $\mathcal{N}^m$, then $\mathcal{N}^m$ is totally umbilical or satisfies the inequality*

$$\inf \rho \leq m\left(\frac{c}{4} + \zeta^2\right)\left(m - \frac{5}{3}\right) \tag{38}$$

where ρ is the scalar curvature.

Proof. Let $\mathrm{J}\xi$ be normal to $\mathcal{N}^m$, and we can consider e_{m+1} such that it is parallel to ξ, so we have

$$tr\zeta_{m+1} = m\zeta, \quad tr\zeta_i = 0, \quad i \neq m+1 \tag{39}$$

From (19), we have

$$\begin{aligned}\frac{1}{2}\Delta\Pi = &\sum_{iijk} (\mathcal{B}^i_{abc})^2 + \sum_{iabc} \mathcal{B}^i_{ab}\mathcal{B}^i_{ccab} \\ &+ \sum_{iabcl} \mathcal{B}^i_{ab}\left(\mathcal{B}^i_{cl}\mathrm{R}_{labc} + \mathcal{B}^i_{la}\mathrm{R}_{lcbc}\right) - \sum_{ijabc} \mathcal{B}^i_{ab}\mathcal{B}^j_{ca}\mathrm{R}_{ijbc}.\end{aligned} \tag{40}$$

From (10), (39) and $\mathcal{N}^m$ are totally umbilical, and we obtain

$$\begin{aligned}\sum_{iabcl} \mathcal{B}^i_{ab}\left(\mathcal{B}^i_{cl}\mathrm{R}_{labc} + \mathcal{B}^i_{la}\mathrm{R}_{lcbc}\right) =& m\left(\frac{c}{4} + \zeta^2\right)\Pi - m^2\zeta^2 \\ &+ \sum_{ij}\left\{\operatorname{tr}(\zeta_i\zeta_j)^2 - \operatorname{tr}\left(\zeta_i^2\zeta_j^2\right)\right\} \\ &- \sum_{ij}\left(\operatorname{tr}(\zeta_i\zeta_j)\right)^2,\end{aligned} \tag{41}$$

$$\sum_{ijabc} \mathcal{B}^i_{ab}\mathcal{B}^j_{ca}\mathrm{R}_{ijbc} = -\sum_{ij}\left\{\mathrm{tr}(\zeta_i\zeta_j)^2 - \mathrm{tr}(\zeta_i^2\zeta_j^2)\right\} - \sum_a \mathrm{tr}\zeta^2_{a^*}. \tag{42}$$

In view of (39) and the pseudo-umbilical such that $\mathcal{B}^{m+1}_{ab} = \zeta\delta_{ab}$, we derive

$$\sum_{iabc} \mathcal{B}^i_{ab}\mathcal{B}^i_{ccab} = m\zeta\Delta\zeta, \tag{43}$$

$$\sum_{iabc}(\mathcal{B}^i_{abc})^2 \geq \sum_{ac}(\mathcal{B}^{m+1}_{aac})^2 = m\sum_a(\nabla_a\zeta)^2, \tag{44}$$

$$\frac{1}{2}\Delta\zeta^2 = \zeta\Delta\zeta + \sum_a(\nabla_a\zeta)^2. \tag{45}$$

According to the pseudo-umbilical condition with Lemma 2, $\mathcal{B}^{m+1}_{ab} = \zeta\delta_{ab}$, we have

$$\begin{aligned}
2\sum_{ij}\left\{\mathrm{tr}(\mathrm{1_i1_j})^2 - \mathrm{tr}\left(\zeta_i^2\zeta_j^2\right)\right\} - \sum_{ij}\mathrm{tr}(\zeta_i\zeta_j)^2
&= 2\sum_{ij\neq m+1}\left\{\mathrm{tr}(\zeta_i\zeta_j)^2 - \mathrm{tr}\left(\zeta_i^2\zeta_j^2\right)\right\} \\
&\quad - \sum_{ij\neq m+1} tr\left(\zeta_i\zeta_j\right)^2 - \left(\mathrm{tr}H^2_{m+1}\right)^2 \\
&\geq -\frac{3}{2}\tau^2 - m^2\zeta^4 \\
&= -\frac{3}{2}(\Pi - m\zeta^2)^2 - m^2\zeta^4.
\end{aligned} \tag{46}$$

Substituting (41)–(46) into (40), we have

$$\begin{aligned}
\frac{1}{2}\Delta\Pi &\geq \frac{1}{2}m\Delta\zeta^2 + m\left(\frac{c}{4} + \zeta^2\right)\Pi - \frac{3}{2}\left(\Pi - m\zeta^2\right)^2 - m^2\zeta^4 - m^2\zeta^2 \\
&= \frac{1}{2}m\Delta\zeta^2 + \left(\Pi - m\zeta^2\right)\left\{m\left(\frac{c}{4} + \zeta^2\right) - \frac{3}{2}\left(\Pi - m\zeta^2\right)\right\} \\
&= \frac{1}{2}m\Delta\zeta^2 + \tau\left\{m\left(\frac{c}{4} + \zeta^2\right) - \frac{3}{2}\tau\right\}.
\end{aligned} \tag{47}$$

Applying a similar argument as that in [10], we deduce that either $\mathcal{N}^m$ is totally umbilical or

$$\mathrm{inf}\rho \leq m\left(\frac{c}{4} + \zeta^2\right)\left(m - \frac{5}{3}\right).$$

This completes the proof of the theorem. □

Theorem 3. *Let* Jξ *be normal to an m-dimensional compact totally real submanifold* $\mathcal{N}^m$ *in a quaternionic space form* $\mathcal{M}^{4m+h}(c)$*. Then, the following inequality holds*

$$\int\left\{2m\left(\frac{c}{4} + \zeta^2\right)\Pi - 3\Pi^2 - 5m^2\zeta^4 - 4m^2\zeta^2 + 2m\zeta^2\right\}\mathrm{dV} \leq 0 \tag{48}$$

Proof. Assume that Jξ is normal to the submanifold $\mathcal{N}^m$, and, without loss of generality, let the vector e_{1^*} be chosen to be parallel to the mean curvature vector ξ. This choice implies that $tr\zeta_{1^*} = m\zeta$. Furthermore, for all indices $i \neq 1^*$, we have $tr\zeta_i = 0$. Under these assumptions, and in view of Equation (10), we obtain the following:

$$\begin{aligned}\sum_{ijabc} \mathcal{B}^i_{ab}\mathcal{B}^j_{ca}\mathrm{R}_{ijbc} =& m^2\zeta^2 - \sum_a \mathrm{tr}\zeta^2_{a^*} - \sum_{ij}\left\{\mathrm{tr}(\zeta_i\zeta_j)^2 - \mathrm{tr}(\zeta_i^2\zeta_j^2)\right\}\\ \leq& m^2\zeta^2 - \mathrm{tr}\zeta^2_{1^*} - \sum_{ij}\left\{\mathrm{tr}(\zeta_i\zeta_j)^2 - \mathrm{tr}(\zeta_i^2\zeta_j^2)\right\}\\ =& m^2\zeta^2 - m\zeta^2 - \sum_{ij}\left\{\mathrm{tr}(\zeta_i\zeta_j)^2 - \mathrm{tr}(\zeta_i^2\zeta_j^2)\right\}. \end{aligned} \tag{49}$$

Applying a similar argument as in Theorem 2, we deduce that

$$\begin{aligned}\frac{1}{2}\Delta\Pi \geq& \frac{1}{2}m\Delta\zeta^2 + m\left(\frac{c}{4}+\zeta^2\right)\Pi\\ &-\frac{3}{2}(\Pi - m\zeta^2)^2 - m^2\zeta^4 - 2m^2\zeta^2 + m\zeta^2.\end{aligned}$$

Since the boundary of $\mathcal{N}^m$ is compact, it follows from Stokes' theorem that

$$\int\left\{m\left(\frac{c}{4}+\zeta^2\right)\Pi - \frac{3}{2}(\Pi - m\zeta^2)^2 - m^2\zeta^4 - 2m^2\zeta^2 + m\zeta^2\right\} \leq 0$$

which implies (48). This completes the proof of the theorem. □

Remark 1. *It has been noted that Theorems 1–3 are extended versions of Theorems 1 and 2 in [29].*

4. Conclusions

The study of rigidity theorems in space forms, particularly for totally real submanifolds, has been an area of limited exploration. This work extends previous studies on totally real submanifolds by investigating scalar curvature pinching conditions in quaternionic space forms. We derived refined inequalities involving the scalar curvature, mean curvature, and the second fundamental form. These results enhance our understanding of the rigidity and geometry of such submanifolds and provide a basis for further exploration of their topological and analytical properties.

Author Contributions: Conceptualization, F.A. and A.A.; methodology, F.A. and A.A.; investigation, F.A. and A.A.; writing—original draft preparation, A.A.; writing—review and editing, F.A. and A.A.; funding acquisition, A.A. All authors have read and agreed to the published version of the manuscript.

Funding: This work was funded by the Deanship of Research and Graduate Studies at King Khalid University through a Large Research Project under grant number RGP2/22/45.

Data Availability Statement: No data were used for this study.

Acknowledgments: Author Akram Ali extends his appreciation to the Deanship of Research and Graduate Studies at King Khalid University for funding this work through a Large Research Project under grant number RGP2/22/45. The authors express their sincere thanks to the referee for providing valuable suggestions which helped to improve this paper.

Conflicts of Interest: The authors declare no conflicts of interest.

ferences

Simons, J. Minimal varieties in Riemannian manifolds. *Ann. Math.* **1968**, *88*, 62–105. [CrossRef]
Lawson, H.B. Local rigidity theorems for minimal hypersurfaces. *Ann. Math.* **1969**, *89*, 187–197. [CrossRef]
Chern, S.-S.; do Carmo, M.; Kobayashi, S. Minimal submanifolds of a sphere with the second fundamental form of constant length. In *Functional Analysis and Related Fields*; Browder, F.E., Ed.; Springer: Berlin/Heidelberg, Germany, 1970; pp. 59–75.
Li, A.-M.; Li, J. An intrinsic rigidity theorem for minimal submanifolds in a sphere. *Arch. Math.* **1992**, *58*, 582–594.

5. Chen, Q.; Xu, S. Rigidity of compact minimal submanifolds in a unit sphere. *Geom. Dedicata* **1993**, *45*, 83–88. [CrossRef]
6. Leung, P.-F. Minimal submanifolds in a sphere. *Math. Z.* **1983**, *183*, 75–86. [CrossRef]
7. Song, W.; Shao, W. Totally real submanifolds with constant scalar curvature in a complex space form. *J. Math. (PRC)* **2013**, *33*, 20–26.
8. Xu, H.-W. A rigidity theorem for a submanifold with parallel mean curvature in a sphere. *Arch. Math.* **1993**, *61*, 489–496. [CrossRef]
9. Xu, H.-W.; Gu, J.-R. A general gap theorem for submanifolds with parallel mean curvature in $\mathbb{R}^{n+p}$. *Commun. Anal. Geom.* **20** *15*, 175–293.
10. Liu, M.; Song, W. Complete totally real pseudo-umbilical submanifolds in a complex projective space. *J. Math. Res. Expo.* **2011**, *31*, 946–9
11. Gu, J.R.; Xu, H.W. On Yao rigidity theorem for minimal submanifolds in spheres. *Math. Res. Anal.* **2012**, *19*, 511–523.
12. Zhao, G.-S. An intrinsic rigidity theorem for totally real minimal submanifolds in a complex projective space. *J. Sichuan Un* **1992**, *29*, 174–178.
13. Bektaş, M. Totally real submanifolds in a quaternion space form. *Czechoslov. Math. J.* **2004**, *54*, 341–346. [CrossRef]
14. Carriazo, A.; Kim, Y.H.; Yoon, D.W. Some inequalities on totally real submanifolds in quaternionic space forms. *J. Korean Ma Soc.* **2004**, *41*, 795–808.
15. Mutlu, P.; Sentürk, Z. On quaternionic space forms. *Filomat* **2015**, *29*, 593–597. [CrossRef]
16. Lee, J.; Vîlcu, G.-E. Inequalities for generalized normalized δ-Casorati curvatures of slant submanifolds in quaternionic spa forms. *Taiwan. J. Math.* **2015**, *19*, 691–702. [CrossRef]
17. Alghamdi, F. Characterizations of pointwise hemi-slant warped product submanifolds in *LCK* manifolds. *Symmetry* **2024**, *16*, 281. [CrossR
18. Alhouiti, N. Pointwise hemislant submanifolds in a complex space form. *J. Math.* **2023**, 8940238. [CrossRef]
19. Alghamdi, F. On warped product pointwise pseudo-slant submanifolds of *LCK*-manifolds and their applications. *Axioms* **20** *13*, 807. [CrossRef]
20. Chen, B.-Y.; Ogiue, K. On totally real submanifolds. *Trans. Am. Math. Soc.* **1974**, *193*, 257–266. [CrossRef]
21. Ali, A.; Alkhaldi, A.H.; Laurian-Ioan, P.; Ali, R. Eigenvalue inequalities for the p-Laplacian operator on C-totally real submanifol in Sasakian space forms. *Appl. Anal.* **2022**, *101*, 702–713. [CrossRef]
22. Ali, A.; Lee, J.W.; Alkhaldi, A.H. The first eigenvalue for the p-Laplacian on Lagrangian submanifolds in complex space forr *Int. J. Math.* **2022**, *33*, 2250016. [CrossRef]
23. Li, Y.L.; Mofarreh, F.; Agrawal, R.P.; Ali, A. Reilly-type inequality for the Φ-Laplace operator on semi-slant submanifolds Sasakian space forms. *J. Inequal. Appl.* **2022**, *2022*, 102. [CrossRef]
24. Alluhaibi, N.; Ali, A. The eigenvalue estimates of p-Laplacian of totally real submanifolds in generalized complex space forr *Ric. Mat.* **2024**, *73*, 1307–1321. [CrossRef]
25. Ahmad, L.M. Basic inequalities for submanifolds of quaternionic space forms with a quarter-symmetric connection. *J. Geom. Ph* **2021**, *159*, 103927.
26. Siddiqi, M.D.; Siddiqui, A.N.; Ahmad, K. A generalized Wintgen inequality in quaternion Kähler geometry. *Int. J. Geom. Methc Mod. Phys.* **2024**, *21*, 2450185. [CrossRef]
27. Ge, J.Q.; Tang, Z.Z. A proof of the DDVV conjecture and its equality case. *Pac. J. Math.* **2008**, *237*, 87–95. [CrossRef]
28. Lu, Z. Normal scalar curvature conjecture and its applications. *J. Funct. Anal.* **2011**, *261*, 1284–1308. [CrossRef]
29. Zhou, J.-D.; Xu, C.-Y.; Song, W.-D. The rigidity of total real submanifolds in a complex projective space. *J. Math. (PRC)* **2015**, *35*, 1139–11

mathematics

ticle

air of Associated η-Ricci–Bourguignon Almost Solitons with ;eneralized Conformal Killing Potential on Sasaki-like Almost ontact Complex Riemannian Manifolds

ıncho Manev

Department of Algebra and Geometry, Faculty of Mathematics and Informatics, University of Plovdiv Paisii Hilendarski, 24 Tzar Asen St., 4000 Plovdiv, Bulgaria; mmanev@uni-plovdiv.bg; Tel.: +359-889-521-244

Abstract

The subject of this study is almost contact complex Riemannian manifolds, also known as almost contact B-metric manifolds. The considerations are restricted to a special class of these manifolds, namely those of the Sasaki-like type, because of their geometric construction and the explicit expression of their classification tensor by the pair of B-metrics. Here, each of the two B-metrics is considered as an η-Ricci–Bourguignon almost soliton, where η is the contact form. The soliton potential is chosen to be a conformal Killing vector field (in particular, concircular or concurrent) and then a generalization of the notion of conformality using contact conformal transformations of B-metrics. The resulting manifolds, equipped with the introduced almost solitons, are geometrically characterized. In the five-dimensional case, an explicit example on a Lie group depending on two real parameters is constructed, and the properties obtained in the theoretical part are confirmed.

Keywords: η-Ricci–Bourguignon almost soliton; almost contact B-metric manifold; almost contact complex Riemannian manifold; Sasaki-like manifold; conformal Killing vector field

MSC: 53C25; 53D15; 53C50; 53C44; 53D35; 70G45

1. Introduction

Let $\mathcal{M}$ be a smooth manifold equipped with a time-dependent family $g(t)$ of (pseudo-) Riemannian metrics. Suppose that the corresponding Ricci tensor $\rho(t)$ and scalar curvature $\tau(t)$ satisfy together with $g(t)$ the following flow equation

$$\frac{\partial}{\partial t}g = -2(\rho - \ell\tau g), \qquad g(0) = g_0,$$

where ℓ is a real constant. The evolution equation defined as described above is known as the Ricci–Bourguignon flow (or RB flow for short) and was introduced by J. P. Bourguignon in 1981 [1] to generalize other well-known geometric flows by choosing specific values of ℓ. Namely, these are the Ricci flow for $\ell = 0$, the Einstein flow for $\ell = \frac{1}{2}$, the traceless Ricci flow for $\ell = \frac{1}{m}$, and the Schouten flow for $\ell = \frac{1}{2(m-1)}$, where m is the dimension of the manifold [2,3].

It is known that the solitons of an internal geometric flow on $\mathcal{M}$ are its fixed points or self-similar solutions. The corresponding soliton of the Ricci–Bourguignon flow is called the Ricci–Bourguignon soliton (abbreviated RB soliton) and is defined by [4]

 https://doi.org/10.3390/math13132165

$$\rho + \frac{1}{2}\mathcal{L}_{\vartheta}g + (\lambda + \ell\tau)g = 0, \tag{1}$$

where $\mathcal{L}_{\vartheta}g$ denotes the Lie derivative of g with respect to the vector field ϑ, called the soliton potential, and λ is called the soliton constant.

An RB soliton is said to be expanding, steady, or shrinking if $\lambda > 0$, $\lambda = 0$, or $\lambda < 0$, respectively.

In the case where λ is not a constant, but a differentiable function on $\mathcal{M}$, then (1) defines an RB almost soliton [5].

Other recent studies by other authors on RB (almost) solitons are published in refs. [6–11].

An RB (almost) soliton is called trivial if the soliton potential is a Killing vector field. Such a vector field is an infinitesimal generator of isometry, which means that the flow of a Killing vector field ϑ preserves the given metric, i.e., $\mathcal{L}_{\vartheta}g = 0$ holds.

A slightly more general concept than the latter is that of conformal Killing vector fields; i.e., vector fields with a flow preserving a given conformal class of metrics. A conformal Killing vector field on a manifold with (pseudo-)Riemannian metric g is a vector field ϑ whose (locally defined) flow defines conformal transformations; i.e., preserve g up to scale and preserve the conformal structure. In other words, the Lie derivative of g with respect to ϑ is a multiple of g by a coefficient of some function on the manifold. This refers to the classical conformal transformation of the metric, which performs a change on the metric tensor by multiplying it by a scalar function known as the conformal factor. Conformal vector fields appear in various physical theories where angle-preserving symmetries are important, such as conformal field theory and its applications in condensed matter physics, statistical mechanics, quantum statistical mechanics, and string theory [12]. In particular, if the conformal factor is a constant, such vector fields are known as homothetic and they find application in the study of singularities in general relativity [13].

We focus our research on almost contact complex Riemannian manifolds (abbreviated accR manifolds), also known as almost contact B-metric manifolds. What is special about them is that they are equipped with a pair of pseudo-Riemannian metrics (known as B-metrics) g and $\tilde{g}$, which are interrelated by the almost contact structure. Some studies on such manifolds by other authors that have been published in recent years are [14–17].

One approach to using both B-metrics is given in ref. [18]. There, a generalization of the RB almost soliton of (1) is studied, defined as follows

$$\rho + \frac{1}{2}\mathcal{L}_{\vartheta}g + \frac{1}{2}\mathcal{L}_{\vartheta}\tilde{g} + (\lambda + \ell\tau)g + (\tilde{\lambda} + \ell\tilde{\tau})\tilde{g} = 0, \tag{2}$$

where $\tilde{\lambda}$ is also a function on $\mathcal{M}$ and $\tilde{\tau}$ is the scalar curvature of $\tilde{g}$.

In ref. [19], we launched another idea, different from that of (2), to include both B-metrics in the definition of RB almost solitons. Namely, to generate an η-RB almost soliton from each of the two B-metrics via the contact form η.

In this paper, we specialize accR manifolds into a type called Sasaki-like accR manifolds. Thus, we not only determine the Lie derivative of the metric, but we are also motivated by the rich geometric properties of manifolds of Sasaki type, their connections to physics, and their role as a generalization of Riemannian geometry. The Sasaki-like accR manifolds are defined geometrically by the condition that the complex cone of such a manifold is a holomorphic complex Riemannian manifold (also called a Kähler–Norden manifold).

The soliton potential is usually chosen to be at some special positions relative to the structures. A popular possibility is that the soliton potential is a vertical vector field, i.e., pointwise collinear with the Reeb vector field, as this case is studied in ref. [19]. Another substantial case, which is the subject of study here, is when the soliton potential is a conformal Killing vector field, i.e., the Lie derivative of any B-metric with respect to the potential is the same metric multiplied by a function. We also study in this paper a more general case, when the Lie derivative of any B-metric with respect to the potential is obtained by a contact conformal transformation inherent to accR manifolds. In this case, we said that the potential is a generalized conformal Killing vector field.

We also consider the soliton potential to be conformal Killing of a special kind, namely a concircular vector field. Such type of vector fields on Riemannian manifolds have been introduced in ref. [20]. A concircular vector field is defined by its covariant derivative being parallel to the identity tensor. It is well known that concircular vector fields play an important role in the theory of projective and conformal transformations and have applications in general relativity and mathematical physics [21].

An even narrower specialization is made for the soliton potential, with its covariant derivative coinciding with the identity tensor. Then, the vector field is called concurrent [22]. Such vector fields can describe flows in several areas of theoretical physics. For example, in general relativity, they are used to model the trajectories of galaxies and to characterize space–time structures.

2. accR Manifolds

Let $\mathcal{M}$ be a smooth $(2n+1)$-dimensional manifold equipped with an almost contact structure (φ, ξ, η) and a B-metric g. In more detail, φ is an endomorphism in the Lie algebra $\mathfrak{X}(\mathcal{M})$ of vector fields on $\mathcal{M}$, ξ is a Reeb vector field, η is its dual contact form, and g is a pseudo-Riemannian metric of signature $(n+1, n)$ for which the following identities hold:

$$\begin{gathered}\varphi\xi = 0, \qquad \varphi^2 = -\iota + \eta \otimes \xi, \qquad \eta \circ \varphi = 0, \qquad \eta(\xi) = 1, \\ g(\varphi x, \varphi y) = -g(x, y) + \eta(x)\eta(y),\end{gathered} \tag{3}$$

where ι is the identity in $\mathfrak{X}(\mathcal{M})$ [23]. In the last equality and further on, by x, y, z we denote arbitrary elements of $\mathfrak{X}(\mathcal{M})$ or vectors in the tangent space $T_p\mathcal{M}$ of $\mathcal{M}$ at an arbitrary point p of $\mathcal{M}$.

It is known that on $\mathcal{M}$ there exists an associated metric $\tilde{g}$ of g, which is also a B-metric and is defined by

$$\tilde{g}(x, y) = g(x, \varphi y) + \eta(x)\eta(y).$$

Such a manifold is called an almost contact B-metric manifold or an almost contact complex Riemannian manifold (abbreviated accR manifold) and we use $(\mathcal{M}, \varphi, \xi, \eta, g, \tilde{g})$ to denote it. The well-known Ganchev–Mihova–Gribachev classification of accR manifolds is introduced in ref. [23]. It consists of 11 basic classes $\mathcal{F}_i$, $i \in \{1, 2, \ldots, 11\}$, determined by conditions for the (0,3)-tensor F defined by

$$F(x, y, z) = g\big((\nabla_x \varphi)y, z\big),$$

where ∇ denotes the Levi-Civita connection of g. It has the following basic properties:

$$\begin{gathered}F(x, y, z) = F(x, z, y) = F(x, \varphi y, \varphi z) + \eta(y)F(x, \xi, z) + \eta(z)F(x, y, \xi), \\ F(x, \varphi y, \xi) = (\nabla_x \eta)(y) = g(\nabla_x \xi, y).\end{gathered}$$

The Lee forms of $(\mathcal{M}, \varphi, \xi, \eta, g, \tilde{g})$ are the following 1-forms associated with F:

$$\theta(z) = g^{ij}F(e_i, e_j, z), \quad \theta^*(z) = g^{ij}F(e_i, \varphi e_j, z), \quad \omega(z) = F(\xi, \xi, z),$$

where (g^{ij}) is the inverse matrix of the matrix (g_{ij}) of g with respect to a basis $\{e_i;\xi\}$ $(i = 1, 2, \ldots, 2n)$ of T_pM.

2.1. Sasaki-like accR Manifolds

A Sasaki-like accR manifold, according to the definition in ref. [24], is such that its warped product manifold with the real line, called the complex cone of such a manifold, is a Kähler–Norden manifold. An accR manifold, which is Sasaki-like, is defined using F by the condition

$$F(x, y, z) = g(\varphi x, \varphi y)\eta(z) + g(\varphi x, \varphi z)\eta(y).$$

Therefore, Sasaki-like accR manifolds have the following Lee forms

$$\theta = -2n\,\eta, \qquad \theta^* = \omega = 0 \tag{4}$$

and the following identities hold [24]

$$\nabla_x\xi = -\varphi x, \qquad (\nabla_x\eta)(y) = -g(x, \varphi y), \tag{5}$$

$$\rho(x, \xi) = 2n\,\eta(x), \qquad \rho(\xi, \xi) = 2n, \tag{6}$$

where ρ denotes the Ricci tensor for g.

Since the contact form η is closed on these manifolds, taking into account the second equality in (5), then θ in (4) is also closed.

Let $\tilde{\rho}$ denote the Ricci tensor defined by $\tilde{g}$. In ref. [19], for every Sasaki-like accR manifold it is proved that ρ and $\tilde{\rho}$ coincide; i.e.,

$$\tilde{\rho}(y, z) = \rho(y, z). \tag{7}$$

As a consequence of (6) and (7) we obtain the following for a Sasaki-like accR manifold:

$$\tilde{\rho}(x, \xi) = 2n\,\eta(x), \qquad \tilde{\rho}(\xi, \xi) = 2n. \tag{8}$$

Contracting (7), we obtain a dependence between the scalar curvature $\tilde{\tau}$ concerning $\tilde{g}$ and the associated quantity of τ regarding φ, defined by $\tau^* = g^{ij}\rho(e_i, \varphi e_j)$. This formula has the form $\tilde{\tau} = -\tau^* + \rho(\xi, \xi)$, which combining with the last result in (6) implies

$$\tilde{\tau} = -\tau^* + 2n. \tag{9}$$

2.2. Almost Einstein-like accR Manifolds

For accR manifolds, the notion of an Einstein-like manifold was introduced in ref. [25] and studied for Ricci-like solitons, but is applicable to any $(\mathcal{M}, \varphi, \xi, \eta, g, \tilde{g})$. Such a manifold is called Einstein-like if its Ricci tensor ρ satisfies the condition

$$\rho = a\,g + b\,\tilde{g} + c\,\eta \otimes \eta \tag{10}$$

for some triplet of constants (a, b, c). In particular, $(\mathcal{M}, \varphi, \xi, \eta, g, \tilde{g})$ is called an η-Einstein manifold if $b = 0$, or an Einstein manifold if $b = c = 0$.

In ref. [26], another, more specialized notion of Einstein-like accR manifolds was introduced and studied, namely the so-called v-Einstein manifold (v is an abbreviation for vertically), if $a = b = 0$ and $c \neq 0$. Manifolds with such a Ricci tensor of rank 1 have been known for a long time (see, e.g., [27,28]) and are also called Ricci simple manifolds (e.g., [29,30]).

If a, b, and c in (10) are functions on $\mathcal{M}$, then the manifold is called almost Einstein-like, almost η-Einstein, and almost Einstein, respectively [31].

A corresponding definition for an Einstein-like accR manifold using Ricci tensor $\tilde{\rho}$ is appropriate, and the corresponding condition of (10) is the following

$$\tilde{\rho} = \tilde{a}\,\tilde{g} + \tilde{b}\,g + \tilde{c}\,\eta \otimes \eta$$

for some triplet of constants $(\tilde{a}, \tilde{b}, \tilde{c})$. The notions arising from this idea, such as those for ρ above, are also relevant.

3. η-RB Almost Solitons

Many publications have studied more general solitons, for which an additional 1-form η is used in the definition, the so-called η-solitons. For example, some recent publications on η-RB solitons are [6,10]. This approach makes particular sense when the 1-form is part of the tensor structure of the manifold. In our case, this is the contact form η, and we use this idea for the RB almost solitons defined by (1).

This gives us an η-Ricci–Bourguignon almost soliton (in short, η-RB almost soliton), induced by the B-metric g in the following way:

$$\rho + \frac{1}{2}\mathcal{L}_{\vartheta} g + (\lambda + \ell\tau)g + \mu\,\eta \otimes \eta = 0, \tag{11}$$

where μ is also a function on $\mathcal{M}$ [6]. Hereafter, we denote this almost soliton by $(g; \vartheta; \lambda, \mu, \ell)$.

Recall that when λ and μ are constants on the manifold, $(g; \vartheta; \lambda, \mu, \ell)$ is called an η-RB soliton.

In this paper, we exploit the idea of including both B-metrics in the definition of η-RB almost solitons, but with exactly the same roles, unlike the so-called RB-like almost soliton defined by (2) in ref. [18]. Recall that both B-metrics act on the vertical distribution span(ξ) as $\eta \otimes \eta$.

Analogously to (11), we can also consider an η-RB almost soliton induced by the other B-metric $\tilde{g}$, denoted by $(\tilde{g}; \vartheta; \tilde{\lambda}, \tilde{\mu}, \tilde{\ell})$, and defined as follows

$$\tilde{\rho} + \frac{1}{2}\mathcal{L}_{\vartheta}\tilde{g} + (\tilde{\lambda} + \tilde{\ell}\tilde{\tau})\tilde{g} + \tilde{\mu}\,\eta \otimes \eta = 0, \tag{12}$$

where $\tilde{\lambda}$ and $\tilde{\mu}$ are functions on $\mathcal{M}$, and $\tilde{\ell}$ is a constant.

Clearly, an η-RB almost soliton $(\tilde{g}; \vartheta; \tilde{\lambda}, \tilde{\mu}, \tilde{\ell})$ with $\tilde{\mu} = 0$ is simply an RB almost soliton with respect to $\tilde{g}$.

Similarly, if $\tilde{\lambda}$ and $\tilde{\mu}$ are constants on $\mathcal{M}$, then the solitons of the corresponding type are said to be given.

Further in this paper, we study manifolds for which we use the following

Definition 1 ([19]). *An accR manifold $(\mathcal{M}, \varphi, \xi, \eta, g, \tilde{g})$ is said to be equipped with a pair of associated η-RB almost solitons with potential vector field ϑ if the corresponding Ricci tensors $\rho, \tilde{\rho}$ and scalar curvatures $\tau, \tilde{\tau}$ satisfy* (11) *and* (12)*, respectively.*

3.1. *The Potential Is a Conformal Killing Vector Field*

A vector field on $\mathcal{M}$, e.g., the potential ϑ, is called conformal Killing with respect to g if there exists a function ψ on $\mathcal{M}$ such that [5]

$$\mathcal{L}_{\vartheta} g = 2\psi g, \tag{13}$$

where $\psi = \operatorname{div}\vartheta / \dim\mathcal{M}$. Particular cases of conformal vector fields are homothetic vector fields, for which $\psi =$ const, and isometric vector fields, also called Killing vector fields, for which $\psi = 0$. The conformal Killing vector field is non-trivial if $\psi \neq 0$; otherwise, ϑ is called a Killing vector field with respect to g.

The counterpart notion regarding the other metric can also be considered. Namely, a vector field ϑ is called a conformal with respect to $\tilde{g}$ if there exists a function $\tilde{\psi}$ on $\mathcal{M}$ such that

$$\mathcal{L}_{\vartheta}\tilde{g} = 2\tilde{\psi}\tilde{g}. \tag{14}$$

Similarly, it is determined by $\tilde{\psi}$ whether a conformal vector field is non-trivial or Killing with respect to $\tilde{g}$.

Theorem 1. *Let $(\mathcal{M}, \varphi, \xi, \eta, g, \tilde{g})$ be a $(2n+1)$-dimensional Sasaki-like accR manifold that admits a pair of associated η-RB almost solitons $(g; \vartheta; \lambda, \mu, \ell)$ and $(\tilde{g}; \vartheta; \tilde{\lambda}, \tilde{\mu}, \tilde{\ell})$, where the potential ϑ is a conformal Killing vector field with respect to g and $\tilde{g}$ with functions ψ and $\tilde{\psi}$, respectively.*

Then, the manifold is v-Einstein regarding ρ and $\tilde{\rho}$ with one and the same triplet of constants $(a, b, c) = (\tilde{a}, \tilde{b}, \tilde{c}) = (0, 0, 2n)$, i.e., $\mathcal{M}$ has the following Ricci tensors with respect to the pair of B-metrics:

$$\rho = \tilde{\rho} = 2n\,\eta \otimes \eta, \tag{15}$$

and equal scalar curvatures with respect to g and $\tilde{g}$ as follows:

$$\tau = \tilde{\tau} = 2n. \tag{16}$$

Furthermore, the following conditions for the used functions are valid:

$$\lambda + \psi = -2n\ell, \qquad \tilde{\lambda} + \tilde{\psi} = -2n\tilde{\ell}, \tag{17}$$

$$\mu = \tilde{\mu} = -2n. \tag{18}$$

Proof. Taking into account (11) and (13), we obtain the form of the Ricci tensor for g as follows

$$\rho = -(\lambda + \ell\tau + \psi)g - \mu\,\eta \otimes \eta. \tag{19}$$

Therefore, the manifold is almost η-Einstein. Similarly, (14) and (12) imply the following form of the Ricci tensor for $\tilde{g}$:

$$\tilde{\rho} = -(\tilde{\lambda} + \tilde{\ell}\tilde{\tau} + \tilde{\psi})\tilde{g} - \tilde{\mu}\,\eta \otimes \eta. \tag{20}$$

We use (6) and (8) with (19) and (20) to get the following conditions:

$$\lambda + \ell\tau + \psi = -(\mu + 2n), \tag{21}$$

$$\tilde{\lambda} + \tilde{\ell}\tilde{\tau} + \tilde{\psi} = -(\tilde{\mu} + 2n). \tag{22}$$

After that, we substitute (21) and (22) into (19) and (20), respectively, to obtain

$$\rho = (\mu + 2n)g - \mu\,\eta \otimes \eta, \tag{23}$$

$$\tilde{\rho} = (\tilde{\mu} + 2n)\tilde{g} - \tilde{\mu}\,\eta \otimes \eta. \tag{24}$$

Then, we take the appropriate traces of (23) and (24) to obtain τ and $\tilde{\tau}$ for g and $\tilde{g}$, respectively. The results are the following

$$\tau = 2n(\mu + 2n + 1), \tag{25}$$

$$\tilde{\tau} = 2n(\tilde{\mu} + 2n + 1). \tag{26}$$

Taking into account (7); then, (23) and (24) imply

$$(\mu + 2n)g - (\tilde{\mu} + 2n)\tilde{g} - (\mu - \tilde{\mu})\eta \otimes \eta = 0,$$

which is equivalent to the following equality

$$(\mu + 2n)g(\varphi x, \varphi y) + (\tilde{\mu} + 2n)g(x, \varphi y) = 0. \tag{27}$$

Replacing y with φy in (27) gives the following equation:

$$(\tilde{\mu} + 2n)g(\varphi x, \varphi y) - (\mu + 2n)g(x, \varphi y) = 0. \tag{28}$$

Therefore, there exists a solution to the system of Equations (27) and (28) for arbitrary x and y if and only if the conditions in (18) are satisfied. The latter result is taken into account in (25) and (26). Thus, the corresponding scalar curvatures are obtained as in (16).

Similarly, (18), (23), and (24) imply the form of the two Ricci tensors given in (15).

Next, we substitute (16) and (18) into (21) and (22) and get the equalities in (17). □

3.2. The Potential Is a Generalized Conformal Killing Vector Field

A vector field ϑ is called a generalized conformal Killing vector field with respect to g if the following condition is satisfied:

$$\frac{1}{2}(\mathcal{L}_\vartheta g) = e^{2u}\cos 2v\, g + e^{2u}\sin 2v\, \tilde{g} + \left(e^{2w} - e^{2u}\cos 2v - e^{2u}\sin 2v\right)\eta \otimes \eta,$$

where $u, v, w \in \mathcal{F}(\mathcal{M})$. For brevity, we can denote $\psi = e^{2u}\cos 2v$, $\chi = e^{2u}\sin 2v$, $\omega = e^{2w}$ and the above definition condition takes the following form:

$$\frac{1}{2}(\mathcal{L}_\vartheta g) = \psi\, g + \chi\, \tilde{g} + (\omega - \psi - \chi)\eta \otimes \eta, \tag{29}$$

where $\psi, \chi, \omega \in \mathcal{F}(\mathcal{M})$.

The definition of the generalized conformal Killing vector field is inspired by the most general type of conformal transformation inherent to an accR manifold. In ref. [32], a general type of contact conformal transformation of an accR structure $(\varphi, \xi, \eta, g, \tilde{g})$ into another accR structure $(\varphi, \bar{\xi}, \bar{\eta}, \bar{g}, \bar{\tilde{g}})$ has been introduced and it can be written in terms of ψ, χ, ω in the following form:

$$\bar{\xi} = \omega^{-\frac{1}{2}}\xi, \qquad \bar{g} = \psi\, g + \chi\, \tilde{g} + (\omega - \psi - \chi)\eta \otimes \eta,$$
$$\bar{\eta} = \omega^{\frac{1}{2}}\eta, \qquad \bar{\tilde{g}} = \psi\, \tilde{g} - \chi\, g + (\omega - \psi - \chi)\eta \otimes \eta.$$

In ref. [24] (Proposition 7), it is proved that the class of Sasaki-like accR manifolds remains closed under contact conformal transformations of general type if and only if the following conditions are satisfied

$$\mathrm{d}w \circ \varphi = 0, \qquad \mathrm{d}u - \mathrm{d}v \circ \varphi = 0, \qquad \mathrm{d}u \circ \varphi + \mathrm{d}v = (1 - e^{w})\eta.$$

If $\chi = 0$ and $\omega = \psi$ are valid, then (29) implies (13), the condition for a usual conformal Killing vector field regarding g.

Similarly, ϑ is called a generalized conformal Killing vector field with respect to $\tilde{g}$ if it satisfies:

$$\frac{1}{2}(\mathcal{L}_\vartheta \tilde{g}) = \tilde{\psi}\, \tilde{g} + \tilde{\chi}\, g + (\tilde{\omega} - \tilde{\psi} - \tilde{\chi})\eta \otimes \eta, \tag{30}$$

where $\tilde{\psi}, \tilde{\chi}, \tilde{\omega} \in \mathcal{F}(\mathcal{M})$.

In this case, the specialization $\tilde{\chi} = 0$ and $\tilde{\omega} = \tilde{\psi}$ gives us (14), the condition for a usual conformal Killing vector field regarding $\tilde{g}$.

Theorem 2. *Let $(\mathcal{M}, \varphi, \xi, \eta, g, \tilde{g})$ be a $(2n+1)$-dimensional Sasaki-like accR manifold that admits a pair of associated η-RB almost solitons $(g; \vartheta; \lambda, \mu, \ell)$ and $(\tilde{g}; \vartheta; \tilde{\lambda}, \tilde{\mu}, \tilde{\ell})$, where the potential ϑ is a generalized conformal Killing vector field with respect to g and $\tilde{g}$ with functions (ψ, χ, ω) and $(\tilde{\psi}, \tilde{\chi}, \tilde{\omega})$, respectively.*

Then, the manifold is almost Einstein-like regarding ρ and $\tilde{\rho}$ with triplets of functions $(a, b, c) = (-\tilde{\chi}, -\chi, \chi + \tilde{\chi} + 2n)$ and $(\tilde{a}, \tilde{b}, \tilde{c}) = (-\chi, -\tilde{\chi}, \chi + \tilde{\chi} + 2n)$; i.e., $\mathcal{M}$ has the following Ricci tensors with respect to each of the B-metrics:

$$\rho = \tilde{\rho} = -\tilde{\chi} g - \chi \tilde{g} + (\chi + \tilde{\chi} + 2n)\eta \otimes \eta \tag{31}$$

and the scalar curvatures with respect to g and $\tilde{g}$ are as follows:

$$\tau = 2n(1 - \tilde{\chi}), \qquad \tilde{\tau} = 2n(1 - \chi). \tag{32}$$

Furthermore, the following conditions for the used functions are valid:

$$\omega - \psi + \mu + 2n = -\tilde{\chi}, \qquad \tilde{\omega} - \tilde{\psi} + \tilde{\mu} + 2n = -\chi, \tag{33}$$

$$\lambda - 2n\ell\psi + (1 + 2n\ell)(\omega + \mu) = -2n\{1 + (2n+1)\ell\}, \tag{34}$$

$$\tilde{\lambda} - 2n\tilde{\ell}\tilde{\psi} + (1 + 2n\tilde{\ell})(\tilde{\omega} + \tilde{\mu}) = -2n\{1 + (2n+1)\tilde{\ell}\}. \tag{35}$$

In particular, we get that the manifold is:

(i) almost η-Einstein regarding g if and only if χ is zero.
(ii) almost η-Einstein regarding $\tilde{g}$ if and only if $\tilde{\chi}$ is zero.
(iii) v-Einstein if and only if $\chi = \tilde{\chi} = 0$ holds.
(iv) scalar flat regarding g ($\tilde{g}$, respectively) if and only if $\tilde{\chi} = 1$ ($\chi = 1$, respectively) holds.

Proof. Combining (11) and (29), we obtain the following

$$\rho = -(\lambda + \ell\tau + \psi)g - \chi \tilde{g} - (\omega - \psi - \chi + \mu)\eta \otimes \eta, \tag{36}$$

We apply the last expression of ρ for (ξ, ξ) and, due to the last equality in (6), we obtain

$$\ell\tau = -(\lambda + \omega + \mu + 2n). \tag{37}$$

Then, we put (37) into (36) and the form of the Ricci tensor for g is as follows

$$\rho = (\omega - \psi + \mu + 2n)g - \chi \tilde{g} - (\omega - \psi + \mu - \chi)\eta \otimes \eta. \tag{38}$$

Taking the trace of (38), we get an expression of the scalar curvature for g

$$\tau = 2n(\omega - \psi + \mu + 2n + 1). \tag{39}$$

Now, we take the trace of ρ_{ij} from (38) with $\varphi^j_s g^{is}$ to obtain the associated quantity of τ as follows

$$\tau^* = 2n\chi. \tag{40}$$

We proceed similarly with the quantities for the other B-metric $\tilde{g}$. The equalities in (8), (12) and (30) imply the following expressions for the Ricci tensor and the scalar curvature regarding $\tilde{g}$:

$$\tilde{\rho} = (\tilde{\omega} - \tilde{\psi} + \tilde{\mu} + 2n)\tilde{g} - \tilde{\chi} g - (\tilde{\omega} - \tilde{\psi} + \tilde{\mu} - \tilde{\chi})\eta \otimes \eta, \tag{41}$$

$$\tilde{\tau} = 2n(\tilde{\omega} - \tilde{\psi} + \tilde{\mu} + 2n + 1). \tag{42}$$

Taking into account (9), (40), and (42), we obtain the relation in the second equality of (33), which is between some of the used functions. The equality in (38) simplifies the form of $\tilde{\rho}$ from (41) to the corresponding form in (31).

Similarly, (42) and the second equality in (33) give the expression of $\tilde{\tau}$ in (32).

The equalities in (7), (31), and (38), together with the last identity in (3), imply

$$(\omega - \psi + \mu + \tilde{\chi} + 2n)g(\varphi x, \varphi y) = 0,$$

which is true if and only if the first equality in (33) holds. Considering the same equality, the expressions for ρ and τ from (38) and (39) take the forms in (31) and (32), respectively.

Let us return to (37) and use (39). Then, we get the relation in (34) between the functions λ, μ and the constant ℓ for the η-RB almost soliton concerning g and the functions ψ and ω for its potential ϑ.

In a similar way, the dependence in (35) is obtained for the corresponding functions of the η-RB almost soliton concerning $\tilde{g}$.

The conclusions in (i)–(iii) follow directly from (31), and those in (iv) are elementary consequences of (32). □

3.3. *The Potential Is Concircular or Concurrent*

Recall that if for a vector field, e.g., ϑ, there exists a function β on the manifold such that $\nabla_x\vartheta = \beta x$ is valid for any vector field x, then ϑ is called a concircular vector field. In particular, if β is a constant, then ϑ is called a concurrent vector field.

Since it is known that $(\mathcal{L}_\vartheta g)(x,y) = g(\nabla_x\vartheta, y) + g(x, \nabla_y\vartheta)$ is true, then for a concircular vector field ϑ with respect to g it follows $\mathcal{L}_\vartheta g = 2\beta g$, i.e., every concircular vector field ϑ is conformal regarding g. A similar statement is also true regarding $\tilde{g}$, i.e., $\mathcal{L}_\vartheta \tilde{g} = 2\tilde{\beta}\tilde{g}$ is valid if ϑ is concircular with respect to $\tilde{g}$ with function $\tilde{\beta}$.

Corollary 1. *Let the conventions of Theorem 1 be given. A special case arises when the potential ϑ is concurrent with respect to each of the two B-metrics. Then, $(g; \vartheta; \lambda, \mu, \ell)$ and $(\tilde{g}; \vartheta; \tilde{\lambda}, \tilde{\mu}, \tilde{\ell})$ are η-RB solitons.*

Proof. In the case under consideration, since $\psi = \beta$ and $\tilde{\psi} = \tilde{\beta}$ are constants and the results in (17) are known, then the functions λ and $\tilde{\lambda}$ become constants. Then, λ and $\tilde{\lambda}$, together with the constants μ and $\tilde{\mu}$ from (17), determine $(g; \vartheta; \lambda, \mu, \ell)$ and $(\tilde{g}; \vartheta; \tilde{\lambda}, \tilde{\mu}, \tilde{\ell})$ as η-RB solitons, respectively. □

Corollary 2. *Let $(\mathcal{M}, \varphi, \xi, \eta, g, \tilde{g})$ satisfy the conditions of Theorem 2. In particular, if the potential ϑ is:*

(i) *concircular with respect to g and $\tilde{g}$ for functions β and $\tilde{\beta}$, respectively, then the manifold is v-Einstein with η-RB almost solitons $(g; \vartheta; \lambda, \mu, \ell)$ and $(\tilde{g}; \vartheta; \tilde{\lambda}, \tilde{\mu}, \tilde{\ell})$, so that*

$$\rho = \tilde{\rho} = 2n\,\eta \otimes \eta, \qquad \tau = \tilde{\tau} = 2n, \tag{43}$$

$$\lambda + 2n\ell = -\beta, \qquad \tilde{\lambda} + 2n\tilde{\ell} = -\tilde{\beta}, \qquad \mu = \tilde{\mu} = -2n. \tag{44}$$

(ii) *concurrent with respect to each of the two B-metrics, then $(g; \vartheta; \lambda, \mu, \ell)$ and $(\tilde{g}; \vartheta; \tilde{\lambda}, \tilde{\mu}, \tilde{\ell})$ are η-RB solitons.*

Proof. Under the assumption about ϑ in (i), it follows that $\psi = \omega = \beta$, $\tilde{\psi} = \tilde{\omega} = \tilde{\beta}$, and $\chi = \tilde{\chi} = 0$ hold. Then, (31)–(35) is specialized in the corresponding form in (43) and (44).

The assumption in (ii) is a special case of that in (i), with β and $\tilde{\beta}$ being constants. Then, the results in (44) show that the considered almost solitons become solitons of this type. □

3.4. An Example of an η-RB Almost Soliton with a Generalized Conformal Killing Potential

We now consider an explicit example of a Sasaki-like accR manifold given as Example 2 in ref. [24]. It is on a Lie group G of dimension 5, i.e., for $n = 2$, with a basis of left-invariant vector fields $\{e_0, \ldots, e_4\}$. The corresponding Lie algebra is defined by the commutators

$$\begin{array}{ll}[e_0, e_1] = pe_2 + e_3 + qe_4, & [e_0, e_2] = -pe_1 - qe_3 + e_4, \\ [e_0, e_3] = -e_1 - qe_2 + pe_4, & [e_0, e_4] = qe_1 - e_2 - pe_3, \qquad p, q \in \mathbb{R}.\end{array} \tag{45}$$

An accR structure $(\varphi, \xi, \eta, g, \tilde{g})$ is defined as follows

$$\begin{array}{l} g(e_0, e_0) = g(e_1, e_1) = g(e_2, e_2) = -g(e_3, e_3) = -g(e_4, e_4) = 1, \\ g(e_i, e_j) = 0, \quad i, j \in \{0, 1, \ldots, 4\},\ i \neq j, \\ \xi = e_0, \quad \varphi e_1 = e_3, \quad \varphi e_2 = e_4, \quad \varphi e_3 = -e_1, \quad \varphi e_4 = -e_2. \end{array}$$

After that, in ref. [25], the components of the curvature tensor $R_{ijkl} = R(e_i, e_j, e_k, e_l)$ and those of the Ricci tensor $\rho_{ij} = \rho(e_i, e_j)$ are calculated. Thus, the form of its Ricci tensor is established as $\rho = 4\,\eta \otimes \eta$, and the manifold is v-Einstein. The scalar curvatures for g and $\tilde{g}$ are obtained as $\tau = \tilde{\tau} = 4$ and the constructed manifold is $*$-scalar flat, i.e.$\tau^* = 0$. These calculations confirm the results in (15) and (16) and the related conclusions of Theorem 1.

Using (11) and (12), we construct a pair of associated η-RB almost solitons $(g; \vartheta; \lambda, \mu, \ell)$ and $(\tilde{g}; \vartheta; \tilde{\lambda}, \tilde{\mu}, \tilde{\ell})$ on $(G, \varphi, \xi, \eta, g, \tilde{g})$ with the following functions

$$\lambda = 2(t - 2\ell), \quad \tilde{\lambda} = -2(t + 2\tilde{\ell}), \qquad \mu = \tilde{\mu} = -4, \tag{46}$$

assuming that $\eta = dt$. We then choose the potential ϑ to be a generalized conformal Killing vector field with respect to each of the two B-metrics for a corresponding set of functions (ψ, χ, ω) and $(\tilde{\psi}, \tilde{\chi}, \tilde{\omega})$, i.e., (29) and (30) are valid. We choose the involved functions as follows

$$\psi = \omega = -\tilde{\psi} = -\tilde{\omega} = -2t, \qquad \chi = \tilde{\chi} = 0. \tag{47}$$

We verify that (46) and (47) satisfy (33), (34) and (35) according to Theorem 2 and (44) in Corollary 2.

The same choice of functions as in (46) also satisfies the requirement that ϑ be concircular with respect to g and $\tilde{g}$ for functions $\beta = -2t$ and $\tilde{\beta} = 2t$, respectively.

As another option, we choose β and $\tilde{\beta}$ to be constants, e.g., $\beta = -2$ and $\tilde{\beta} = 2$. Therefore, we determine that ϑ is concurrent with respect to g and $\tilde{g}$, respectively, as well as determining that $\lambda = 2 - 4\ell$ and $\tilde{\lambda} = -2 - 4\tilde{\ell}$ are constants. The last conclusion means that $(g; \vartheta; \lambda, \mu, \ell)$ and $(\tilde{g}; \vartheta; \tilde{\lambda}, \tilde{\mu}, \tilde{\ell})$ are η-RB solitons in this case.

We establish that the constructed η-RB soliton $(g; \vartheta; \lambda, \mu, \ell)$ is expanding, steady, or shrinking if and only if we choose $\ell < \frac{1}{2}$, $\ell = \frac{1}{2}$, or $\ell > \frac{1}{2}$, respectively. Analogously, $(\tilde{g}; \vartheta; \tilde{\lambda}, \tilde{\mu}, \tilde{\ell})$ is expanding, steady, or shrinking if and only if we set $\tilde{\ell} < -\frac{1}{2}$, $\tilde{\ell} = -\frac{1}{2}$, or $\tilde{\ell} > -\frac{1}{2}$, respectively.

We need to find whether such a vector field ϑ exists. Let it be $\vartheta = \vartheta^s e_s$ with respect to the considered basis for $s = 0, 1, 2, 3, 4$. By virtue of the well-known Koszul formula and (45), we obtain the condition $p^4 + q^4 + 2p^2q^2 - 2p^2 + 2q^2 + 1 = 0$ for parameters p and q, which has the solutions $p_{1,2} = 1 \pm qi$, $p_{3,4} = -1 \pm qi$. Since p and q are real, then we have $q = 0$ and $p = 1$ or $p = -1$. If we choose $q = 0$ and $p = 1$, we determine the potential

$\vartheta = c_1 e_1 + c_2 e_2 - c_2 e_3 + c_1 e_4$ for the arbitrary constants c_1 and c_2. We verify that this ϑ is also concurrent with respect to $\tilde{g}$.

Funding: The research is partially supported by project FP25-FMI-010 "Innovative Interdisciplinary Research in Informatics, Mathematics, and Pedagogy of Education", of the Scientific Fund of the Paisii Hilendarski University of Plovdiv, Bulgaria.

Data Availability Statement: The original contributions presented in this study are included in the article. Further inquiries can be directed to the corresponding author.

Conflicts of Interest: The funders had no role in the design of the study; in the collection, analyses, or interpretation of data; in the writing of the manuscript; or in the decision to publish the results. The author declares no conflict of interest.

ferences

Bourguignon, J.-P. Ricci Curvature and Einstein Metrics. In *Global Differential Geometry and Global Analysis (Berlin, 1979), Proceedings of the Colloquium, Berlin, Germany, 21–24 November 1979*; Lecture Notes in Mathematics; Springer: Berlin/Heidelberg, Germany, 1981; Volume 838, pp. 42–63.
Siddiqui, A.N.; Siddiqi, M.D. Almost Ricci–Bourguignon solitons and geometrical structure in a relativistic perfect fluid spacetime. *Balk. J. Geom. Appl.* **2021**, *26*, 126–138.
Catino, G.; Mazzieri, L.; Mongodi, S. Rigidity of gradient Einstein shrinkers. *Commun. Contemp. Math.* **2015**, *17*, 1550046. [CrossRef]
Catino, G.; Cremaschi, L.; Djadli, Z.; Mantegazza, C.; Mazzieri, L. The Ricci–Bourguignon Flow. *Pac. J. Math.* **2017**, *287*, 337–370. [CrossRef]
Dwivedi, S. Some results on Ricci–Bourguignon solitons and almost solitons. *Can. Math. Bull.* **2021**, *64*, 591–604. [CrossRef]
Blaga, A.M.; Taştan, H.M. Some results on almost η-Ricci–Bourguignon solitons. *J. Geom. Phys.* **2021**, *168*, 104316. [CrossRef]
Mi, R. Remarks on scalar curvature of gradient Ricci–Bourguignon sollitons. *Bull. Sci. Math.* **2021**, *171*, 103034. [CrossRef]
Soylu, Y. Ricci–Bourguignon solitons and almost solitons with concurrent vector field. *Differ. Geom. Dyn. Syst.* **2022**, *24*, 191–200.
Cunha, A.W.; Lemos, R.; Roing, F. On Ricci–Bourguignon solitons: Triviality, uniqueness and scalar curvature estimates. *J. Math. Anal. Appl.* **2023**, *526*, 127333. [CrossRef]
Dogru, Y. η-Ricci–Bourguignon solitons with a semi-symmetric metric and semi-symmetric non-metric connection. *AIMS Math.* **2023**, *8*, 11943–11952. [CrossRef]
Al-Sodais, H.; Bin Turki, N.; Deshmukh, S. Some new characterizations of trivial Ricci–Bourguignon solitons. *J. Math.* **2025**, *1*, 7917018. [CrossRef]
Ginsparg, P. Applied Conformal Field Theory. In *Champs, Cordes et Phénomènes Critiques/Fields, Strings and Critical Phenomena (Les Houches)*; Brézin, E., Zinn-Justin, J., Eds.; Elsevier Science Publishers B.V.: Amsterdam, The Netherlands, 1989.
Stephani, H.; Kramer, D.; MacCallum, M.; Hoenselaers, C.; Herlt, E. *Exact Solutions of Einstein's Field Equations*; Cambridge University Press: Cambridge, UK, 2003; p. 163, ISBN 978-0-521-46136-8.
Bulut, Ş. A quarter-symmetric metric connection on almost contact B-metric manifolds. *Filomat* **2019**, *33*, 5181–5190. [CrossRef]
Bulut, Ş. A semisymmetric metric connection on almost contact B-metric manifolds. *Turk. J. Math.* **2021**, *45*, 2455–2465. [CrossRef]
Özdemir, N.; Eren, E. Some results on almost contact manifolds with B-metric. *Fundam. Contemp. Math. Sci.* **2025**, *6*, 81–92. [CrossRef]
Özdemir, N.; Aktay, Ş.; Solgun, M. Some results on normal almost contact manifolds with B-metric. *Kragujev. J. Math.* **2026**, *50*, 597–611.
Manev, M. Ricci–Bourguignon almost solitons with special potential on Sasaki-like almost contact complex Riemannian manifolds. *Mathematics* **2024**, *12*, 2100. [CrossRef]
Manev, M. Pairs of associated η-Ricci–Bourguignon almost solitons with vertical potential on Sasaki-like almost contact complex Riemannian manifolds. *Mathematics* **2025**, *13*, 1863. [CrossRef]
Fialkow, A. Conformal geodesics. *Trans. Amer. Math. Soc.* **1939**, *45*, 443–473. [CrossRef]
Takeno, H. Concircular scalar field in spherically symmetric space-times, I. *Tensor* **1967**, *20*, 167–176.
Yano, K.; Chen, B.-Y. On the concurrent vector fields of immersed manifolds. *Kodai Math. Sem. Rep.* **1971**, *23*, 343–350. [CrossRef]
Ganchev, G.; Mihova, V.; Gribachev, K. Almost contact manifolds with B-metric. *Math. BalKa (N.S.)* **1993**, *7*, 261–276.
Ivanov, S.; Manev, H.; Manev, M. Sasaki-like almost contact complex Riemannian manifolds. *J. Geom. Phys.* **2016**, *105*, 136–148. [CrossRef]
Manev, M. Ricci-like solitons on almost contact B-metric manifolds. *J. Geom. Phys.* **2020**, *154*, 103734. [CrossRef]

26. Manev, M.; Nakova, G. Curvature properties of some three-dimensional almost contact B-metric manifolds. *Plovdiv Univ. S Work.* **2004**, *34*, 51–60.
27. Brinkmann, H.W. Einstein spaces which are mapped conformally on each other. *Math. Ann.* **1925**, *94*, 119–145. [CrossRef]
28. Walker, A.G. Canonical form for a Riemannian space with a parallel field of null planes. *Quart. J. Math.* **1949**, *20*, 135–1 [CrossRef]
29. Sharma, R. On a class of contact metric manifolds. *Ill. J. Math.* **1989**, *33*, 207–218.
30. Deprez, J.; Roter, W.; Verstraelen, L. Conditions on the projective curvature tensor of conformally flat Riemannian manifol *Kyungpook Math. J.* **1989**, *29*, 153–165.
31. Manev, M. Ricci-like solitons with arbitrary potential and gradient almost Ricci-like solitons on Sasaki-like almost contact B-met manifolds. *Results Math.* **2020**, *77*, 149. [CrossRef]
32. Manev, M. Contactly conformal transformations of general type of almost contact manifolds with B-metric. Applications. *Ma Balk. (N.S.)* **1997**, *11*, 347–357.

mathematics MDPI

ticle

On Surfaces of Exceptional Lorentzian Lie Groups with a Four-Dimensional Isometry Group

ovanni Calvaruso * and Lorenzo Pellegrino

Dipartimento di Matematica e Fisica "E. De Giorgi", Università del Salento, Prov. Lecce-Arnesano, 73100 Lecce, Italy; lorenzo.pellegrino@unisalento.it
* Correspondence: giovanni.calvaruso@unisalento.it

Abstract

In total, geodesic surfaces and their generalizations, namely totally umbilical and parallel surfaces, are well-known topics in Submanifold Theory and have been intensively studied in three-dimensional ambient spaces, both Riemannian and Lorentzian. In this paper, we prove the non-existence of parallel and totally umbilical (in particular, totally geodesic) surfaces for three-dimensional Lorentzian Lie groups, which admit a four-dimensional isometry group, but are neither of Bianchi–Cartan–Vranceanu-type nor homogeneous plane waves. Consequently, the results of the present paper complete the investigation of these fundamental types of surfaces in all homogeneous Lorentzian manifolds, whose isometry group is four-dimensional. As a byproduct, we describe a large class of flat surfaces of constant mean curvature in these ambient spaces and exhibit a family of examples.

Keywords: Lorentzian Lie groups; parallel surfaces; totally umbilical surfaces; CMC surfaces

MSC: 53C50; 53C25

1. Introduction

The study of three-dimensional homogeneous manifolds is a very active area of research in differential geometry, having drawn the attention of many scholars, due to the richness of their geometric properties. A nondegenerate metric in a three-dimensional pseudo-Riemannian manifold must be (up to reversing its sign [1]) either Riemannian or Lorentzian. Notably, Lorentzian settings tend to exhibit a broader spectrum of geometric behaviors, many of which have no analogue in the Riemannian context.

A key aspect in the analysis of these spaces lies in understanding their isometry groups. For a three-dimensional pseudo-Riemannian manifold (M,g), the group of isometries $\mathrm{Iso}(M,g)$ has a dimension of at most six. The case where $\dim(\mathrm{Iso}(M,g)) = 6$ characterizes manifolds of constant sectional curvature. It is also known that there are no three-dimensional pseudo-Riemannian manifolds whose isometry group has a dimension of five.

This naturally motivates the study of pseudo-Riemannian three-manifolds (M,g) admitting a four-dimensional isometry group. In such cases, the action of $\mathrm{Iso}(M,g)$ is transitive, which implies that (M,g) is a homogeneous manifold.

The Riemannian setting has been completely studied and is now well understood. Three-dimensional Riemannian manifolds with a four-dimensional isometry group are referred to as *Bianchi–Cartan–Vranceanu spaces* (often abbreviated as *BCV spaces*), named after the three mathematicians who made pioneering works on these spaces from distinct points of view (see [2–5]).

As it is often the case, the Lorentzian case offers more possibilities than its Riemannian counterpart. Indeed, some different types of homogeneous Lorentzian three-manifolds possess a four-dimensional isometry group. This becomes evident when one compares the classification of non-isometric homogeneous structures in the Riemannian setting [6] with the one in the Lorentzian context [7]. Specifically, as shown in [7], three-dimensional Lorentzian Lie groups whose isometry group is four-dimensional form three mutually exclusive classes, which exist for some distinct forms of the self-adjoint structure operator L associated with their Lie algebra:

(I) *Lorentzian Bianchi–Cartan–Vranceanu spaces*, for which L is diagonal, show a structure similar to Riemannian BCV spaces (although the Lorentzian class presents more cases than its Riemannian analogue) and exhibit several interesting geometric properties. In particular, they are naturally reductive spaces and are defined by submersions over a pseudo-Riemannian surface of constant curvature [8].

(II) An *exceptional example* $\mathcal{E}$, for which the minimal polynomial of L has a non-zero double root. At the Lie algebra level, this Lorentzian Lie group is described by

$$[u_1, u_2] = \mu u_3, \quad [u_2, u_3] = \mu u_2, \quad [u_3, u_1] = \mu u_1 + \varepsilon u_2, \qquad \mu \neq 0,\ \varepsilon = \pm 1, \tag{1}$$

where $\{u_1, u_2, u_3\}$ is a pseudo-orthonormal basis with u_3 being spacelike, i.e., such that the Lorentzian inner product $\langle\ ,\ \rangle$ is completely determined by the non-vanishing components

$$\langle u_1, u_2 \rangle = \langle u_3, u_3 \rangle = 1. \tag{2}$$

Equations (1) and (2) describe a one-parameter family of left-invariant Lorentzian metrics on $G = \widetilde{SL}(2, \mathbb{R})$ (we may refer either to [7] or to Section 3 for more details). These examples are still naturally reductive but do not have a structure similar to Lorentzian BCV spaces.

(III) Some *homogeneous plane waves*, for which the minimal polynomial of L has 0 as a triple root. Homogeneous plane waves are a well-known topic, whose study which dates back to the work by [9] in the framework of theoretical physics. These examples are not naturally reductive.

We refer to the above case (Case II) as "exceptional" with respect to Case (I), which admits a Riemannian analogue, and Case (III), which is well-known in the literature.

In this framework, the investigation of immersed surfaces plays a fundamental role. Exploring the geometric features of surfaces in homogeneous spaces provides deeper insights into the ambient manifold's global structure and enables the classification of surfaces through intrinsic and extrinsic criteria, such as total umbilicity, the parallelism of the second fundamental form, and the constancy of mean curvature. The investigation of such surfaces often reveals essential features of the ambient homogeneous space, serving as a powerful geometric tool to distinguish between different types of ambient spaces. This is true both for the cases where examples occur and classifications are achieved, and for the ones where non-existence results of submanifolds with the required properties are proved (see, for example, [10,11]). An exhaustive list of references concerning the geometry of surfaces in homogeneous Riemannian three-manifolds will dive into an investigation beyond the possibilities and purposes of the actual work. Besides the ones we already cited, a few works in this context are given by [12–29] and the references therein.

In Lorentzian settings, several authors have studied some remarkable families of surfaces in some ambient spaces that have a correspondence with the Riemannian cases, such as Lorentzian Berger spheres [30] and the Heisenberg group endowed with some left-invariant Lorentzian metrics [31–33]. In recent works ([8,34]), the authors of this paper described parallel, totally geodesic and totally umbilical surfaces of the non-exceptional cases—Cases (I) and (III)—of three-dimensional Lorentzian manifold admitting a group of isometry with a dimension of four, providing some examples of minimal and CMC surfaces. Together with M. Castrillon-Lopez [31], we investigated totally umbilical and minimal surfaces of the Lorentzian Heisenberg groups, proving the existence of a useful relation between minimality and total umbilicity in the Lorentzian context, which involves the conformality of the Gauss map of the surface. Moreover, the complete characterization of totally umbilical surfaces of Lorentzian reducible spaces was provided in [35].

Taking into account the results of [8,34,35], in order to complete the investigation of remarkable surfaces in Lorentzian homogeneous three-manifolds with a four-dimensional isometry group, one needs to consider these surfaces in the exceptional example $\mathcal{E}$. This research is the aim of this paper.

The paper is organized in the following way. In Section 2, we report some basic information concerning the geometric properties of $\mathcal{E}$ and the geometry of surfaces in a Lorentzian ambient space. In Section 3, we prove the non-existence of parallel (in particular, totally geodesic) surfaces of $\mathcal{E}$. On the other hand, the study of the case of parallel surfaces is a special case of surfaces with a Codazzi second fundamental form. We investigate the broader class of Codazzi surfaces, which turn out to be flat, and CMC surfaces, which exhibit a family that can be used as an example. In Section 4, we prove the non-existence of totally umbilical surfaces in $\mathcal{E}$.

2. Preliminaries

2.1. The Geometry of the Exceptional Example

For any choice of the real parameter $\mu \neq 0$, the unimodular Lie group $G = \widetilde{SL}(2, \mathbb{R})$ with the left-invariant Lorentzian metric determined by (1) and (2) is isometric to $\mathbb{R}^3$, equipped with the metric

$$g = \frac{\varepsilon}{\mu}(e^{-2\mu x_3} - 1)dx_1^2 + 2dx_1dx_2 + 2\mu x_2 dx_1 dx_3 + dx_3^2, \qquad \varepsilon = \pm 1. \tag{3}$$

In order to simplify the notation, throughout this paper, we shall use $\mathcal{E}$ to denote the exceptional example with four-dimensional isometry group, i.e., (an open subset of) $\mathbb{R}^3$ equipped with the Lorentzian metric (3).

We shall use the notation $\partial_i = \partial/\partial x_i$, for all indices i. As ∂_i, $i = 1, 2$, and 3 are global coordinate vector fields on $\mathcal{E}$, the well-known *Koszul formula* simplifies as

$$2g(\nabla_{\partial_i}\partial_j, \partial_k) = \partial_i(g(\partial_j, \partial_k)) + \partial_j(g(\partial_i, \partial_k)) - \partial_k(g(\partial_j, \partial_i)). \tag{4}$$

and allows us to determine $\nabla_{\partial_i}\partial_j$ for all indices i, j. For example, starting from (3) and setting $i = 2$ and $j = 3$ in (4), we get

$$\begin{cases} 2g(\nabla_{\partial_2}\partial_3, \partial_1) = \mu, \\ 2g(\nabla_{\partial_2}\partial_3, \partial_2) = 0, \\ 2g(\nabla_{\partial_2}\partial_3, \partial_3) = 0 \end{cases}$$

whence, $\nabla_{\partial_2}\partial_3 = \nabla_{\partial_3}\partial_2 = \frac{1}{2}\mu\partial_2$. Similar calculations allow us to describe the Levi-Civita connection ∇ of g, which is then explicitly determined by the following possibly non-vanishing components:

$$\begin{aligned}
\nabla_{\partial_1}\partial_1 &= -\varepsilon\mu x_2 e^{-2\mu x_3}\partial_2 + \varepsilon e^{-2\mu x_3}\partial_3, \\
\nabla_{\partial_1}\partial_2 &= \nabla_{\partial_2}\partial_1 = -\tfrac{1}{2}\mu^2 x_2\partial_2 + \tfrac{1}{2}\mu\partial_3, \\
\nabla_{\partial_1}\partial_3 &= \nabla_{\partial_3}\partial_1 = -\tfrac{1}{2}\mu\partial_1 - \tfrac{1}{2}(\varepsilon e^{-2\mu x_3} + \varepsilon + \mu^3 x_2^2)\partial_2 + \tfrac{1}{2}\mu^2 x_2\partial_3, \\
\nabla_{\partial_2}\partial_3 &= \nabla_{\partial_3}\partial_2 = \tfrac{1}{2}\mu\partial_2.
\end{aligned} \tag{5}$$

Next, we consider the curvature tensor R, taken with the sign convention $R(X,Y) = \nabla_{[X,Y]} - [\nabla_X, \nabla_Y]$. A direct calculation yields that, for example,

$$R(\partial_2,\partial_3)\partial_3 = \nabla_{\partial_2}\nabla_{\partial_3}\partial_3 - \nabla_{\partial_3}\nabla_{\partial_2}\partial_3 = -\frac{1}{2}\mu\nabla_{\partial_3}\partial_2 = \frac{1}{4}\mu^2\partial_2.$$

With similar calculations, we find that the curvature tensor R is completely determined, up to symmetries, by the following possibly non-vanishing components:

$$\begin{aligned}
R(\partial_1,\partial_2)\partial_1 &= \tfrac{1}{4}\mu^2\partial_1 - \tfrac{1}{4}\varepsilon\mu(e^{-2\mu x_3} - 1)\partial_2, \\
R(\partial_1,\partial_2)\partial_2 &= -\tfrac{1}{4}\mu^2\partial_2, \\
R(\partial_1,\partial_2)\partial_3 &= -\tfrac{1}{4}\mu^3 x_2\partial_2, \\
R(\partial_1,\partial_3)\partial_1 &= \tfrac{1}{4}\mu^3 x_2\partial_1 + \varepsilon\mu^2 x_2 e^{-2\mu x_3}\partial_2 - \tfrac{1}{4}\varepsilon\mu(5e^{-2\mu x_3} - 1)\partial_3, \\
R(\partial_1,\partial_3)\partial_2 &= -\tfrac{1}{4}\mu^2\partial_3, \\
R(\partial_1,\partial_3)\partial_1 &= \tfrac{1}{4}\mu^2\partial_1 + \varepsilon\mu e^{-2\mu x_3}\partial_2 - \tfrac{1}{4}\mu^3 x_2\partial_3, \\
R(\partial_2,\partial_3)\partial_1 &= \tfrac{1}{4}\mu^3 x_2\partial_2 - \tfrac{1}{4}\mu^2\partial_3, \\
R(\partial_2,\partial_3)\partial_3 &= \tfrac{1}{4}\mu^2\partial_2.
\end{aligned} \tag{6}$$

In terms of the coordinate vector fields, we have the following explicit expression for the pseudo-orthonormal basis $\{u_i\}$:

$$\begin{cases}
u_1 = e^{\mu x_3}\partial_1 + \frac{1}{2\mu}\left((\mu^3 x_2^2 + \varepsilon)e^{\mu x_3} - \varepsilon e^{-\mu x_3}\right)\partial_2 - \mu x_2 e^{\mu x_3}\partial_3, \\
u_2 = e^{-\mu x_3}\partial_2, \\
u_3 = \partial_3.
\end{cases} \tag{7}$$

Starting from (5) and (7), we can also determine the Levi-Civita connection as follows:

$$\begin{aligned}
&\nabla_{u_1}u_1 = \varepsilon u_3, && \nabla_{u_2}u_1 = -\tfrac{\mu}{2}u_3, && \nabla_{u_3}u_1 = \tfrac{\mu}{2}u_1, \\
&\nabla_{u_1}u_2 = \tfrac{\mu}{2}u_3, && \nabla_{u_2}u_2 = 0, && \nabla_{u_3}u_2 = -\tfrac{\mu}{2}u_2, \\
&\nabla_{u_1}u_3 = -\tfrac{\mu}{2}u_1 - \varepsilon u_2, && \nabla_{u_2}u_3 = \tfrac{\mu}{2}u_2, && \nabla_{u_3}u_3 = 0.
\end{aligned} \tag{8}$$

Moreover, we observe that the Killing vector field u_2 satisfies

$$\nabla_X u_2 = -\frac{\mu}{2}(X \wedge u_2), \tag{9}$$

for all $X \in T\mathcal{E}$, where the wedge product $\wedge$ is completely determined by

$$u_1 \wedge u_2 = -u_3, \quad u_2 \wedge u_3 = -u_2, \quad u_1 \wedge u_3 = u_1 \tag{10}$$

(see [36]).

From (6) and (7), we deduce that the curvature tensor is completely determined by

$$\begin{aligned} &R(u_1,u_2)u_1 = \tfrac{\mu^2}{4}u_1, && R(u_1,u_2)u_2 = -\tfrac{\mu^2}{4}u_2, && \\ &R(u_1,u_3)u_1 = -\varepsilon\mu u_3, && R(u_1,u_3)u_2 = -\tfrac{\mu^2}{4}u_3, && R(u_1,u_3)u_3 = \tfrac{\mu^2}{4}u_1 + \varepsilon\mu u_2, \\ &R(u_2,u_3)u_1 = -\tfrac{\mu^2}{4}u_3, && R(u_2,u_3)u_3 = \tfrac{\mu^2}{4}u_2. && \end{aligned} \tag{11}$$

Next, we can prove the following result.

Proposition 1. *The curvature tensor of $\mathcal{E}$ is given by*

$$R(X,Y)Z = -\frac{\mu^2}{4}(g(X,Z)Y - g(Y,Z)X) + \varepsilon\mu g(X\wedge Y,u_2)Z\wedge u_2$$

for all tangent vector fields X, Y, Z.

Proof. Let X, Y, Z be three arbitrary vector fields on $\mathcal{E}$. We write them as

$$X = \bar{X} + xu_3, \quad Y = \bar{Y} + yu_3, \quad Z = \bar{Z} + zu_3,$$

with $\bar{X}, \bar{Y}, \bar{Z}$ orthogonal to u_3 and $x = g(X,u_3)$, and so on. Observe that from (11), it follows that all terms of $g(R(X,Y)Z,W)$, where u_3 occurs either one, three or four times, necessarily vanish. Therefore, we have

$$\begin{aligned} g(R(X,Y)Z,W) = g(R(\bar{X},\bar{Y})\bar{Z},\bar{W}) \\ &+ xzg(R(u_3,\bar{Y})u_3,\bar{W}) + yzg(R(\bar{X},u_3)u_3,\bar{W}) \\ &+ ywg(R(\bar{X},u_3)\bar{Z},u_3) + xwg(R(u_3,\bar{Y})\bar{Z},u_3). \end{aligned}$$

From the decompositions of $\bar{X}$, $\bar{Y}$, $\bar{Z}$ and $\bar{W}$, it follows that

$$g(R(\bar{X},\bar{Y})\bar{Z},\bar{W}) = -\frac{\mu^2}{4}(g(\bar{X},\bar{Z})g(\bar{Y},\bar{W}) - g(\bar{X},\bar{W})g(\bar{Y},\bar{Z})).$$

Moreover, we have

$$R(\bar{X},u_3)u_3 = \frac{\mu^2}{4}\bar{X} + \varepsilon\mu g(\bar{X},u_2)u_2, \qquad R(u_3,\bar{Y})u_3 = -\frac{\mu^2}{4}\bar{Y} - \varepsilon\mu g(\bar{Y},u_2)u_2.$$

Hence, we conclude that

$$\begin{aligned} g(R(X,Y)Z,W) &= -\frac{\mu^2}{4}(g(\bar{X},\bar{Z})g(\bar{Y},\bar{W}) - g(\bar{X},\bar{W})g(\bar{Y},\bar{Z})) \\ &\quad - \frac{\mu^2}{4}(xzg(\bar{Y},\bar{W}) - yzg(\bar{X},\bar{W})) + \varepsilon\mu z(yg(\bar{X},u_2) - xg(\bar{Y},u_2))g(u_2,\bar{W}) \\ &\quad - \frac{\mu^2}{4}(ywg(\bar{X},\bar{Z}) - xwg(\bar{Y},\bar{Z})) + \varepsilon\mu w(xg(\bar{Y},u_2) - yg(\bar{X},u_2))g(u_2,\bar{Z}) \\ &= -\frac{\mu^2}{4}(g(X,Z)g(Y,W) - g(X,W)g(Y,Z)) \\ &\quad + \varepsilon\mu(yg(X,u_2) - xg(\bar{Y},u_2))(g(u_3,Z)g(u_2,W) - g(u_2,Z)g(u_3,W)) \\ &= g\left(-\frac{\mu^2}{4}(g(X,Z)Y - g(Y,Z)X) + \varepsilon\mu g(X\wedge Y,u_2)Z\wedge u_2, W\right), \end{aligned}$$

which ends the proof, as W is an arbitrary vector field. □

2.2. Geometry of Surfaces in a Three-Dimensional Lorentzian Ambient Space

Let $F : \Sigma \to M$ be an isometric immersion, where (Σ, g^{Σ}) denotes an oriented pseudo-Riemannian surface of a three-dimensional pseudo-Riemannian manifold (M, g), and the metric g^{Σ} on Σ is the pullback of g via the immersion F.

The normal unit vector field N on Σ satisfies $g(N, N) = \delta$, where $\delta \in \{-1, 1\}$. If $\delta = 1$, then N is spacelike, implying that Σ is a timelike surface and the immersion is Lorentzian. On the other hand, if $\delta = -1$, then N is timelike and, therefore, Σ is spacelike and the immersion is Riemannian.

Let ∇^{Σ} and ∇ denote the Levi-Civita connections of g^{Σ} and g, respectively. Given any vector fields X and Y tangent to Σ, these connections are related via the classical *Gauss equation*:

$$\nabla_X Y = \nabla^{\Sigma}_X Y + h(X, Y)N,$$

where h is a symmetric bilinear form on the tangent bundle of Σ, known as the *second fundamental form* associated with the immersion F.

Next, the formula $SX = -\nabla_X N$ defines the *shape operator* S of Σ for every tangent vector X. The shape operator S is related to the second fundamental form h by

$$g(SX, Y) = \delta h(X, Y),$$

for all X and Y tangent to Σ.

The behavior of h (equivalently, of S) at various points provides insights into the geometry of the submanifold. Specifically, a point $p \in \Sigma$ is referred to as *totally umbilical* if there exists some constant $\lambda(p)$, such that

$$h(X_p, Y_p) = \lambda(p)\, g^{\Sigma}_p(X_p, Y_p)$$

for all tangent vectors $X_p, Y_p \in T_p\Sigma$. When this condition holds throughout the entire surface, we say that Σ is *totally umbilical*. This means that around each point, there exists an open set $U \subseteq \Sigma$ and a smooth function $\lambda : U \to \mathbb{R}$, for which

$$h(X, Y) = \lambda\, g^{\Sigma}(X, Y)$$

for all vector fields X, Y defined on U. This geometric condition is equivalent to requiring that the shape operator S is a multiple of the identity (namely $S = \lambda\,\mathrm{Id}$) at each point. A special case arises when $\lambda = 0$. In this case, $h \equiv 0$, meaning that Σ is *totally geodesic*.

Another geometric property connected to some condition on the second fundamental form of the immersion is given by *parallel surfaces*. A surface is said to be parallel if its second fundamental form is covariantly constant, i.e.,

$$\nabla h = 0.$$

A weaker version of this requirement is the *semi-parallel* condition. Let R^{Σ} and R be the curvature tensors of Σ and M, respectively. The immersion is called *semi-parallel* if it satisfies the following:

$$R^{\Sigma} \cdot h = 0,$$

that is, more explicitly,

$$(R^{\Sigma} \cdot h)(X, Y, Z, W) = -h(R^{\Sigma}(X, Y)Z, W) - h(Z, R^{\Sigma}(X, Y)W) = 0,$$

for all vector fields X, Y, Z, W tangent to Σ.

Any parallel surface is also semi-parallel. Moreover, in the case where the ambient manifold is three-dimensional, the first author and Van der Veken in [37] proved the following relation between semi-parallel and totally umbilical surfaces:

Lemma 1 ([37]). *A surface Σ in a three-dimensional Lorentzian manifold (M,g) is semi-parallel if and only if it is either flat or totally umbilical.*

Further geometric constraints are encoded in the *Gauss and Codazzi equations*, which relate the intrinsic and extrinsic geometry of Σ:

$$g(R(X,Y)Z,W) = g(R^{\Sigma}(X,Y)Z,W) + \delta(h(X,Z)h(Y,W) - h(X,W)h(Y,Z)), \quad (12)$$
$$g(R(X,Y)Z,N) = \delta((\nabla h)(X,Y,Z) - (\nabla h)(Y,X,Z)), \quad (13)$$

where X,Y,Z,W are vector fields tangent to Σ. The *Codazzi condition* for the second fundamental form refers to the total symmetry of the covariant derivative ∇h. From Equation (13), one sees that this symmetry condition holds for parallel hyper-surfaces. Moreover, it is equivalent to the following condition:

$$g(R(X,Y)N,Z) = 0 \quad (14)$$

for all vector fields X,Y,Z tangent to Σ. Thus, a surface satisfying Equation (14) is said to have a *Codazzi second fundamental form* (or, briefly, a *Codazzi surface*).

Among submanifolds, *minimal* and *constant mean curvature (CMC) surfaces* have a special interest. The *mean curvature* H of a surface Σ in a three-dimensional ambient space $\tilde{M}$ is given by

$$H = \frac{1}{2}\mathrm{tr}_{g^{\Sigma}} h = \frac{1}{2}\sum \left(g^{\Sigma}\right)^{ij} h_{ij},$$

where $\left(g^{\Sigma}\right)^{ij}$ are the coefficients of the inverse metric $(g^{\Sigma})^{-1}$, expressed in a local frame. The surface is said to be *minimal* if $H = 0$. (In the context of Lorentzian geometry, for timelike surfaces, the vanishing of the mean curvature is often referred to as *maximality*). If the mean curvature is constant on Σ, the surface is called *CMC*. Clearly, totally geodesic surfaces are minimal, and minimal surfaces are necessarily CMC.

3. Codazzi Surfaces in $\mathcal{E}$

We shall now investigate surfaces Σ of $\mathcal{E}$ having a Codazzi second fundamental form. As we already mentioned in the Introduction Section, $\mathcal{E}$ is the class of three-dimensional Lorentzian Lie groups with a four-dimensional isometry group, such that the minimal polynomial of the structure operator L of its Lie algebra has a non-vanishing double root.

In fact, using (1) and (10), we easily obtain that L, completely determined by $L(u_i \wedge u_j) = [u_i, u_j]$, has eigenvalues $\lambda_1 = \lambda_2 = -\mu$ (with a one-dimensional associated eigenspace) and $\lambda_3 = \mu$.

The general form of the Lie algebra of a three-dimensional Lorentzian Lie group, with the structure operator L admitting a double root, is the following: its Lie algebra admits an orthonormal basis $\{e_1, e_2, e_3\}$, with e_3 being timelike, such that

$$(\mathfrak{g}_4): \quad \begin{aligned} [e_1,e_2] &= -e_2 + (2\eta - \beta)e_3, \qquad \eta = \pm 1, \\ [e_1,e_3] &= -\beta e_2 + e_3, \\ [e_2,e_3] &= \alpha e_1. \end{aligned} \quad (15)$$

The eigenvalues of L are given by $\lambda_1 = \lambda_2 = \beta - \eta$ (with a one-dimensional associated eigenspace) and $\lambda_3 = \alpha$. Therefore, $\mathcal{E}$ corresponds to the special case of (15), where $\alpha = \mu \neq 0$ and $\beta - \eta = -\mu \neq 0$, yielding a family of left-invariant metrics on $\widetilde{SL}(2, \mathbb{R})$ [36].

Parallel surfaces of all three-dimensional Lorentzian Lie groups, whose structure operator L has a double root, were already classified in the following result of [37].

Theorem 1 ([37])**.** *Let M be a parallel surface in a non-symmetric three-dimensional Lorentzian Lie group with Lie algebra* (15)*. Then, one of the following statements holds:*

(a) *M is an integral surface of the distribution spanned by $\{e_2, e_3\}$. This case only occurs if $\alpha = 0$. M is parallel, flat and minimal, but not totally geodesic.*

(b) *M is an integral surface of the distribution spanned by $\{e_1, ce_2 + be_3\}$, where b and c are constants satisfying $b^2 - c^2 = \delta$ and $\beta b^2 + 2bc + (\beta - 2\eta)c^2 = 0$. M is totally geodesic and has constant Gaussian curvature $K = -\delta(\beta - \eta)$.*

We now consider a pseudo-Riemannian-oriented surface Σ immersed into $\mathcal{E}$, whose normal vector field N satisfies $g(N, N) = \delta = \pm 1$. Using the notation we introduced in the previous section, the following result easily follows from Proposition 1.

Corollary 1. *For any X, Y tangent to Σ, we have the following:*

$$R(X,Y)N = \varepsilon\mu g(X \wedge Y, u_2)(N \wedge u_2). \tag{16}$$

We now prove the following.

Theorem 2. *Let Σ denote a surface in the exceptional example $\mathcal{E}$ with a Codazzi second fundamental form. Then,*

(I) *Every point of Σ admits an open neighborhood $U \subseteq \Sigma$, where $N = \phi u_2 + u_3$, for some smooth function $\phi : U \to \mathbb{R}$. In particular, Σ is timelike.*

(II) *Σ is flat.*

(III) *Σ is CMC but never minimal.*

(IV) *Σ is never parallel (in particular, totally geodesic).*

Proof. Starting from (7) and (2), we can easily deduce an orthonormal basis for $\mathcal{E}$, namely

$$E_1 = \frac{u_1 + u_2}{\sqrt{2}}, \quad E_2 = \frac{u_1 - u_2}{\sqrt{2}}, \quad E_3 = u_3, \tag{17}$$

where $g(E_1, E_1) = g(E_3, E_3) = -g(E_2, E_2) = 1$.

In a neighborhood U of some point $p \in \Sigma$, we consider $N = AE_1 + BE_2 + CE_3$, for some functions $A, B, C : U \to \mathbb{R}$ such that $g(N, N) = A^2 - B^2 + C^2 = \delta = \pm 1 \neq 0$. Then, the following vector fields are tangent to the surface:

$$X_1 = BE_1 + AE_2, \quad X_2 = CE_1 - AE_3, \quad X_3 = CE_2 + BE_3.$$

If h is Codazzi, then Equation (13) yields that $g(R(X_i, X_j)N, X_k) = 0$ for all indices $i, j, k \in \{1, 2, 3\}$. Hence, using the formula we introduced in Proposition 1, we have

$$
\begin{aligned}
0 &= g(R(X_1,X_2)N,X_k) = \varepsilon\mu A g(N,u_2) g(N\wedge u_2, X_k) = \varepsilon\mu \tfrac{A}{\sqrt{2}}(A+B) g(N\wedge u_2, X_k), \\
0 &= g(R(X_1,X_3)N,X_k) = \varepsilon\mu B g(N,u_2) g(N\wedge u_2, X_k) = \varepsilon\mu \tfrac{B}{\sqrt{2}}(A+B) g(N\wedge u_2, X_k), \\
0 &= g(R(X_2,X_3)N,X_k) = \varepsilon\mu C g(N,u_2) g(N\wedge u_2, X_k) = \varepsilon\mu \tfrac{C}{\sqrt{2}}(A+B) g(N\wedge u_2, X_k).
\end{aligned} \tag{18}
$$

It is easy to check that $g(N \wedge u_2, X_k) = 0$ for every $k \in \{1,2,3\}$ if and only if either N vanishes or it is lightlike. As a consequence, taking into account the fact that A, B, C cannot all vanish, conditions (18) hold if and only if $A + B = 0$; whence, $N = A(E_1 - E_2) + CE_3 = \sqrt{2}Au_2 + Cu_3$. Since $\delta = C^2$, then $\delta = 1$. Thus, Σ is timelike. Without the loss of generality, we can fix $C = 1$. By setting $\phi = \sqrt{2}A$, we obtain the result (I) given in the statement.

Next, let Σ be any Codazzi surface in $\mathcal{E}$, as described in (I). Thus, the vector fields

$$
X = u_2, \qquad Y = u_1 + u_2 - \phi u_3 \tag{19}
$$

span the tangent space to Σ at every point.

We observe that this distribution is integrable if and only if $X(\phi) = -\mu$. In fact, we have

$$
[X,Y] = -(\mu + X(\phi))N + \phi X(\phi) X.
$$

Moreover, a direct calculation, using the Gauss equation, (8) and (19), allows us to obtain

$$
\nabla^{\Sigma}_X X = \nabla^{\Sigma}_Y X = 0, \quad \nabla^{\Sigma}_X Y = -\mu\phi X, \quad \nabla^{\Sigma}_Y Y = \phi Y(\phi) X \tag{20}
$$

and the second fundamental form h of the immersion is determined by

$$
\begin{aligned}
&h(X,X) = 0, \quad h(X,Y) = -X(\phi) - \tfrac{\mu}{2}, \\
&h(Y,X) = \tfrac{\mu}{2}, \quad h(Y,Y) = \varepsilon - Y(\phi).
\end{aligned} \tag{21}
$$

It easily follows from (20) that $R^{\Sigma} = 0$; therefore, any Codazzi surface is flat, as stated in point (II).

Next, Equation (21) yields that $h(Y,X) = \frac{\mu}{2} \neq 0$ (hence, Σ cannot be totally geodesic) and the symmetry condition for h is equivalent to the integrability condition $X(\phi) = -\mu$ for the distribution $\{X,Y\}$. Since $\mu \neq 0$, it excludes that $X(\phi)$ vanishes. In particular, we have $\phi \neq 0$.

From (21), a standard calculation shows that the mean curvature of Σ is given by

$$
H = \frac{1}{2}\mathrm{tr}_{g^{\Sigma}} h = \frac{1}{2}\sum \left(g^{\Sigma}\right)^{ij} h_{ij} = \frac{\mu}{2},
$$

which proves the point (III) of the statement, i.e., the fact that any Codazzi surface is CMC but never minimal.

Finally, in order to consider the special case of parallel surfaces, we shall make use of the results given in Theorem 1. As previously emphasized, $\mathcal{E}$ is the special case corresponding to $\alpha = \mu \neq 0$ and $\beta - \eta = -\mu \neq 0$. As $\alpha \neq 0$, this excludes case (a) of Theorem 1. On the other hand, case (b) cannot occur for $\mathcal{E}$, either. In fact, in such a case, we have a totally geodesic surface; however, we already proved that totally geodesic surfaces do not exist in $\mathcal{E}$. Therefore, as stated in point (IV), $\mathcal{E}$ does not admit any parallel surfaces, and this completes the proof. □

Next, we shall exhibit a class of examples of Codazzi surfaces, which, by the above Theorem 2, are CMC and flat.

Example 1. *From Theorem 2, the normal vector field associated with any Codazzi surface is of the form* $N = \phi u_2 + u_3$. *In general,* $\phi = \phi(u, v)$ *is a smooth function defined on the surface* Σ, *which depends on some local coordinates* (u, v) *defined on* Σ, *if they exist.*

Using the tangent vector fields described in (19), *we find that*

$$\partial_u = \beta(u,v)X = \beta(u,v)u_2, \qquad \partial_v = Y = u_1 + u_2 - \phi(u,v)u_3, \tag{22}$$

are coordinate vector fields, provided that the smooth function β *satisfies*

$$\beta_v(u,v) = -\mu\phi(u,v)\beta(u,v), \quad \beta(u,v) = -\frac{\phi_u(u,v)}{\mu}. \tag{23}$$

We now show that the plane Σ, *defined by the equation* $x_3 = 0$, *admits some coordinates as described in* (22) *and* (23). *Therefore,* Σ *provides an example of a Codazzi surface (whence, a flat, CMC surface) for any choice of* ϕ.

Let $F : \Sigma \to \mathcal{E}$, $(u, v) \mapsto (x_1(u, v), x_2(u, v), x_3(u, v))$ *be the immersion of the surface* Σ *in the local coordinates in* (22). *By combining* (7) *and* (22) *with* $x_3 = 0$, *we obtain*

$$\begin{aligned} &\partial_u x_1(u,v) = 0, && \partial_v x_1(u,v) = 1, \\ &\partial_u x_2(u,v) = \beta(u,v), && \partial_v x_2(u,v) = \tfrac{\mu^2 x_2^2(u,v)}{2} + 1, \\ &\partial_u x_3(u,v) = 0, && \partial_v x_3(u,v) = -\mu x_2(u,v) - \phi(u,v). \end{aligned} \tag{24}$$

By integration of (24), *we deduce*

$$x_1(u,v) = v + c_1, \quad x_2(u,v) = \frac{\sqrt{2}}{\mu}\tan\left(\frac{\mu}{\sqrt{2}}v + f(u)\right), \quad x_3(u,v) = 0$$

for some smooth function f, *which depends only on* u.

Moreover, as $x_3 = 0$, *we have* $\partial_v x_3(u, v) = 0$; *therefore, using the last equation in* (24),

$$x_2(u,v) = -\frac{\phi(u,v)}{\mu}. \tag{25}$$

It is now easy to check that condition (23) *is satisfied for any suitable choice of* $f(u)$ *that determines* ϕ *by* (25) *and then* β *as* $\beta(u,v) = -\frac{\phi_u(u,v)}{\mu}$. *Observe that since* $\beta \neq 0$, *we have* $\phi_u \neq 0$, *i.e.,* $f'(u) \neq 0$, *so that* f *is not constant.*

Finally, since translations in x_1 *are isometries of the ambient space, we conclude that the plane* Σ *defined by* $x_3 = 0$ *can be parametrized by the following CMC immersion:*

$$F(u,v) = \left(v, \frac{\sqrt{2}}{\mu}\tan\left(\frac{\mu}{\sqrt{2}}v + f(u)\right), 0\right)$$

for any arbitrary non-constant smooth function $f(u)$.

4. Totally Umbilical Surfaces in $\mathcal{E}$

In this section, we will investigate the existence of totally umbilical surfaces in the exceptional example $\mathcal{E}$. Let Σ denote any oriented surface immersed into $\mathcal{E}$, and let N be its $\delta-$unit normal vector field ($g(N, N) = \delta$). The vector field u_2 decomposes as

$$u_2 = T + \nu N,$$

where $\nu := \delta g(N, u_2)$, and T is tangent to Σ; whence,

$$g(T, T) = -\delta\nu^2.$$

Remark 1. *We observe that $\nu = 0$ if and only if $g(N, u_2) = 0$, i.e., the surface Σ is Codazzi (see Theorem 2). Then, in the rest of this section, we will consider two cases separately, depending on whether ν vanishes. We explicitly observe that from Theorem 2, totally umbilical surfaces Σ cannot be totally geodesic.*

We now prove the following:

Proposition 2. *Any totally umbilical surface Σ immersed in $\mathcal{E}$ has a Codazzi second fundamental form.*

Proof. If Σ is a surface, immersed in $\mathcal{E}$, which is not Codazzi, then $\nu \neq 0$. Assume now that Σ is totally umbilical, with a unit normal N such that $g(N, N) = \delta$.

Observe that if T vanishes identically on a nonempty open subset, we can choose $N = u_2$ on this set, which contradicts the fact that, as $g(N, N) = \delta$, N is not lightlike. Therefore, we can assume that T does not vanish on any nonempty open subset.

Next, we define the vector field $JX := N \wedge X$, as some sort of rotation of X around an axis that follows the direction of N and satisfies

$$g(JX, JY) = -\delta\, g(X, Y), \qquad J^2X = -\delta\, X,$$

for any vector fields X, Y tangent to Σ. As a consequence, $JT = N \wedge (u_2 - \nu N) = N \wedge u_2$ is tangent to Σ and horizontal.

We recall that Σ is totally umbilical if and only if $\nabla_X N = \lambda X$ for every X tangent to Σ. We can now calculate

$$\begin{aligned}
\nabla_T JT &= \nabla_T(N \wedge T) \\
&= (\nabla_T N) \wedge T + N \wedge (\nabla_T T) \\
&= -\lambda T \wedge T + N \wedge (\nabla_T(u_2 - \nu N)) \\
&= -\tfrac{\mu}{2} N \wedge (T \wedge u_2) + \lambda\nu N \wedge T \\
&= -\tfrac{\mu}{2}\delta\nu\, T + \lambda\nu\, JT,
\end{aligned}$$

$$\begin{aligned}
\nabla_{JT} T &= \nabla_{JT}(u_2 - \nu N) \\
&= -\tfrac{\mu}{2}\,(JT \wedge u_2) - JT(\nu)N + \nu\lambda\, JT \\
&= -\tfrac{\mu}{2}\,(JT \wedge T) - \tfrac{\mu}{2}\nu\,(JT \wedge N) - JT(\nu)N + \nu\lambda\, JT \\
&= -\tfrac{\mu}{2}\delta\nu\, T + \lambda\nu\, JT,
\end{aligned}$$

where we used

$$\begin{aligned}
JT(\nu)N &= \delta JT(g(N, u_2))N \\
&= \delta\big(g(\nabla_{JT}N, u_2) + g(N, \nabla_{JT}u_2)\big)N \\
&= \delta\big(-\lambda g(JT, T + \nu N) - \tfrac{\mu}{2} g(N, JT \wedge (T + \nu N))\big)N \\
&= -\delta\tfrac{\mu}{2} g(N, JT \wedge T)N = -\tfrac{\mu}{2}\, JT \wedge T.
\end{aligned}$$

Moreover, we obtain

$$\begin{aligned}\nabla_{JT}JT = &\ \nabla_{JT}(N \wedge T) \\ = &\ (\nabla_{JT}N) \wedge T + N \wedge (\nabla_{JT}T) \\ = &\ -\lambda(JT \wedge T) - \delta\tfrac{\mu}{2}\nu\, JT - \delta\nu\lambda\, T.\end{aligned}$$

Therefore, $h(JT, JT) = -\delta\lambda g(JT \wedge T, N)$. If we require total umbilicity, we then get

$$JT(\nu)\frac{2\lambda}{\mu} = -\delta\lambda g(JT \wedge T, N) = \lambda g(JT, JT) = \lambda\nu^2.$$

However, $\lambda \neq 0$ because Σ cannot be totally geodesic according to Theorem 2. Therefore, the above equation yields

$$JT(\nu) = \frac{\mu}{2}\nu^2. \tag{26}$$

Observe that $JT(\nu) \neq 0$ as $\nu \neq 0$. Moreover, as $\nabla_T JT = \nabla_{JT}T$, we obtain that $[T, JT] = 0$. Therefore,

$$\begin{aligned}R(T, JT)N &= -\nabla_T\nabla_{JT}N + \nabla_{JT}\nabla_T N \\ &= \nabla_T\lambda JT - \nabla_{JT}\lambda T \\ &= T(\lambda)JT - JT(\lambda)T.\end{aligned} \tag{27}$$

On the other hand, by (16), we obtain

$$R(T, JT)N = \varepsilon\mu\nu g(T \wedge JT, N)JT = 2\varepsilon\mu\delta JT(\nu)JT. \tag{28}$$

We now compare (27) and (28), and we get

$$T(\lambda) = 2\varepsilon\mu\delta JT(\nu), \quad JT(\lambda) = 0,$$

whence, using $[T, JT](\lambda) = 0$, we deduce

$$2\varepsilon\mu\delta JT(JT(\nu)) = JT(T(\lambda)) = T(JT(\lambda)) = 0.$$

Consequently, Equation (26) yields

$$0 = JT(JT(\nu)) = \mu\nu JT(\nu),$$

which cannot occur, since $\nu \neq 0$ and $JT(\nu) \neq 0$. □

Next, we are left to investigate the existence of totally umbilical surfaces in the class of Codazzi surfaces.

Let Σ be a Codazzi surface immersed in $\mathcal{E}$. We know from Theorem 2 that its normal vector field is given by $N = \phi u_2 + u_3$. We recall that in this case, vector fields X, Y described in (19) span the tangent space to Σ.

From (21), we deduce that Σ is totally umbilical if and only if

$$\begin{cases}\frac{\mu}{2} = h(X, Y) = \lambda g(X, Y) = \lambda \\ \varepsilon - Y(\phi) = h(Y, Y) = \lambda g(Y, Y) = \lambda(2 + \phi^2).\end{cases}$$

Next, we find

$$h(X, X) = 0, \quad h(X, Y) = \frac{\mu}{2}, \quad h(Y, Y) = \frac{\mu}{2}\phi^2 + \mu. \tag{29}$$

Therefore, using (20) and (29), we obtain

$$\nabla^{\Sigma}_{X}h(X,X) = \nabla^{\Sigma}_{Y}h(X,X) = 0,$$
$$\nabla^{\Sigma}_{X}h(X,Y) = \nabla^{\Sigma}_{Y}h(X,Y) = 0,$$
$$\nabla^{\Sigma}_{X}h(Y,Y) = \mu\phi X(\phi) + 2\mu\phi h(X,Y) = 0,$$
$$\nabla^{\Sigma}_{Y}h(Y,Y) = \mu\phi Y(\phi) - 2\phi Y(\phi)h(X,Y) = 0.$$

This proves that a totally umbilical surface in $\mathcal{E}$ is necessarily parallel. Then, taking into account Theorem 2 and Proposition 2, we obtain the following conclusion.

Theorem 3. *The exceptional examples $\mathcal{E}$ do not admit any totally umbilical surface.*

As we already mentioned in the Introduction Section, totally umbilical and parallel (in particular, totally geodesic) surfaces were investigated in the Lorentzian Heisenberg groups [31], Lorentzian BCV spaces $\mathbb{M}^3_1(\kappa,\tau)$ with $\tau \neq 0$ [8], Lorentzian reducible spaces [35] (which in the homogeneous case correspond to $\mathbb{M}^3_1(\kappa,0)$)-spaces), and homogeneous plane waves $\mathcal{H}$ [34]. Therefore, the present work completed the investigation of totally umbilical and parallel surfaces of three-dimensional homogeneous Lorentzian manifolds with a four-dimensional isometry group. The results concerning the existence of these surfaces in all such homogeneous Lorentzian manifolds are summarized in Table 1 below.

Table 1. Surfaces in 3D Lorentzian spaces with 4D isometry group.

	$\mathbb{M}^3_1(\kappa,\tau \neq 0)$	$\mathbb{M}^3_1(\kappa,0)$	$\mathcal{E}$	$\mathcal{H}$
Totally Umbilical	✗	✓	✗	✗
Totally Geodesic	✗	✓	✗	✓
Parallel	✓	✓	✗	✓

Author Contributions: Conceptualization, G.C. and L.P.; methodology, G.C. and L.P.; validation, G.C. and L.P.; formal analysis, G.C. and L.P.; investigation, G.C. and L.P.; writing—original draft preparation, G.C. and L.P.; writing—review and editing, G.C. and L.P. All authors have read and agreed to the published version of the manuscript.

Funding: This research received no external funding.

Data Availability Statement: The original contributions presented in this study are included in the article. Further inquiries can be directed to the corresponding author.

Conflicts of Interest: The authors declare no conflicts of interest.

ferences

O'Neill, B. *Semi-Riemannian Geometry;* Academic Press: New York, NY, USA, 1983.
Bianchi, L. *Lezioni di Geometria Differenziale;* Enrico Spoerri Libraio-Editore: Pisa, Italy, 1894.
Bianchi, L. *Lezioni Sulla Teoria dei Gruppi Continui e Finiti di Trasformazioni;* Ed. Zanichelli: Bologna, Italy, 1928.
Cartan, E. *Leçons sur la Géométrie des Espaces de Riemann;* Gauthier-Villars: Paris, France, 1928.
Vranceanu, G. *Leçons de Géométrie Différentielle I;* Education Academic Representative of Romania: Bucharest, Romania, 1947.
Calviño-Louzao, E.; Ferreiro-Subrido, M.; Garcia-Rio, E.; Vazquez-Lorenzo, R. Homogeneous Riemannian structures in dimension three. *RACSAM Ser. A-Mat.* **2023**, *70*, 11. [CrossRef]
Calvaruso, G.; Zaeim, A. Three-dimensional homogeneous Lorentzian structures. *RACSAM Ser. A-Mat.* **2025**, *119*, 24. [CrossRef]
Calvaruso, G.; Pellegrino, L. Lorentzian BCV spaces: Properties and geometry of their surfaces. *J. Austral. Math. Soc.* 2025, *in press.* [CrossRef]
Blau, M.; O'Loughlin, M. Homogeneous plane waves. *Nucl. Phys. B* **2003**, *654*, 135–176. [CrossRef]
D'haene, M.; Inoguchi, J.; Van der Veken, J. Parallel and totally umbilical hypersurfaces of the four dimensional Thurston geometry Sol^4_0. *Math. Nachr.* **2024**, *297*, 1879–1891. [CrossRef]

11. Erjavec, Z.; Inoguchi, J. Codazzi and totally umbilical hypersurfaces in Sol_1^4. *Glasg. J. Math.* **2025**, *67*, 487–493. [CrossRef]
12. Belkhelfa, M.; Dillen, F.; Inoguchi, J. Surfaces with parallel second fundamental form in Bianchi-Cartan-Vranceanu spaces. In *PDE's, Submanifolds and Affine Differential Geometry*; Banach Center Publications, Polish Academy of Sciences: Warsaw, Poland, 2002; Volume 57, pp. 67–87.
13. Daniel, B. Isometric immersions into 3-dimensional homogeneous manifolds. *Comment. Math. Helv.* **2007**, *82*, 87–131. [CrossRef]
14. Dillen, F.; Fastenakels, J.; der Veken, J.V.; Vrancken, L. Constant angle surfaces in $\mathbb{S}^2 \times \mathbb{R}$. *Monatsh. Math.* **2007**, *152*, 89–9 [CrossRef]
15. Dillen, F.; Munteanu, M.I. Constant angle surfaces in $\mathbb{H}^2 \times \mathbb{R}$. *Bull. Braz. Math. Soc.* **2009**, *40*, 85–97. [CrossRef]
16. Fastenakels, J.; Munteanu, M.I.; Van Der Veken, J. Constant angle surfaces in the Heisenberg group. *Acta Math. Sin.* **201** *27*, 747–756. [CrossRef]
17. Inoguchi, J.; Kumamoto, T.; Ohsugi, N.; Suyama, Y. Differential geometry of curves and surfaces in 3-dimensional homogeneous spaces I. *Fukuoka Univ. Sci. Rep.* **1999**, *29*, 155–182
18. Inoguchi, J.; Kumamoto, T.; Ohsugi, N.; Suyama, Y. Differential geometry of curves and surfaces in 3-dimensional homogeneous spaces II. *Fukuoka Univ. Sci. Rep.* **2000**, *30*, 17–47.
19. Inoguchi, J.; Kumamoto, T.; Ohsugi, N.; Suyama, Y. Differential geometry of curves and surfaces in 3-dimensional homogeneous spaces III. *Fukuoka Univ. Sci. Rep.* **2000**, *30*, 131–160.
20. Inoguchi, J.; Kumamoto, T.; Ohsugi, N.; Suyama, Y. Differential geometry of curves and surfaces in 3-dimensional homogeneous spaces IV. *Fukuoka Univ. Sci. Rep.* **2000**, *30*, 161–168.
21. Inoguchi, J.; Van der Veken, J. A complete classification of parallel surfaces in three-dimensional homogeneous spaces. *Geom. Dedicata* **2008**, *131*, 159–172. [CrossRef]
22. Lopez, R.; Munteanu, M.I. On the geometry of constant angle surfaces in Sol_3. *Kyushu J. Math.* **2011**, *65*, 237–249. [CrossRef]
23. Manzano, J.M.; Souam, R. The classification of totally umbilical surfaces in homogeneous 3-manifolds. *Math. Z.* **2015**, *27* 557–576. [CrossRef]
24. Montaldo, S.; Onnis, I.I. Helix surfaces in the Berger sphere. *Israel J. Math.* **2014**, *201*, 949–966. [CrossRef]
25. Montaldo, S.; Onnis, I.I.; Passamani, A.P. Helix surfaces in the special linear group. *Ann. Mat. Pura Appl.* **2016**, *195*, 59–7 [CrossRef]
26. Souam, R.; Toubiana, E. Totally umbilic surfaces in homogeneous 3-manifolds. *Comment. Math. Helv.* **2009**, *84*, 673–704. [CrossRef]
27. Torralbo, F.; Urbano, F. Compact stable constant mean curvature surfaces in homogeneous 3-manifolds. *Indiana Univ. Math.* **2012**, *61*, 1129–1156. [CrossRef]
28. Torralbo, F.; Van der Veken, J. Rotationally invariant constant Gauss curvature surfaces in Berger spheres. *J. Math. Anal. Ap* **2020**, *489*, 124183. [CrossRef]
29. Van der Veken, J. Higher order parallel surfaces in Bianchi-Cartan-Vranceanu spaces. *Results Math.* **2008**, *51*, 339–359. [CrossRef]
30. Onnis, I.I.; Passamani, A.P.; Piu, P. Constant angle surfaces in Lorentzian Berger spheres. *J. Geom. Anal.* **2019**, *29*, 1456–14 [CrossRef]
31. Calvaruso, G.; Castrillón-Lopez, M.; Pellegrino, L. On totally umbilical and minimal surfaces of the Lorentzian Heisenberg groups. *Math. Nachr.* **2025**, *298*, 1922–1942. [CrossRef]
32. Onnis, I.I.; Piu, P. Constant angle surfaces in the Lorentzian Heisenberg group. *Arch. Math.* **2017**, *109*, 575–589. [CrossRef]
33. Turhan, E.; Altay, G. Minimal surfaces in three dimensional Lorentzian Heisenberg group. *Beiträge Alg. Geom.* **2014**, *55*, 1–2 [CrossRef]
34. Calvaruso, G.; Pellegrino, L. Surfaces of 3*D* homogeneous plane waves. *J. Geom. Phys.* 2025, *in press*. [CrossRef]
35. Calvaruso, G.; Pellegrino, L. Totally umbilical surfaces of Lorentzian reducible spaces. *J. Geom. Anal.* **2025**, *35*, 22. [CrossRef]
36. Rahmani, S. Métriques de Lorentz sur les groupes de Lie unimodulaires de dimension trois. *J. Geom. Phys.* **1992**, *9*, 295–3 [CrossRef]
37. Calvaruso, G.; Van der Veken, J. Parallel surfaces in three-dimensional Lorentzian Lie groups. *Taiwan. J. Math.* **2010**, *14*, 223–2 [CrossRef]

mathematics

icle

olutions of Da Rios Vortex Filament Equation of Cartan Null urves with Combescure Transformation

ılin Li [1], Osman Keçilioğlu [2], Kazım İlarslan [2] and Qingyou Sun [1,*]

1 School of Mathematics, Hangzhou Normal University, Hangzhou 311121, China; liyl@hznu.edu.cn
2 Department of Mathematics, Faculty of Engineering and Natural Sciences, Kırıkkale University, Kırıkkale 71450, Turkey; kecilioglu@kku.edu.tr (O.K.); kilarslan@kku.edu.tr (K.İ.)
* Correspondence: qysun@hznu.edu.cn

Abstract

In this study, Cartan null curves connected via the Combescure transformation are investigated within the framework of Minkowski 3-space, and the necessary conditions for establishing such connections are derived. The relationships between the Frenet vectors and curvatures of these curve pairs are also analyzed. Furthermore, when a ruled surface generated by a Cartan null curve provides a solution to the Da Rios equation, the conditions under which the ruled surface generated was by the corresponding Cartan null curve, related through the Combescure transformation, also satisfies the equation. All obtained results are supported with illustrative examples.

Keywords: Cartan null curves; Da Rios vortex filament equations; Combescure related curves

MSC: 53C50; 53C40

1. Introduction

A central topic in the study of space curves in Euclidean 3-space is the investigation of pairs of curves defined through relationships among their respective Frenet vector fields [1]. From such geometric correspondences, several well-known families of curves, such as Bertrand, Mannheim, evolute, and involute curves, naturally arise as special cases [2,3]. These curve classes have attracted significant attention, and ongoing research continues to explore their properties and generalizations [4–7].

Another important consequence of these geometric relations is the appearance of space curves connected via the Combescure transformation. Two curves in Euclidean 3-space that correspond pointwise and have parallel tangent vectors at corresponding points are said to be related by this transformation [8]. Under such a correspondence, the two curves share a common Frenet frame.

In [9], various geometric properties of a curve β, obtained from a given curve α through a Combescure transformation in $\mathbb{E}^3$, were analyzed. By exploiting the fact that these curves possess a common Frenet frame, an equivalence relation was introduced, and the corresponding equivalence classes were examined for specific families of curves. It was shown that all members of the equivalence class of a helix are themselves helices, and the same result holds for k-slant helices. The Combescure transformation also finds applications in the theory of surfaces, manifold theory, and Riemannian spaces [10–14]. However, since these topics lie beyond the scope of the present study, they will not be discussed here.

The theoretical framework developed for curves in Euclidean space has also been extended to spaces equipped with different metric structures. The most notable examples include three-dimensional Minkowski space and Minkowski spacetime. The differential geometry of curves in Minkowski space (or, more generally, in semi-Riemannian manifolds) has been extensively studied by both mathematicians and theoretical physicists.

In Minkowski 3-space, curves are classified as spacelike, timelike, or null (lightlike) depending on the causal character of their velocity vectors. It is well known that timelike and spacelike curves exhibit many analogous geometric properties [15,16]. However, the degeneracy of the induced metric along a null curve makes their study significantly more involved than in the non-degenerate cases. In Minkowski 3-space, non-null and pseudo-null curves associated with the Combescure transformation have been investigated in two separate studies [17].

Since it constitutes a part of our study, let us provide a brief literature overview on the Vortex Filament and Da Rios equations and their solutions. The vortex filament equation is an evolution equation for space curves in $\mathbb{E}^3$, first introduced by L. S. Da Rios [18] to describe the motion of a one-dimensional vortex filament in an incompressible, inviscid fluid. If $x(s,t)$ denotes the position vector of the filament, it satisfies

$$x_t = x_s \times x_{ss}, \tag{1}$$

which is known as the vortex filament equation. This relation was later rediscovered by Betchov, Arms, and Hama [19,20] as a local approximation of vortex tube evolution derived from the Biot–Savart law.

The vortex filament can also be regarded as a dynamical system on the space of curves in Minkowski 3-space [21]. Motions that preserve the filament's form correspond to travelling wave solutions of the nonlinear Schrödinger (NLS) equation [22], and the associated soliton surface is called the Hasimoto surface (or NLS surface).

Geometrically, if $x(s,t)$ is a spacelike curve with a timelike normal (or binormal) vector field, the motion governed by the vortex filament equation generates a spacelike (or timelike) Hasimoto surface. These situations are related to the nonlinear heat system

$$q_t = q_{ss} + q^2 x, \qquad x_t = -x_{ss} - x^2 q,$$

as discussed in [23]. Moreover, if $x(s,t)$ is timelike, the motion produces a timelike Hasimoto surface governed by the repulsive-type Schrödinger equation

$$ix_t = -q_{ss} + 2|q|^2 q,$$

which was analyzed in detail in [24]. The vortex filament equation for null Cartan curves has been studied by Grbović and Nešović [25]. The following significant results were obtained.

Theorem 1. *Let α be a null Cartan curve in $\mathbb{E}^3_1$ with the Cartan frame $\{T, N, B\}$, and let S be a ruled surface defined by*

$$x(s,t) = T(s) + t\big(a(s)T(s) + b(s)N(s) + c(s)B(s)\big),$$

where $a(s)$, $b(s)$, and $c(s)$ are differentiable functions of the pseudo-arc parameter s of α.

Then S is a solution of the vortex filament Equation (1) *if and only if:*

1. *α is a null Cartan helix with constant nonzero torsion $\tau_0 \neq 0$, and S is a non-degenerate cylindrical ruled surface with spacelike or timelike rulings, given by*

$$x(s,t) = T(s) + t(-\tau_0 T(s) + B(s));$$

2. *α is a null Cartan cubic, and S is a lightlike cylindrical ruled surface with null rulings, expressed as*

$$x(s,t) = T(s) + tB(s).$$

Let us now summarize what has been accomplished in this study.

This paper is organized as follows. In Section 2, Minkowski 3-space is first introduced, and the Frenet frame and curvatures of a Cartan null curve in this space are presented. In Section 3, the Combescure mate of a Cartan null curve is defined, and the parametric equation of the Combescure mate is obtained by means of a differentiable function C (Theorem 2). Furthermore, the relations between the Frenet vectors and curvatures of a Cartan null curve and its Combescure mate are derived (Theorem 3). Appropriate examples along with their graphical representations are also provided to illustrate the obtained results. In Section 4, considering Theorem 1 mentioned in the introduction, the necessary and sufficient conditions for the differentiable function C appearing in the parametrization of the conjugate curve are determined so that the ruled surface S^*, generated by the Cartan null curve associated with the Combescure transformation, also becomes a solution of the Da Rios vortex filament equation, provided that the ruled surface S, generated by a given Cartan null curve, satisfies the same equation (Theorem 4). Appropriate examples together with their graphical representations are also presented. It should be noted that only the parametric equations of the obtained surfaces are derived, and no discussion is made regarding their differential geometric properties.

2. Preliminaries

Minkowski space $\mathbb{E}^3_1$ is a three-dimensional affine space endowed with an indefinite flat metric $g(,)$ with signature $(-,+,+)$. This means that metric bilinear form can be written as

$$g(u,v) = -u_1v_1 + u_2v_2 + u_3v_3,$$

for any two vectors $u = (u_1, u_2, u_3)$ and $v = (v_1, v_2, v_3)$ in $\mathbb{E}^3_1$. Recall that a vector $u \in \mathbb{E}^3_1$ is called spacelike, if $g(u,u) > 0$ or $u = 0$, timelike if $g(u,u) < 0$, and null (lightlike) if $g(u,u) = 0$ and $u \neq 0$. The norm of a vector u is given by $||u|| = \sqrt{|g(u,u)|}$, and two vectors u and v are said to be orthogonal if $g(u,v) = 0$. An arbitrary curve $\varphi(s)$ in $\mathbb{E}^3_1$ can locally be spacelike, timelike or null (lightlike), if all its velocity vectors $\varphi'(s)$ are respectively spacelike, timelike or null. A null curve φ is parameterized by pseudo-arc s if $g(\varphi''(s), \varphi''(s)) = 1$. A spacelike or a timelike curve $\varphi(s)$ has unit speed, if $g(\varphi'(s), \varphi'(s)) = \pm 1$ [15,16]. The Lorentzian vector product of two vectors u and v is given by

$$u \times v = (u_3v_2 - u_2v_3, u_3v_1 - u_1v_3, u_1v_2 - u_2v_1).$$

A null curve $\varphi = \varphi(s)$ is called a null Cartan curve if it is parameterized by the pseudo-arc function s defined by

$$s(t) = \int_0^t ||\varphi''(u)||\, du.$$

Let $\{T, N, B\}$ denote the moving Frenet frame along a curve φ in $\mathbb{E}^3_1$, T, N and B, representing the tangent, principal normal, and binormal vector fields, respectively. Where

$T(s) = \varphi'(s)$, $N(s) = \varphi''(s)$, the vector B is a scalar multiple of the vector $T(s) \times_E N(s)$, (by $\times_E$ we denote Euclidean cross product) satisfying $g(T, B) = 1$ [26].

It is known that there exists a unique Cartan frame $\{T, N, B\}$ along a null Cartan curve α satisfying the Cartan equations [27]:

$$\begin{bmatrix} T' \\ N' \\ B' \end{bmatrix} = \begin{bmatrix} 0 & \kappa & 0 \\ -\tau & 0 & -\kappa \\ 0 & \tau & 0 \end{bmatrix} \begin{bmatrix} T \\ N \\ B \end{bmatrix}. \tag{2}$$

where the first curvature $\kappa(s) = 1$. The second curvature (torsion) $\tau(s)$ is an arbitrary function of the pseudo-arc parameter s. The Cartan frame vectors of α satisfy the relations

$$\begin{aligned} &\langle T, T\rangle = \langle B, B\rangle = 0, \quad \langle N, N\rangle = 1, \quad \langle T, N\rangle = \langle N, B\rangle = 0, \quad \langle T, B\rangle = 1, \\ &T \times B = N, \qquad N \times T = T, \qquad B \times N = B. \end{aligned} \tag{3}$$

3. Combescure-Related Cartan Null Curves in Minkowski 3-Space

In this section, we studied Combescure-related Cartan null curves in Minkowski 3-space. By constructing an explicit parametrization of the associated curve, we derive the relations between the corresponding Frenet frame elements under this transformation. The theoretical results are supported by a representative example, and the graphical illustration of the curves is also provided to visualize the geometric behavior of the transformation.

Definition 1. *Let $\varphi : I \subseteq \mathbb{R} \to \mathbb{E}_1^3$ and $\varphi^* : I^* \subseteq \mathbb{R} \to \mathbb{E}_1^3$ be null curves in $\mathbb{E}_1^3$ with Frenet apparatus $\{T, N, B, \kappa, \tau, s\}$ and $\{T^*, N^*, B^*, \kappa^*, \tau^*, s^*\}$, respectively. If the tangent vectors at the corresponding points of φ and φ^* are parallel, these curves are called related by a transformation of Combescure.*

Theorem 2. *Let $\varphi : I \subseteq \mathbb{R} \to \mathbb{E}_1^3$ and $\varphi^* : I^* \subseteq \mathbb{R} \to \mathbb{E}_1^3$ be null curves in $\mathbb{E}_1^3$ with Frenet apparatus $\{T, N, B, \kappa, \tau\}$ and $\{T^*, N^*, B^*, \kappa^*, \tau^*\}$, respectively. Then φ and φ^* Combescure-related curves if and only if there exists $C : I \subseteq \mathbb{R} \to \mathbb{R}$ differentiable function such that*

$$\varphi^*(f(s)) = \varphi(s) + \left(-\tau(s)C(s) - \frac{d^2C(s)}{ds^2}\right)T(s) + \frac{dC(s)}{ds}N + C(s)B. \tag{4}$$

where

$$f'(s) = \frac{ds^*}{ds} = \sqrt{1 - C(s)\frac{d\tau}{ds} - 2\tau\frac{dC(s)}{ds} - \frac{d^3C(s)}{ds^3}}. \tag{5}$$

Proof. Let $\varphi : I \subseteq \mathbb{R} \to \mathbb{E}_1^3$ and $\varphi^* : I^* \subseteq \mathbb{R} \to \mathbb{E}_1^3$ be null curves in $\mathbb{E}_1^3$ with Frenet apparatus $\{T, N, B, \kappa, \tau\}$ and $\{T^*, N^*, B^*, \kappa^*, \tau^*\}$, respectively. Assume that the vectors T and T^* are parallel, and

$$\varphi^*(s^*) = \varphi^*(f(s)) = \varphi(s) + u(s)T(s) + v(s)N(s) + w(s)B(s) \tag{6}$$

where u, v and w are differentiable functions on $I \subseteq \mathbb{R}$. Then differentiating (6) with respect to s and using the (2), we obtain

$$\frac{d\varphi^*}{ds^*}\frac{ds^*}{ds} = T^*\frac{ds^*}{ds} = (1 + u' - v\tau)T + (u + v' + \tau w)N + (w' - v)B \tag{7}$$

By taking the scalar product of Equation (7) with T, N and B, respectively, and using the fact that T^* and T are parallel, we obtain:

$$\left.\begin{array}{r} w' - v = 0 \\ u + v' + \tau w = 0 \\ 1 + u' - v\tau \neq 0 \end{array}\right\}. \tag{8}$$

Let $C : I \to \mathbb{R}$ be a differentiable function. If we take $w(s) = C(s)$, then from (8), we obtain

$$v = \frac{dC(s)}{ds} \text{ and } u = -\tau C(s) - \frac{d^2C(s)}{ds^2}. \tag{9}$$

Substituting (9) in (6) yields

$$\frac{d\varphi^*}{ds^*}\frac{ds^*}{ds} = T^*\frac{ds^*}{ds} = \left(1 - C(s)\frac{d\tau}{ds} - 2\tau\frac{dC(s)}{ds} - \frac{d^3C(s)}{ds^3}\right)T. \tag{10}$$

Differentiating (10) with respect to s and using (2), we obtain

$$\frac{d^2\varphi^*}{ds^{*2}}\left(\frac{ds^*}{ds}\right)^2 + \frac{d\varphi^*}{ds^*}\frac{d^2s^*}{ds^2} = \left(-C(s)\frac{d^2\tau}{ds^2} - 3\frac{d\tau}{ds}\frac{dC(s)}{ds} - 2\tau\frac{d^2C(s)}{ds^2} - \frac{d^4C(s)}{ds^4}\right)T \tag{11}$$

$$+ \left(1 - C(s)\frac{d\tau}{ds} - 2\tau\frac{dC(s)}{ds} - \frac{d^3C(s)}{ds^3}\right)N.$$

Substituting (10) in (11) and the scalar product with itself, we have

$$\left(\frac{ds^*}{ds}\right)^4 g\left(\frac{d^2\varphi^*}{ds^{*2}}, \frac{d^2\varphi^*}{ds^{*2}}\right) = \left(1 - C(s)\frac{d\tau}{ds} - 2\tau\frac{dC(s)}{ds} - \frac{d^3C(s)}{ds^3}\right)^2.$$

Taking into account that the curve φ^* is a pseudo arc-length parametrized Cartan null curve, it follows that

$$\frac{ds^*}{ds} = \sqrt{1 - C(s)\frac{d\tau}{ds} - 2\tau\frac{dC(s)}{ds} - \frac{d^3C(s)}{ds^3}}. \tag{12}$$

Conversely, we assume that φ^* is given by

$$\varphi^*(f(s)) = \varphi(s) + \left(-\tau(s)C(s) - \frac{d^2C(s)}{ds^2}\right)T(s) + \frac{dC(s)}{ds}N + C(s)B. \tag{13}$$

Differentiating (13) with respect to s and using Frenet frame, we get

$$T^*f' = \left(1 - C(s)\frac{d\tau}{ds} - 2\tau\frac{dC(s)}{ds} - \frac{d^3C(s)}{ds^3}\right)T. \tag{14}$$

We get that $T^*(f(s))$ and $T(s)$ are parallel, thus the Cartan null curves φ^* and φ are related by a transformation of Combescure. This completes the proof. □

Theorem 3. *Let $\varphi : I \subseteq \mathbb{R} \to \mathbb{E}_1^3$ and $\varphi^* : I^* \subseteq \mathbb{R} \to \mathbb{E}_1^3$ be Cartan null curves related by a transformation of Combescure with Frenet apparatus $\{T, N, B, \kappa, \tau\}$ and $\{T^*, N^*, B^*, \kappa^*, \tau^*\}$, respectively. Then, the following relationships hold between their Frenet vectors and curvatures:*

$$T^* = f'T \quad , \quad N^* = \frac{f''}{f'}T + N \quad , \quad B^* = -\frac{(f'')^2}{2(f')^3}T - \frac{f''}{(f')^2}N + \frac{1}{f'}B$$
$$\kappa^* = \kappa = 1 \qquad \text{and} \qquad \tau^* = -\frac{f'''}{(f')^3} + \frac{3(f'')^2}{2(f')^4} + \frac{\tau}{(f')^2} \tag{15}$$

where $f'(s) = \frac{ds^*}{ds} = \sqrt{1 - C(s)\frac{d\tau}{ds} - 2\tau\frac{dC(s)}{ds} - \frac{d^3C(s)}{ds^3}}$.

Proof. Assume that φ and φ^* are pseudo null curves related by a transformation of Combescure and the parameterization of φ^* is given by (4). Differentiating (4) with respect to s and using Frenet equations, we get

$$\frac{d\varphi^*}{ds^*}\frac{ds^*}{ds} = \left(1 - C(s)\frac{d\tau}{ds} - 2\tau\frac{dC(s)}{ds} - \frac{d^3C(s)}{ds^3}\right)T \tag{16}$$

Since $f'(s) = \frac{ds^*}{ds} = \sqrt{1 - C(s)\frac{d\tau}{ds} - 2\tau\frac{dC(s)}{ds} - \frac{d^3C(s)}{ds^3}}$, then:

$$T^* = f'T. \tag{17}$$

Differentiating (17) with respect to s and using Frenet formulae, we have:

$$N^* = \frac{f''}{f'}T + N. \tag{18}$$

Now assume that there exist differentiable functions a, b, c such that the binormal vector B^* can be expressed as:

$$B^* = aT + bN + cB. \tag{19}$$

Taking the scalar product of Equation (19) with T and N, respectively, using (3), we get

$$\begin{aligned} g(B^*, T) &= c \\ g(B^*, N) &= b. \end{aligned} \tag{20}$$

From (17) and (18), we obtain $T = \frac{1}{f'}T^*$ and $N = N^* - \frac{f''}{(f')^2}T^*$. If we use these equations in (20), we have

$$c = \frac{1}{f'} \text{ and } b = -\frac{f''}{(f')^2}. \tag{21}$$

Substituting these results back in (19) yields

$$B^* = aT - \frac{f''}{(f')^2}N + \frac{1}{f'}B. \tag{22}$$

Next, using the property $B^* \times N^* = B^*$ and Equation (18), we derive

$$\left(aT - \frac{f''}{(f')^2}N + \frac{1}{f'}B\right) \times \left(\frac{f''}{f'}T + N\right) = aT - \frac{f''}{(f')^2}N + \frac{1}{f'}B. \tag{23}$$

From (3),

$$-\frac{(f'')^2}{(f')^3}T = 2aT. \tag{24}$$

Therefore

$$a = -\frac{(f'')^2}{2(f')^3}. \tag{25}$$

Thus, the final expression for the binormal vector B^* is

$$B^* = -\frac{(f'')^2}{2(f')^3}T - \frac{f''}{(f')^2}N + \frac{1}{f'}B. \tag{26}$$

Differentiating (18) with respect to s and using Frenet formulae, we have:

$$-\tau^* T^* f' - B^* f' = \left(\left(\frac{f''}{f'}\right)' - \tau\right)T + \frac{f''}{f'}N - B. \tag{27}$$

Taking the inner product of Equations (26) and (27) side by side yields

$$\tau^*(f(s)) = -\frac{f'''}{(f')^3} + \frac{3(f'')^2}{2(f')^4} + \frac{\tau}{(f')^2}.$$

□

Corollary 1. *If $f' = c$, $c \in \mathbb{R}_0$ is taken in Theorem 2, then $N = N^*$ is obtained. In this case, the Combescure-related curve pair (φ, φ^*) is also a Bertrand curve pair.*

Example 1. *Consider the curve $\mathbb{E}_1^3$ given by*

$$\varphi(s) = \left(-\frac{s}{2}, \frac{1}{4}\cos(2s), -\frac{1}{4}\sin(2s)\right)$$

with the curvatures $\kappa = 1, \tau = 2$ and Frenet vectors

$$\begin{aligned} T &= \left(-\frac{1}{2}, -\frac{1}{2}\sin(2s), -\frac{1}{2}\cos(2s)\right) \\ N &= (0, -\cos(2s), \sin(2s)) \\ B &= (1, -\sin(2s), -\cos(2s)) \end{aligned}$$

Since $g(\varphi''(s), \varphi''(s)) = 1$, $\varphi(s)$ is a Cartan null curve, by using Theorem 2 and taking $C(s) = -\frac{s^3}{6}$, we obtain the curve φ^* related by transformation of Combescure as follows

$$\varphi^*(f(s)) = \begin{pmatrix} -s - \frac{s^3}{3}, \\ \frac{1}{4}\left((1+2s^2)\cos(2s) - 2s\sin(2s)\right), \\ -\frac{1}{4}\left((1+2s^2)\sin(2s) + 2s\cos(2s)\right) \end{pmatrix}.$$

Since $g\left((\varphi^*)''(s^*),(\varphi^*)''(s^*)\right) = 1$, $\varphi^*(f(s))$ *is a Cartan null curve with curvatures* $\kappa^*(f(s)) = 1$ *and* $\tau^*(f(s)) = \frac{2+11s^2+4s^4}{4(1+s^2)^3}$. *If the Frenet vectors of* φ^* *are calculated, we get*

$$T(f(s)) = \frac{1}{2}\begin{pmatrix} -\sqrt{2+2s^2}, \\ -\sqrt{2+2s^2}\sin(2s), \\ -\sqrt{2+2s^2}\cos(2s) \end{pmatrix}$$

$$N(f(s)) = \begin{pmatrix} -\frac{s}{2+2s^2}, \\ -\cos(2s) - \frac{s\sin(2s)}{2+2s^2}, \\ \sin(2s) - \frac{s\cos(2s)}{2+2s^2} \end{pmatrix}$$

$$B(f(s)) = \begin{pmatrix} \frac{s^2+(2+2s^2)^2}{(2+2s^2)^{\frac{5}{2}}}, \\ \frac{4(s+s^3)\cos(2s)-(4+7s^2+4s^4)\sin(2s)}{4\sqrt{2}(1+s^2)^{\frac{5}{2}}}, \\ \frac{-(4+7s^2+4s^4)\cos(2s)-4(s+s^3)\sin(2s)}{4\sqrt{2}(1+s^2)^{\frac{5}{2}}} \end{pmatrix}.$$

Since $T^*(f(s)) = \sqrt{2+2s^2}T(s)$, *the vectors* T *and* T^* *are parallel. This implies that the curves* φ *and* φ^* *are Cartan null curves with pseudo arc-length parameter, related by a Combescure transformation (See Figure 1).*

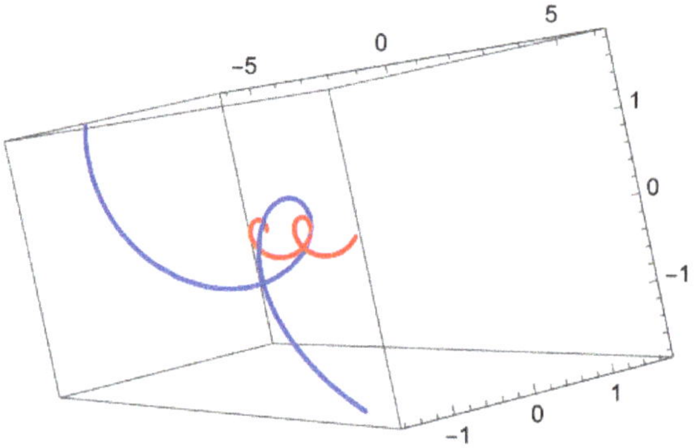

Figure 1. The blue graphic is φ and the red graphic is φ^* in Example 1.

4. Solutions of the Da Rios Vortex Filament Equation Generated by Combescure-Related Cartan Null Curves

In this section, by using a given Cartan null curve and its Combescure-related Cartan null mate curve, the differentiable function $C(s)$ appearing in the parametrization of the mate curve is determined so that the ruled surface S^* generated by the mate curve provides a solution to the Da Rios vortex filament equation. As a result, new solutions to the Da Rios vortex filament equation are obtained.

Theorem 4. *Let* $\varphi : I \subseteq \mathbb{R} \to \mathbb{E}^3_1$ *and* $\varphi^* : I^* \subseteq \mathbb{R} \to \mathbb{E}^3_1$ *be Cartan null curves related by transformation of Combescure with Frenet apparatus* $\{T, N, B, \kappa, \tau, s\}$ *and* $\{T^*, N^*, B^*, \kappa^*, \tau^*, s^*\}$, *respectively. If the ruled surface S generated by the curve φ satisfies the Da Rios vortex filament equation, then a necessary and sufficient condition for the Combescure-related curve φ^* to also satisfy the Da Rios vortex filament equation is that the differentiable function $C(s)$ in the parametrization of φ^* given by Equation (4) is one of the following:*

1. *Let φ be a Cartan null curve with torsion function $\tau(s) = 0$. In this case:*
   (a) *If*
   $$C(s) = \frac{s^3}{6} + \frac{c_2^2}{6(s+2c_1)} + c_3 + c_4 s + c_5 s^2$$
   then the torsion function of φ^* is $\tau^*(s^*) = 0$, where $c_2 > 0$.
   (b) *If*
   $$\begin{aligned} C(s) &= +c_3 + c_4 s + c_5 s^2 + \frac{s^3}{6} - \frac{4c_1(s+c_2)}{8d} \\ &\quad - \frac{\sqrt{2}(8d + c_1^2(s+c_2)^2)}{8d\sqrt{d}} \arctan\left(\frac{c_1(s+c_2)}{2\sqrt{2d}}\right) \end{aligned}$$
   then the torsion function of φ^* is $\tau^*(s^*) = d$, where $c_1, d \in \mathbb{R}^+$.
   (c) *If*
   $$\begin{aligned} C(s) &= c_3 + c_4 s + c_5 s^2 + \frac{3s + ds^3 + 3c_2 + 3dsc_2^2}{6d} \\ &\quad + \frac{(8d - c_1^2(s+c_2)^2)}{4c_1 d\sqrt{2d}} \tanh^{-1}\left(\frac{c_1(s+c_2)}{2\sqrt{2d}}\right) \end{aligned}$$
   then the torsion function of φ^* is $\tau^*(s^*) = -d$, where $c_1, d \in \mathbb{R}^+$.
2. *Let φ be a Cartan null curve with torsion function $\tau(s) = c$. In this case:*
   (a) *If*
   $$C(s) = \frac{1}{12c^{\frac{3}{2}}} \left(\begin{array}{l} 6\sqrt{c}(s + 2c_1 + 2cc_5) - 6c^{\frac{3}{2}} c_4 \cos(\sqrt{2c}(s+2c_1)) \\ +(3\sqrt{2} + 8\sqrt{2}c_2^2 + 6c^{\frac{3}{2}} c_3) \sin(\sqrt{2c}(s+2c_1)) \\ -4\sqrt{2}c_2^2 \tan(\frac{\sqrt{c}(s+2c_1)}{\sqrt{2}}) \end{array} \right)$$
   then the torsion function of φ^* is $\tau^*(s^*) = 0$, where $c > 0$, $c_2 \neq 0$,
   (b) *If*
   $$\begin{aligned} C(s) &= \frac{s - 2c_1}{2c} + \frac{2\sqrt{2}c_1^2 \exp(-\sqrt{-2c}(s-2c_1))}{3c^{\frac{3}{2}}(1 + \exp(\sqrt{-2c}(s-2c_1)))} \\ &\quad + \frac{c_3}{2} \exp(\sqrt{-2c}(s-2c_1)) \\ &\quad - \frac{c_4}{2} \exp(-\sqrt{-2c}(s-2c_1)) + c_5 \end{aligned}$$
   then the torsion function of φ^* is $\tau^*(s^*) = 0$, where $c_2 \neq 0$, $c < 0$.

(c) *If*

$$\begin{aligned} C(s) &= c_3 + c_4\cos(\sqrt{2c}s) + c_5\sin(\sqrt{2c}s) + \frac{s}{2c} \\ &- \int \frac{8c}{(c_1+\sqrt{c_1^2-16dc}\sin(\sqrt{2k}s+c_2)^2}ds \\ &+ \cos(\sqrt{2c}s)\int \frac{8c\cos(\sqrt{2c}s)}{(c_1+\sqrt{c_1^2-16dc}\sin(\sqrt{2k}s+c_2)^2}ds \\ &+ \sin(\sqrt{2c}s)\int \frac{8c\sin(\sqrt{2c}s)}{(c_1+\sqrt{c_1^2-16dc}\sin(\sqrt{2k}s+c_2)^2}ds \end{aligned}$$

then the torsion function of φ^* *is* $\tau^*(s^*) = d$, *where* $c > 0, d \in \mathbb{R}_0$.

(d) *If*

$$\begin{aligned} C(s) &= c_3 + c_4e^{\sqrt{-2c}s} + c_5e^{-\sqrt{-2c}s} + \frac{s}{2c} \\ &- \int \frac{8c}{(c_1+\sqrt{16dc-c_1^2}\sinh(\sqrt{-2c}(s+c_2))^2}ds \\ &+ e^{\sqrt{-2c}s}\int \frac{4ce^{-\sqrt{-2c}s}}{(c_1+\sqrt{16dc-c_1^2}\sinh(\sqrt{-2c}(s+c_2))^2}ds \\ &+ e^{-\sqrt{-2c}s}\int \frac{4ce^{\sqrt{-2c}s}}{(c_1+\sqrt{16dc-c_1^2}\sinh(\sqrt{-2c}(s+c_2))^2}ds \end{aligned}$$

then the torsion function of φ^* *is* $\tau^*(s^*) = d$, *where* $c < 0, d \in \mathbb{R}_0$.

Proof. Let us assume that the curves φ and φ^* are Combescure-related Cartan null curves in the Minkowski 3-space, and that the ruled surface S generated from φ is a solution of the Da Rios vortex filament equation. Then, by Theorem 4, φ must be either a null Cartan cubic or a null Cartan helix. In order for the ruled surface S^*, generated by the Combescure-related curve φ^*, to also be a solution of the Da Rios vortex filament equation, we need to determine the differentiable function $C(s)$ given in the parameterization of φ^*.

We consider the following two cases:

Case 1: Assume that φ is a null Cartan cubic. In this case, the torsion $\tau = 0$. For the ruled surface S^*, generated by φ^*, to be a solution of the Da Rios vortex filament equation, φ^* must also be either a null Cartan cubic or a null Cartan helix.

Suppose φ^* is a null Cartan cubic, then $\tau^* = 0$. Using this in Equation (15), we obtain

$$0 = -\frac{f'''}{(f')^3} + \frac{3(f'')^2}{2(f')^4} \tag{28}$$

Solving differential Equation (28), we find that the general solution is

$$f(s) = -\frac{c_2}{s+2c_1} + c_3 \tag{29}$$

Taking the derivative of (29) with respect to s and substituting into Equation (5) gives

$$\frac{c_2}{(s+2c_1)^2} = \sqrt{1 - \frac{d^3C(s)}{ds^3}}. \tag{30}$$

From (30), it is clear that $c_2 > 0$. Squaring both sides yields

$$\frac{c_2^2}{(s+2c_1)^4} = 1 - \frac{d^3C(s)}{ds^3}. \tag{31}$$

The general solution to this differential equation is

$$C(s) = \frac{s^3}{6} + \frac{c_2^2}{6(s+2c_1)} + c_3 + c_4 s + c_5 s^2 \tag{32}$$

where c_3, c_4 and c_5 are real constants.

Now assume that φ^* is a null Cartan helix. Then, for a non-zero real constant d, we have $\tau^* = d$. Substituting $\tau = 0$ and $\tau^* = d$ into Equation (15) gives

$$d = -\frac{f'''}{(f')^3} + \frac{3(f'')^2}{2(f')^4}. \tag{33}$$

Solving (6), we obtain the following general solutions

For $d > 0$,

$$f(s) = \frac{\sqrt{2}}{\sqrt{d}} \arctan\left(\frac{c_1(s+c_2)}{2\sqrt{2d}}\right) + c_3. \tag{34}$$

For $d < 0$,

$$f(s) = -\frac{\sqrt{2}}{\sqrt{-d}} \arctan\left(\frac{c_1(s+c_2)}{2\sqrt{-2d}}\right) + c_3. \tag{35}$$

Now by using $d > 0$ and differentiating (34) with respect to s and substituting into Equation (15), we get

$$\frac{4c_1}{8d + c_1^2(s+2c_1)^2} = \sqrt{1 - \frac{d^3C(s)}{ds^3}} \tag{36}$$

Here, it is evident that $c_1 > 0$. Squaring both sides of (36) leads to

$$\left(\frac{4c_1}{8d + c_1^2(s+2c_1)^2}\right)^2 = 1 - \frac{d^3C(s)}{ds^3}. \tag{37}$$

The general solution to this equation is

$$C(s) = \frac{s^3}{6} - \frac{4c_1(s+c_2)}{8d} - \frac{\sqrt{2}(8d + c_1^2(s+c_2)^2)}{8d\sqrt{d}} \arctan\left(\frac{c_1(s+c_2)}{2\sqrt{2d}}\right) + c_3 + c_4 s + c_5 s^2$$

For $d < 0$, similar steps yield the following result

$$C(s) = \frac{3s + ds^3 + 3c_2 + 3dsc_2^2}{-6d} + \frac{(8d - c_1^2(s+c_2)^2)}{4c_1 d\sqrt{-2d}} \tanh^{-1}\left(\frac{c_1(s+c_2)}{2\sqrt{-2d}}\right) + c_3 + c_4 s + c_5 s^2$$

Case 2: Now, assume that φ is a null Cartan helix, so its torsion is $\tau = c$, where c is a non-zero real constant. Then, for the ruled surface S^*, generated by φ^*, to be a solution of the Da Rios vortex filament equation, the curve φ^* must again be either a null Cartan cubic or a null Cartan helix. The proof follows similarly to Case 1, and thus the second part of Theorem 4 is proven. □

Example 2. *Consider the null Cartan helix in Minkowski 3-space given by*

$$\varphi(s) = \left(-\frac{s}{2}, \frac{1}{4}\sin 2s, \frac{1}{4}\cos 2s\right)$$

with curvatures $\kappa(s) = 1$, $\tau(s) = 2$ *and Frenet vectors*

$$T(s) = \frac{1}{2}(-1, \cos 2s, -\sin 2s)$$

$$N(s) = (0, -\sin 2s, -\cos 2s)$$

$$B(s) = (1, \cos 2s, -\sin 2s).$$

If we take $c = 2$, $c_2 = 1$ *and* $c_1 = c_3 = c_4 = c_5 = 0$ *in part (2a) of Theorem 4, then it follows that*

$$C_1(s) = \frac{1}{24}(6s + 11 \sin 2s - 4 \tan s). \tag{38}$$

Substituting (38) *in* (4), *the curve* φ_1^*, *which is Combescure-related to* φ, *is obtained as follows*

$$\varphi_1^*(f_1(s)) = \begin{pmatrix} -\frac{1}{6}(2 + \cos 2s) \sec^2 s \tan s, \\ \\ \frac{1}{6} \sec^3 s \sin 3s, \\ \\ -\frac{1}{4}(1 + 2 \sec^2 s) \end{pmatrix}$$

with curvatures $\kappa_1^*(f_1(s)) = 1$, $\tau_1^*(f_1(s)) = 0$. *Where* $f_1(s) = \tan s$.

If we take $c = 2, d = -1$ *and* $c_1 = c_2 = c_3 = c_4 = c_5 = 0$ *in part (2c) of Theorem 4, then it follows that*

$$C_2(s) = \frac{s}{4} + \sin(2s) \ln(\tan s) \tag{39}$$

Substituting (39) *in* (4), *the curve* φ_2^*, *which is Combescure-related to* φ, *is obtained as follows*

$$\varphi_2^*(f(s)) = \left(\frac{1}{4}(\cot s - \tan s), -\frac{1}{4} \csc s \sec s, \frac{1}{s} \ln(\cot s)\right)$$

with curvatures $\kappa_1^*(f_2(s)) = 1$, $\tau_1^*(f_2(s)) = -1$. *Where* $f_2(s) = \frac{1}{\sqrt{2}}(s\sqrt{1+s^2} - \ln(-\sqrt{2}s + \sqrt{2+2s^2}))$.

$$\phi(s,t) = \left(-\frac{1}{2} + 2t, \frac{1}{2} \cos 2s, -\frac{1}{2} \sin 2s\right),$$

$$\phi_1(f_1(s), t)) = \begin{pmatrix} t - \frac{1}{2} \sec^2 s, \\ \\ 1 + t - \frac{1}{2} \sec^2 s, \\ \\ -\tan s \end{pmatrix},$$

$$\phi_2(f_2(s), t)) = \left(-\frac{1}{\sqrt{2}} \csc 2s, \frac{1}{\sqrt{2}} \cot 2s, -\frac{1+2t}{\sqrt{2}}\right).$$

Figures 2 and 3 are shown the example's curves and surfaces.

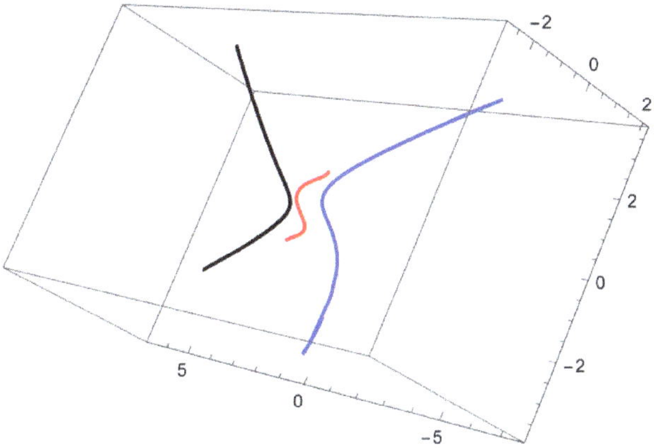

Figure 2. The red graph represents the main curve (φ), while the blue ($\varphi_1^{\star}$) and black ($\varphi_2^{\star}$) graphs illustrate the Combescure mate curves of the main curve in Example 2.

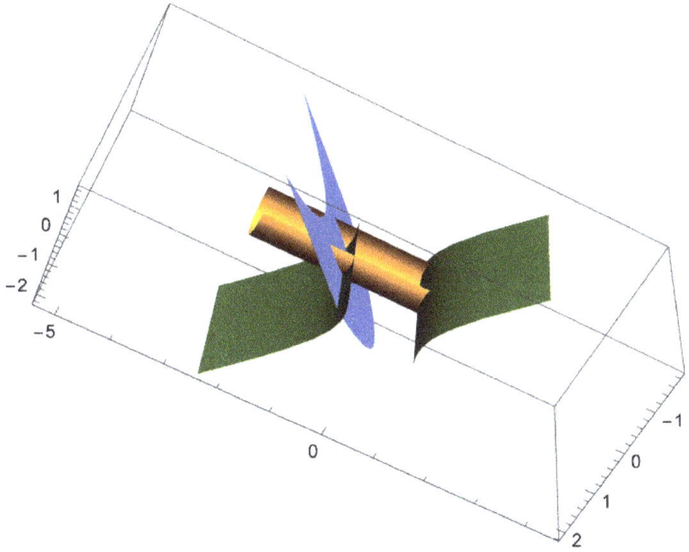

Figure 3. In Example 2, the ruled surface generated by the main curve and satisfying the Da Rios equation is shown in yellow, while the ruled surfaces generated by the Combescure mate curves of the main curve and also satisfying the Da Rios vortex filament equation are depicted in blue and green.

5. Conclusions

In this study, we have investigated Cartan null curves in Minkowski 3-space and their counterparts related via the Combescure transformation. The necessary conditions for establishing such connections were derived, and the relationships between the Frenet vectors and curvatures of these curve pairs were thoroughly analyzed. Additionally, we determined the conditions under which ruled surfaces generated by Combescure-related Cartan null curves satisfy the Da Rios equation. All results were illustrated with concrete examples, demonstrating the applicability and effectiveness of the proposed framework. These findings not only deepen the understanding of the geometric properties of Cartan null curves but also provide a systematic approach for generating new solutions to the

Da Rios vortex filament equation. Future work may explore further generalizations to higher-dimensional Minkowski spaces.

Author Contributions: Conceptualization, Y.L., O.K., K.İ. and Q.S.; software, Y.L., O.K., K.İ. and Q.S. Writing—original draft preparation, Y.L., O.K., K.İ. and Q.S.; Writing—review and editing, Y.L., O.K., K.İ. and Q.S. All authors have read and agreed to the published version of the manuscript.

Funding: This research was funded by the HZNU scientific research and innovation team project (Grant No. TD2025007).

Data Availability Statement: No new data were created or analyzed in this study.

Acknowledgments: The authors would like to express their sincere thanks to the editor and the anonymous reviewers for their helpful comments and suggestions.

Conflicts of Interest: The authors declare no conflicts of interest.

References

1. Miller, J. Note on Tortuous Curves. *Proc. Edinburgh Math. Soc.* **1905**, *24*, 51–55. [CrossRef]
2. Bertrand, J.M. Mémoire sur la théorie des courbes á double courbure. *J. Math. Pures Appl.* **1850**, *15*, 332–350.
3. Hanif, M.; Hou, Z.; Nisar, K. On special kinds of involute and evolute curves in 4-dimensional Minkowski space. *Symmetry* **20** *10*, 317. [CrossRef]
4. Li, Y.; Kecilioglu, O.; İlarslan, K. Generalized Bertrand Curve Pairs in Euclidean Four-Dimensional Space. *Axioms* **2025**, *14*, 2 [CrossRef]
5. Liu, H.; Wang, F. Mannheim partner curves in 3-space. *J. Geom.* **2008**, *88*, 120–126. [CrossRef]
6. Song, X.; Li, E.; Pei, D. Legendrian dualities and evolute-involute curve pairs of spacelike fronts in null sphere. *J. Geom. Ph* **2022**, *178*, 104543. [CrossRef]
7. Zhu, M.; Yang, H.; Li, Y.; Abdel-Baky, R.A.; AL-Jedani, A.; Khalifa, M. Directional developable surfaces and their singularities Euclidean 3-space. *Filomat* **2024**, *38*, 11333–11347. [CrossRef]
8. Graustein, W.C. On two related transformations of space curves. *Am. J. Math.* **1917**, *39*, 233–240. [CrossRef]
9. Camcı, Ç.; Uçum, A.; İlarslan, K. Space curves related by a transformation of Combescure. *J. Dyn. Syst. Geom. Theor.* **2021**, 271–287. [CrossRef]
10. Bataray, B.; Camcı, Ç. Applications of Equivalent Curves to Ruled Surfaces. *Int. Electron. J. Geom.* **2025**, *18*, 135–142. [CrossRe
11. Kilian, M.; Müller, C.; Tervooren, J. Smooth and discrete cone-nets. *Results Math.* **2023**, *78*, 110. [CrossRef]
12. Pirahmad, O.; Pottmann, H.; Skopenkov, M.B. Area preserving Combescure transformations. *Results Math.* **2025**, *80*, 27. [CrossR
13. Pottmann, H.; Wallner, J. The focal geometry of circular and conical meshes. *Adv. Comput. Math.* **2008**, *29*, 249–268. [CrossRef
14. Terng, C.-L. Geometric transformations and soliton equations. In *Handbook of Geometric Analysis*; Advanced Lectures in Mat matics (ALM), International Press: Somerville, MA, USA, 2010; Volume 2, pp. 301–358.
15. Lopez, R. Differential geometry of curves and surfaces in Lorentz-Minkowski space. *Int. Electron. J. Geom.* **2014**, *7*, 44–1 [CrossRef]
16. O'Neill, B. *Semi-Riemannian Geometry. With Applications to Relativity*; Pure and Applied Mathematics, 103; Academic Press, I New York, NY, USA, 1983.
17. Keçilioğlu, O.; İlarslan, K. Combescure related pseudo null curves and their applications to Da Rios vortex filament equati *U.P.B. Sci. Bull. Ser. A* **2025**, *87*, 99–112.
18. Da Rios, L.S. On the motion of an unbounded fluid with a vortex filament of any shape. *Rend. Circ. Mat. Palermo* **1906**, *22*, 117–1 [CrossRef]
19. Arms, R.J.; Hama, F.R. Localized-induction concept on a curved vortex and motion of an elliptic vortex ring. *Phys. Fluids* **1965** 553–559. [CrossRef]
20. Betchov, R. On the curvature and torsion of an isolated vortex filament. *J. Fluid Mech.* **1965**, *22*, 471–479. [CrossRef]
21. Erdoğdu, M.; Özdemir, M. Geometry of Hasimoto surfaces in Minkowski 3-Space. *Math. Phys. Anal. Geom.* **2014**, *17*, 169–1 [CrossRef]
22. Rogers, C.; Schief, W.K. *Bäcklund and Darboux Transformations Geometry and Modern Applications in Soliton Theory*; Cambrid University Press: Cambridge, UK, 2002.
23. Ding, Q.; Inoguchi, J. Schrödinger flows, binormal motion for curves and the second AKNS-hierarchies. *Chaos Solitons Fract* **2004**, *21*, 669–677. [CrossRef]
24. Gürbüz, N. The motion of timelike surfaces in timelike geodesic coordinates. *Int. J. Math. Anal.* **2010**, *4*, 349–356.

Grbović M.; Nešović, E. On Bäcklund transformation and vortex filament equation for null Cartan curve in Minkowski 3-space. *Math. Phys. Anal. Geom.* **2016**, *19*, 23. [CrossRef]
Couto, I.T.; Lymberopoulos, A. *Introduction to Lorentz Geometry Curves and Surfaces*; CRC Press, Taylor and Francis Group: Boca Raton, FL, USA, 2021.
Duggal, K.L.; Jin, D.H. *Null Curves and Hypersurfaces of Semi-Riemannian Manifolds*; World Scientific: Singapore, 2007.

mathematics

MDPI

Article

Intrinsic and Extrinsic Geometry of Kählerian Slant Submanifolds in Complex Space Forms

Andreea Olteanu [1,*,†], Mohammed Mohammed [2,3,†], Ion Mihai [4,†] and Sfundo Cebolenkosi Gumede [2,†]

1 Faculty of Land Reclamation and Environmental Engineering, University of Agronomic Sciences and Veterinary Medicine of Bucharest, 011464 Bucharest, Romania
2 Department of Mathematical Sciences, Mangosuthu University of Technology (MUT), 511 Griffiths Mxenge Highway, Umlazi 4031, South Africa; mohammed.mohammed@mut.ac.za (M.M.); gumedesc@mut.ac.za (S.C.G.)
3 National Institute for Theoretical and Computational Sciences (NITheCS), Stellenbosch 3201, South Africa
4 Faculty of Mathematics and Computer Science, University of Bucharest, 010014 Bucharest, Romania; imihai@fmi.unibuc.ro
* Correspondence: andreea.olteanu@fifim.ro
† These authors contributed equally to this work.

Abstract

The study of complex space forms plays a central role in differential geometry, as these manifolds provide a natural framework for exploring geometric structures endowed with rich symmetries, extending both Riemannian and Kählerian geometries. In this paper, we extend the definition of Kählerian slant submanifolds when the ambient complex space form admits a quarter-symmetric connection. Furthermore, we establish sharp relationships between intrinsic and extrinsic geometric invariants of Kählerian slant submanifolds in complex space forms admitting a quarter-symmetric connection.

Keywords: Chen invariant; Chen inequalities; Kählerian slant; semi-symmetric connection; non-metric connection; quarter symmetric connection

MSC: 53C25; 53C40

1. Introduction

The concept of semi-symmetric connections was first introduced by Friedmann and Schouten [1], marking an important development in affine differential geometry. Later, Hayden [2] independently considered similar notions. The geometric properties were further examined by Yano [3] in Riemannian manifolds. His work resulted in the formulation of semi-symmetric metric connections.

In 1975, Golab [4] extended this idea and introduced the notion of a quarter-symmetric connection, which generalizes both semi-symmetric metric and non-metric connections. By preserving metric compatibility while permitting a specific torsion component, this connection offers a richer framework for geometric analysis than the classical Levi-Civita connection. Subsequent contributions by Agashe and Chafle [5,6] expanded the theory to include non-metric cases and explored its applications in the geometry of submanifolds.

Addressing Chen's open problem, which concerns identifying optimal relationships between the intrinsic and extrinsic invariants of Riemannian submanifolds, central challenge in submanifold geometry, Chen introduced the so-called δ-invariants (or Chen invariants) in [7,8]. These invariants provide a powerful framework for formulating and proving sharp inequalities that relate the curvature quantities of the submanifolds.

https://doi.org/10.3390/math140100

Following Chen's pioneering work, numerous researchers have investigated these invariants in various ambient spaces, and manifolds admitting semi-symmetric metric connections, where the definition of sectional curvature still preserves its skew-symmetric properties.

Several B.-Y. Chen inequalities for submanifolds of generalized complex space forms were established in [9]. A Chen inequality for submanifolds of S-space forms tangent to the structure vector fields, with applications to slant immersions, was obtained in [10]. Sharp inequalities involving δ-invariants for submanifolds of generalized complex and generalized Sasakian space forms with arbitrary codimension were studied in [11], while a B.-Y. Chen inequality for semi-slant submanifolds of T-space forms was derived in [12].

Chen inequalities relating intrinsic and extrinsic curvatures were further established for submanifolds of real, complex, and Sasakian space forms endowed with a semi-symmetric metric connection [13,14], and for submanifolds of Riemannian manifolds of nearly quasi-constant curvature equipped with a semi-symmetric non-metric connection [15].

Recently, in [16], the geometry and topology of warped product Legendrian submanifolds in Kenmotsu space forms were investigated, and the first Chen inequality was established.

When the ambient manifold is endowed with a semi-symmetric non-metric connection, the classical notion of sectional curvature, defined via the symmetric and metric-preserving Levi-Civita connection, becomes inadequate. To overcome this limitation, A. Mihai and the third author proposed in [17] a modified definition of sectional curvature that is compatible with semi-symmetric non-metric connections. This new formulation preserves the essential skew-symmetric properties of curvature and has been successfully applied to the study of various geometric inequalities.

Furthermore, motivated by Golab's introduction of quarter-symmetric connections [4], several researchers have studied submanifolds in ambient manifolds equipped with this more general type of connection. In particular, Wang [18] developed a unified framework for Chen-type inequalities in complex space forms and Sasakian space forms endowed with quarter-symmetric metric connections. By deriving a generalized Gauss equation adapted to quarter-symmetric connections, Wang established sharp inequalities relating intrinsic invariants—such as Ricci and scalar curvature—to extrinsic quantities defined via the second fundamental form. His results demonstrate that the quarter-symmetric setting simultaneously extends both the semi-symmetric metric and semi-symmetric non-metric cases, thereby providing a broader geometric structure for studying curvature inequalities.

2. Preliminaries

Let $\overline{M}$ be an $(n+k)$-dimensional Riemannian manifold endowed with a Riemannian metric g and its associated Levi-Civita connection $\overline{\hat{\nabla}}$. We define a linear connection $\overline{\nabla}$ by

$$\overline{\nabla}_X Y = \overline{\hat{\nabla}}_X Y + \lambda_1 \omega(Y) X - \lambda_2 g(X,Y) P, \tag{1}$$

where P represents a vector field in M defined by $\omega(X) = g(P,X)$.

When $\lambda_1 = \lambda_2 = 1$, the connection $\overline{\nabla}$ reduces to a semi-symmetric metric connection, whereas for $\lambda_1 = 1, \lambda_2 = 0$, it defines a semi-symmetric non-metric connection.

Let

$$\overline{R}(X,Y)Z = \overline{\nabla}_X \overline{\nabla}_Y Z - \overline{\nabla}_Y \overline{\nabla}_X Z - \overline{\nabla}_{[X,Y]} Z$$

denote the curvature tensor of $\overline{\nabla}$. The curvature tensor corresponding to $\overline{\hat{\nabla}}$, denoted by $\overline{\hat{R}}$, is expressed analogously.

https://doi.org/10.3390/math14010071

Define the $(0,2)$-tensors:

$$\alpha(X,Y) = (\hat{\overline{\nabla}}_X \omega)(Y) - \lambda_1 \omega(X)\omega(Y) + \frac{\lambda_2}{2} g(X,Y)\omega(P),$$
$$\beta(X,Y) = \frac{\omega(P)}{2} g(X,Y) + \omega(X)\omega(Y).$$

Then, by [18], we obtain

$$\begin{aligned} \overline{R}(X,Y,Z,W) = \hat{\overline{R}}(X,Y,Z,W) &+ \lambda_1 \alpha(X,Z) g(Y,W) - \lambda_1 \alpha(Y,Z) g(X,W) \\ &+ \lambda_2 g(X,Z)\alpha(Y,W) - \lambda_2 g(Y,Z)\alpha(X,W) \\ &+ \lambda_2(\lambda_1 - \lambda_2) g(X,Z)\beta(Y,W) \\ &- \lambda_2(\lambda_1 - \lambda_2) g(Y,Z)\beta(X,W). \end{aligned} \tag{2}$$

Let M be an n-dimensional submanifold of an $(n+k)$-dimensional Riemannian manifold $\overline{M}$. On the submanifold M, we consider the induced quarter-symmetric connection denoted by ∇ and the induced Levi-Civita connection denoted by $\hat{\nabla}$. Let R and $\hat{R}$ denote the curvature tensors corresponding to ∇ and $\hat{\nabla}$, respectively. The vector field P on M can be uniquely decomposed into its tangent and normal components, $P^\top$ and $P^\perp$, so that $P = P^\top + P^\perp$.

The Gauss formulas corresponding to the connections ∇ and $\hat{\nabla}$ can be expressed as

$$\overline{\nabla}_X Y = \nabla_X Y + h(X,Y), \quad X,Y \in \Gamma(TM),$$
$$\hat{\overline{\nabla}}_X Y = \hat{\nabla}_X Y + \hat{h}(X,Y), \quad X,Y \in \Gamma(TM),$$

where $\hat{h}$ denotes the second fundamental form of M in $\overline{M}$. Moreover, $\hat{h}$ is related to h via

$$\hat{h}(X,Y) = h(X,Y) + \lambda_2 g(X,Y) P^\perp. \tag{3}$$

We choose a local orthonormal frame $e_1, \dots, e_n, e_{n+1}, \dots, e_{n+k}$ on $\overline{M}$ with the property that $e_1, \dots, e_n$ are tangent to M when restricted to M.

We denote $\hat{h}^r_{ij} = g(\hat{h}(e_i,e_j), e_r)$, for all $i,j = 1,\dots,n$, $r = n+1,\dots,n+k$. The squared norm of the second fundamental form h is given by

$$|h|^2 = \sum_{i,j=1}^{n} g(h(e_i,e_j), h(e_i,e_j)),$$

and the mean curvature vector of M with respect to ∇ is

$$H = \frac{1}{n}\sum_{i=1}^{n} h(e_i,e_i).$$

In the same way, the mean curvature vector with respect to $\hat{\nabla}$ is

$$\hat{H} = \frac{1}{n}\sum_{i=1}^{n} \hat{h}(e_i,e_i).$$

By Wang [18], the Gauss equation becomes

$$\begin{aligned} \overline{R}(X,Y,Z,W) = R(X,Y,Z,W) &- g(h(X,W), h(Y,Z)) + g(h(Y,W), h(X,Z)) \\ &+ (\lambda_1 - \lambda_2) g(h(Y,Z), P) g(X,W) \\ &+ (\lambda_2 - \lambda_1) g(h(X,Z), P) g(Y,W). \end{aligned} \tag{4}$$

https://doi.org/10.3390/math1401C

Let $\pi \subset T_pM$ be a 2-plane section at a point $p \in M$, with $\{e_1, e_2\}$ forming an orthonormal basis of π. When the ambient manifold is equipped with a non-Levi-Civita connection, such as a quarter-symmetric connection or a semi-symmetric non-metric connection, the classical definition of sectional curvature is no longer directly applicable due to the lack of symmetry in the curvature tensor. Specifically, we generally have the following.

$$R(X,Y,Z,W) \neq -R(X,Y,W,Z).$$

To address this, several alternative definitions have been proposed in the literature. For example, Wang [18] proposed a consistent approach for defining sectional curvature in manifolds admitting quarter-symmetric connections, while the third author and A. Mihai [17] proposed a similar framework for manifolds endowed with semi-symmetric non-metric connections.

Motivated by these works, we adopt the following symmetrized expression to define the sectional curvature $K(\pi)$ associated with a non-metric or torsion-affected connection.

$$K(\pi) = \frac{1}{2}[R(e_1,e_2,e_2,e_1) + R(e_2,e_1,e_1,e_2)]. \tag{5}$$

This definition ensures that $K(\pi)$ remains real-valued and well-defined, even when the curvature tensor lacks the usual anti-symmetry in the last two arguments.

For any orthonormal basis $\{e_1, \dots, e_n\}$ of T_pM, the scalar curvature τ at p is defined by

$$\tau(p) = \sum_{1\le i<j\le n} K(e_i \wedge e_j) = \frac{1}{2} \sum_{1\le i,j\le n} R(e_i,e_j,e_j,e_i). \tag{6}$$

The study of Riemannian, complex, almost contact, hypercomplex, quaternionic, and Cauchy–Riemann structures falls within the broad framework of G-structures. This area plays a fundamental role in modern differential geometry, global analysis, and mathematical physics.

In this subsection, we provide a brief overview of the basic notions and properties related to the differential geometry of manifolds endowed with complex structures. For further details, the reader is referred to [19].

A complex manifold $(\overline{M}, g)$ of real dimension $2m$ is called a Hermitian manifold if there exists an almost complex structure J, tensor field of type $(1,1)$, satisfying

$$J^2X = -X,$$

such that

$$g(JX, JY) = g(X,Y),$$

for all vector fields X, Y on $\overline{M}$.

If the almost complex structure J is parallel with respect to the Levi-Civita connection $\hat{\overline{\nabla}}$, then the manifold $\overline{M}$ is Kähler.

A 2-plane section invariant by J is called a holomorphic plane section. A complex space form is a Kähler manifold whose holomorphic sectional curvature is constant, equal to c. It is known that complex space forms have the highest degree of homogeneity because their isometries spaces have the maximum dimension.

For a complex space form $\overline{M}(c)$, the curvature tensor satisfies (see, for instance, [20])

$$\begin{aligned}\hat{\overline{R}}(X,Y)Z =& \frac{c}{4}[g(Y,Z)X - g(X,Z)Y \\ &+ g(Z,JY)JX - g(Z,JX)JY + 2g(X,JY)JZ],\end{aligned} \tag{7}$$

https://doi.org/10.3390/math14010071

for all vector fields X, Y, Z on $\overline{M}(c)$.

For a $2m$-dimensional complex space form $\overline{M}(c)$ with constant holomorphic sectional curvature c and a quarter-symmetric connection $\tilde{\nabla}$, the curvature tensor $\overline{R}$ is determined from (2) and (7) as follows:

$$\begin{aligned}\overline{R}(X,Y,Z,W) = {} & \frac{c}{4}[g(Y,Z)g(X,W) - g(X,Z)g(Y,W) \\ & + g(Z,JY)g(JX,W) - g(Z,JX)g(JY,W) + 2g(X,JY)g(JZ,W)] \\ & + \lambda_1\alpha(X,Z)g(Y,W) - \lambda_1\alpha(Y,Z)g(X,W) \\ & + \lambda_2 g(X,Z)\alpha(Y,W) - \lambda_2 g(Y,Z)\alpha(X,W) \\ & + \lambda_2(\lambda_1 - \lambda_2)g(X,Z)\beta(Y,W) \\ & - \lambda_2(\lambda_1 - \lambda_2)g(Y,Z)\beta(X,W). \end{aligned} \tag{8}$$

Let $x : M \longrightarrow \overline{M}$ be an isometric immersion of an n-dimensional Riemannian manifold M in a $2m$-dimensional Hermitian manifold $\overline{M}$, equipped with an almost complex structure J and a compatible Riemannian metric g.

For any tangent vector field $X \in \Gamma(TM)$, the image JX can be decomposed into tangential and normal components as

$$JX = TX + FX,$$

where TX is tangent to M and FX is normal to M. Thus, T acts as an endomorphism of the tangent bundle TM, while F defines a 1-form on TM taking values in the normal bundle $T^{\perp}M$.

For a nonzero vector X tangent to M at a point $p \in M$, denote by $\theta(X)$ the angle between JX and T_pM. This angle $\theta(X)$ is termed the Wirtinger angle of X.

The immersion $x : M \rightarrow \overline{M}$ is said to be a slant immersion if $\theta(X)$ is constant, that is, it is independent of both $p \in M$ and the choice of the tangent vector $X \in T_pM$. The common value of $\theta(X)$ is termed the slant angle of the immersion.

- When $\theta = 0$, the immersion is invariant (and it becomes complex because $\overline{M}$ is Hermitian);
- When $\theta = \frac{\pi}{2}$, the immersion is totally real (also called anti-invariant);
- For $0 < \theta < \frac{\pi}{2}$, the immersion is called a proper slant immersion.

For slant submanifolds, the endomorphism T satisfies the algebraic condition $T^2 = \varepsilon I$, for some constant $\varepsilon \in [-1, 0]$, where I denotes the identity transformation on TM. Accordingly,

$$\varepsilon = \begin{cases} -1, & \text{if } \theta = 0, \quad \text{(invariant case)}, \\ 0, & \text{if } \theta = \frac{\pi}{2}, \quad \text{(totally real case)}, \\ -\cos^2\theta, & \text{if } 0 < \theta < \frac{\pi}{2}, \quad \text{(proper slant case)}. \end{cases}$$

A proper slant submanifold is called a Kählerian slant submanifold when its canonical endomorphism T is parallel with respect to the Levi-Civita connection $\hat{\nabla}$ of M, that is, $\hat{\nabla}T = 0$.

It is known [21] that this condition is equivalent to

$$A_{FX}Y = A_{FY}X, \qquad \forall X, Y \in \Gamma(TM), \tag{9}$$

where A_N denotes the shape operator corresponding to the normal vector field N.

https://doi.org/10.3390/math140100

If $\{e_1, \ldots, e_n\}$ is an orthonormal basis of T_pM at $p \in M$, then we set

$$e_i^* = \frac{1}{\sin\theta} Fe_i, \quad i = 1, \ldots, n.$$

We denote $\hat{h}_{ij}^k = g(\hat{h}(e_i, e_j), e_k^*), \forall i, j, k = 1, \ldots, n$. Because the submanifold M is a Kählerian slant submanifold, Equation (9) implies that the components of the second fundamental form are symmetric, that is,

$$\hat{h}_{ij}^k = \hat{h}_{ik}^j = \hat{h}_{jk}^i, \ \forall i, j, k = 1, \ldots, n.$$

In this situation, M itself inherits a Kähler structure; it becomes a Kähler manifold with respect to the induced metric and the almost complex structure

$$J' = (\sec\theta)T,$$

where θ denotes the slant angle.

Explicit constructions and examples of proper slant and Kählerian slant submanifolds can be found in classical references (see, for instance, Chen [21]).

Lemma 1. *Let $f_1(x_1, x_2, \ldots, x_n)$ be a function of $\mathbb{R}^n$ defined by*

$$f_1(x_1, x_2, \ldots, x_n) = x_1 \sum_{j=2}^{n} x_j - \sum_{j=2}^{n} x_j^2.$$

If $x_1 + x_2 + \ldots + x_n = 2na$, then we have

$$f_1(x_1, x_2, \ldots, x_n) \leq \frac{n-1}{4n}(x_1 + x_2 + \ldots + x_n)^2,$$

with the equality sign holding if and only if $\frac{1}{n+1}x_1 = x_2 = \ldots = x_n = a$.

Lemma 2. *Let $f_2(x_1, x_2, \ldots, x_n)$ be a function on $\mathbb{R}^n$ defined by*

$$f_2(x_1, x_2, \ldots, x_n) = x_1 \sum_{j=2}^{n} x_j - x_1^2.$$

If $x_1 + x_2 + \ldots + x_n = 4a$, then we have

$$f_2(x_1, x_2, \ldots, x_n) \leq \frac{1}{8}(x_1 + x_2 + \ldots + x_n)^2,$$

with the equality sign holding if and only if $x_1 = a$ and $x_2 + \ldots + x_n = 3a$.

3. First Chen Like Inequality

In this section, we derive a first Chen-type inequality for Kählerian slant submanifolds of a complex space form equipped with a quarter-metric connection.

Theorem 1. *Let M be an $n = 2s$-dimensional $(n \geq 3)$ Kählerian slant submanifold isometrically immersed in a $2n$-dimensional complex space form $\overline{M}(c)$ endowed with quarter-symmetric connection $\overline{\nabla}$. We assume that the vector field P is tangent to M. Then, for any $p \in M$ and any 2-plane section $\pi \subset T_pM$, we have*

https://doi.org/10.3390/math14010071

$$\tau(p) - K(\pi) \leq \frac{2n-3}{2(2n+3)}\|H\|^2 + \frac{c}{8}[(n-2)(n+1) + 3(n\cos^2\theta - 2\Theta^2(\pi))] - \frac{(\lambda_1+\lambda_2)(n-1)\lambda}{2} - \frac{\lambda_2(\lambda_1-\lambda_2)(n-1)\mu}{2} + \frac{\lambda_1+\lambda_2}{2}\mathtt{tr}(\alpha_{|\pi}) + \frac{\lambda_2(\lambda_1-\lambda_2)}{2}\mathtt{tr}(\beta_{|\pi}), \tag{10}$$

where $\Theta^2(\pi) = g^2(Je_1, e_2)$, *for any orthonormal vectors* $e_1, e_2 \in \pi$.

The equality at a point $p \in M$ *holds if and only if, with respect to a suitable orthonormal basis* $\{e_1, \ldots, e_n\}$, *the second fundamental form has the following expressions:*

$$\begin{aligned} &h(e_1,e_1) = aFe_1 + 3bFe_3, && h(e_1,e_3) = 3bFe_1, && h(e_3,e_j) = 4bFe_j, \\ &h(e_2,e_2) = -aFe1 + 3bFe_3, && h(e_2,e_3) = 3bFe_2, && h(e_j,e_k) = 4bFe_3\delta_{jk}, \\ &h(e_1,e_2) = -aFe_2, && h(e_3,e_3) = 12bFe_3, && h(e_1,e_j) = h(e_2,e_j) = 0, \end{aligned}$$

for some numbers a and b, and $j, k \in \{4, \ldots, n\}$.

Proof. Let $p \in M$ and let $\pi \subset T_pM$ be a 2-plane section. We choose an orthonormal basis $\{e_1, \ldots, e_n\}$ of T_pM such that $e_1, e_2 \in \pi$ and an orthonormal basis $\{e_{n+1}, \ldots, e_{4s}\}$ of the normal space $T_p^{\perp}M$.

Since $\overline{M}(c)$ is a complex space form equipped with a quarter-symmetric connection, it follows from the amended Gauss Equation (4) that

$$2\tau = \sum_{i,j=1}^{n} R(e_i,e_j,e_j,e_i) = \sum_{i,j=1}^{n} [\overline{R}(e_i,e_j,e_j,e_i) - g(h(e_i,e_j),h(e_j,e_i)) + g(h(e_i,e_i),h(e_j,e_j)) - (\lambda_1-\lambda_2)g(h(e_j,e_j),P)g(e_i,e_i) - (\lambda_2-\lambda_1)g(h(e_i,e_j),P)g(e_j,e_i)]. \tag{11}$$

Using the Formula (8) for the curvature tensor, we calculate $\overline{R}(e_i,e_j,e_j,e_i)$ by taking $X = W = e_i$ and $Y = Z = e_j$, for all $i, j = 1, \ldots, 2s$.

$$\begin{aligned} \overline{R}(e_i,e_j,e_j,e_i) = {} & \frac{c}{4}\{g(e_j,e_j)g(e_i,e_i) - g(e_i,e_j)g(e_j,e_i) \\ & + [g(e_j,Je_j)g(Je_i,e_i) \\ & - g(e_j,Je_i)g(Je_j,e_i) + 2g(e_i,Je_j)g(Je_j,e_i)]\} \\ & + \lambda_1\alpha(e_i,e_j)g(e_j,e_i) - \lambda_1\alpha(e_j,e_j)g(e_i,e_i) \\ & + \lambda_2 g(e_i,e_j)\alpha(e_j,e_i) - \lambda_2 g(e_j,e_j)\alpha(e_i,e_i) \\ & + \lambda_2(\lambda_1-\lambda_2)g(e_i,e_j)\beta(e_j,e_i) \\ & - \lambda_2(\lambda_1-\lambda_2)g(e_j,e_j)\beta(e_i,e_i). \end{aligned} \tag{12}$$

This equation yields

https://doi.org/10.3390/math1401OC

$$\sum_{i,j=1}^{n} \overline{R}(e_i, e_j, e_j, e_i) = \frac{c}{4}[n(n-1) + 3\sum_{i,j=1}^{n} g^2(Te_i, e_j)]$$
$$- (\lambda_1 + \lambda_2)(n-1)\sum_{i=1}^{n} \alpha(e_i, e_i)$$
$$- \lambda_2(\lambda_1 - \lambda_2)(n-1)\sum_{i=1}^{n} \beta(e_i, e_i). \quad (13)$$

We denote $\mathrm{tr}(\alpha) = \lambda$ and $\mathrm{tr}(\beta) = \mu$, and $\sum_{i,j=1}^{n} g^2(Te_i, e_j) = \|T\|^2$. We can rewrite (13) as

$$\sum_{i,j=1}^{n} \overline{R}(e_i, e_j, e_j, e_i) = \frac{c}{4}\Big[n(n-1) + 3\|T\|^2\Big]$$
$$- (\lambda_1 + \lambda_2)(n-1)\lambda - \lambda_2(\lambda_1 - \lambda_2)(n-1)\mu. \quad (14)$$

Now, substituting (14) in (11), we get

$$2\tau = \frac{c}{4}\Big[n(n-1) + 3n\cos^2\theta\Big]$$
$$- (\lambda_1 + \lambda_2)(n-1)\lambda - \lambda_2(\lambda_1 - \lambda_2)(n-1)\mu$$
$$+ \sum_{r=n+1}^{4s} \sum_{1\le i,j\le n} [h^r_{ii}h^r_{jj} - (h^r_{ij})^2]. \quad (15)$$

Let π be the plane spanned by e_1, e_2. Then, by the Gauss equation, taking $X = W = e_1$ and $Y = Z = e_2$, we obtain

$$R(e_1, e_2, e_2, e_1) = \overline{R}(e_1, e_2, e_2, e_1) - g(h(e_1, e_2), h(e_2, e_1)) + g(h(e_1, e_1), h(e_2, e_2)). \quad (16)$$

Similarly, by substituting $X = W = e_1$ and $Y = Z = e_2$ into the formula for constant holomorphic sectional curvature, we have

$$\overline{R}(e_1, e_2, e_2, e_1) = \frac{c}{4}\{g(e_2, e_2)g(e_1, e_1) - g(e_1, e_2)g(e_2, e_1)$$
$$+ [g(e_2, Je_2)g(Je_1, e_1)$$
$$- g(e_2, Je_1)g(Je_2, e_1) + 2g(e_1, Je_2)g(Je_2, e_1)]\}$$
$$+ \lambda_1\alpha(e_1, e_2)g(e_2, e_1) - \lambda_1\alpha(e_2, e_2)g(e_1, e_1)$$
$$+ \lambda_2 g(e_1, e_2)\alpha(e_2, e_1) - \lambda_2 g(e_2, e_2)\alpha(e_1, e_1)$$
$$+ \lambda_2(\lambda_1 - \lambda_2)g(e_1, e_2)\beta(e_2, e_1)$$
$$- \lambda_2(\lambda_1 - \lambda_2)g(e_2, e_2)\beta(e_1, e_1), \quad (17)$$

which simplifies to

$$\overline{R}(e_1, e_2, e_2, e_1) = \frac{c}{4}[1 + 3g^2(Je_1, e_2)] - \lambda_1\alpha(e_2, e_2) - \lambda_2\alpha(e_1, e_1)$$
$$- \lambda_2(\lambda_1 - \lambda_2)\beta(e_1, e_1). \quad (18)$$

Substituting (18) in (16) yields

$$R(e_1, e_2, e_2, e_1) = \frac{c}{4}[1 + 3g^2(Je_1, e_2)] - \lambda_1\alpha(e_2, e_2) - \lambda_2\alpha(e_1, e_1)$$
$$- \lambda_2(\lambda_1 - \lambda_2)\beta(e_1, e_1) + \sum_{r=n+1}^{2n} [h^r_{11}h^r_{22} - (h^r_{12})^2]. \quad (19)$$

https://doi.org/10.3390/math14010071

Also, by taking $X = W = e_2$ and $Y = Z = e_1$ in the Gauss equation, we obtain

$$R(e_2, e_1, e_1, e_2) = \frac{c}{4}[1 + 3g^2(Je_1, e_2)] - \lambda_1\alpha(e_1, e_1) - \lambda_2\alpha(e_2, e_2) - \lambda_2(\lambda_1 - \lambda_2)\beta(e_2, e_2) + \sum_{r=n+1}^{2n} [h^r_{11}h^r_{22} - (h^r_{12})^2]. \tag{20}$$

Then by the formula of the sectional curvature

$$K(\pi) = \frac{1}{2}[R(e_1, e_2, e_2, e_1) + R(e_2, e_1, e_1, e_2)],$$

we have

$$K(\pi) = \frac{c}{4} + 3\frac{c}{4}g^2(Je_1, e_2) - \frac{\lambda_1 + \lambda_2}{2}\mathtt{tr}(\alpha_{|\pi}) - \frac{\lambda_2(\lambda_1 - \lambda_2)}{2}\mathtt{tr}(\beta_{|\pi}) + \sum_{r=n+1}^{2n} [h^r_{11}h^r_{22} - (h^r_{12})^2]. \tag{21}$$

For a plane section π =span$\{e_1, e_2\}$, we denote $\Theta^2(\pi) = g^2(Je_1, e_2)$. Then

$$K(\pi) = \frac{c}{4}[1 + 3\Theta^2(\pi)] - \frac{\lambda_1 + \lambda_2}{2}\mathtt{tr}(\alpha_{|\pi}) - \frac{\lambda_2(\lambda_1 - \lambda_2)}{2}\mathtt{tr}(\beta_{|\pi}) + \sum_{r=n+1}^{2n} [h^r_{11}h^r_{22} - (h^r_{12})^2]. \tag{22}$$

Now from (15),

$$\tau = \frac{c}{8}\left[n(n-1) + 3n\cos^2\theta\right] - \frac{(\lambda_1 + \lambda_2)(n-1)\lambda}{2} - \frac{\lambda_2(\lambda_1 - \lambda_2)(n-1)\mu}{2} + \sum_{r=n+1}^{4s} \sum_{1\le i<j\le n} [h^r_{ii}h^r_{jj} - (h^r_{ij})^2], \tag{23}$$

and, in addition, using Equations (23) and (22), we have

$$\tau(p) - K(\pi) = \frac{c}{8}[(n-2)(n+1) + 3(n\cos^2\theta - 2\Theta^2(\pi))] - \frac{(\lambda_1 + \lambda_2)(n-1)\lambda}{2} - \frac{\lambda_2(\lambda_1 - \lambda_2)(n-1)\mu}{2} + \frac{\lambda_1 + \lambda_2}{2}\mathtt{tr}(\alpha_{|\pi}) + \frac{\lambda_2(\lambda_1 - \lambda_2)}{2}\mathtt{tr}(\beta_{|\pi}) + \sum_{r=n+1}^{4s} \sum_{1\le i<j\le n} \left[h^r_{ii}h^r_{jj} - (h^r_{ij})^2\right] - \sum_{r=n+1}^{4s} (h^r_{11}h^r_{22} - (h^r_{12})^2), \tag{24}$$

or equivalently,

$$\begin{aligned}\tau(p)-K(\pi)=&\frac{c}{8}[(n-2)(n+1)+3(n\cos^2\theta-2\Theta^2(\pi))]\\&-\frac{(\lambda_1+\lambda_2)(n-1)\lambda}{2}-\frac{\lambda_2(\lambda_1-\lambda_2)(n-1)\mu}{2}\\&+\frac{\lambda_1+\lambda_2}{2}\mathtt{tr}(\alpha_{|\pi})+\frac{\lambda_2(\lambda_1-\lambda_2)}{2}\mathtt{tr}(\beta_{|\pi})\\&+\sum_{r=n+1}^{2n}\{\sum_{j=3}^{n}(h^r_{11}+h^r_{22})h^r_{jj}+\sum_{3\le i<j\le n}h^r_{ii}h^r_{jj}\\&-\sum_{j=3}^{n}(h^r_{1j})^2-\sum_{2\le i<j\le n}(h^r_{ij})^2\}.\end{aligned}\tag{25}$$

It follows that

$$\begin{aligned}\tau(p)-K(\pi)\le&\frac{c}{8}[(n-2)(n+1)+3(n\cos^2\theta-2\Theta^2(\pi))]\\&-\frac{(\lambda_1+\lambda_2)(n-1)\lambda}{2}-\frac{\lambda_2(\lambda_1-\lambda_2)(n-1)\mu}{2}\\&+\frac{\lambda_1+\lambda_2}{2}\mathtt{tr}(\alpha_{|\pi})+\frac{\lambda_2(\lambda_1-\lambda_2)}{2}\mathtt{tr}(\beta_{|\pi})\\&+\sum_{r=n+1}^{2n}\{\sum_{j=3}^{n}(h^r_{11}+h^r_{22})h^r_{jj}+\sum_{3\le i<j\le n}h^r_{ii}h^r_{jj}\\&-\sum_{j=3}^{n}(h^r_{1j})^2-\sum_{j=3}^{n}(h^{n+1}_{jj})^2-\sum_{i,j\in\{2,3,\ldots,n\},i\neq j}(h^{n+i}_{jj})^2\}.\end{aligned}\tag{26}$$

For any $i=1,\ldots,n$, we denote $e_i^*=\frac{1}{\sin\theta}Fe_i$. Then $\{e_1^*,\ldots,e_n^*\}$ is an orthonormal basis of $T_p^\perp M$.

Simply, we put $h_{ij}^k=g(h(e_i,e_j),e_k^*)$, $i,j,k=1,\ldots,n$.

Since M is a Kählerian slant submanifold, the components of the second fundamental form are symmetric, that is,

$$\hat{h}^k_{ij}=\hat{h}^j_{ik}=\hat{h}^i_{jk},\ \forall i,j,k=1,\ldots,n.$$

We assume that the vector field P is tangent to M. Then, $\hat{h}=h$.

In our inequality we use $h^1_{1j}=h^j_{11}$, $h^j_{1j}=h^1_{jj}$, for $3\le j\le n$, and $h^j_{ij}=h^i_{jj}$, for $2\le i\neq j\le n$.

The above inequality implies that

$$\begin{aligned}\tau(p)-K(\pi)=&\sum_{r=1}^{n}\{\sum_{j=1}^{n}(h^r_{11}+h^r_{22})h^r_{jj}+\sum_{3\le i<j\le n}h^r_{ii}h^r_{jj}\\&-\sum_{j=3}^{n}(h^j_{11})^2-\sum_{j=3}^{n}(h^1_{jj})^2-\sum_{2\le i\neq j\le n}(h^i_{jj})^2\}\\&+\frac{c}{8}[(n-2)(n+1)+3(n\cos^2\theta-2\Theta^2(\pi))]\\&-\frac{(\lambda_1+\lambda_2)(n-1)\lambda}{2}-\frac{\lambda_2(\lambda_1-\lambda_2)(n-1)\mu}{2}\\&+\frac{\lambda_1+\lambda_2}{2}\mathtt{tr}(\alpha_{|\pi})+\frac{\lambda_2(\lambda_1-\lambda_2)}{2}\mathtt{tr}(\beta_{|\pi}).\end{aligned}\tag{27}$$

In order to achieve the proof, we will use ideas from [22]. In this respect, recall the following inequalities

https://doi.org/10.3390/math14010071

$$\begin{aligned}
&\sum_{j=1}^{n}(h_{11}^{r}+h_{22}^{r})h_{jj}^{r}+\sum_{3\leq i<j\leq n}h_{ii}^{r}h_{jj}^{r}-\sum_{j=3}^{n}(h_{jj}^{r})^{2}\\
&\leq\frac{n-2}{2(n+1)}(h_{11}^{r}+\ldots+h_{nn}^{r})^{2}\\
&\leq\frac{2n-3}{2(2n+3)}(h_{11}^{r}+\ldots+h_{nn}^{r})^{2} \qquad (28)
\end{aligned}$$

for $r=1,2$. The first inequality in (28) can be rewritten as

$$\sum_{j=3}^{n}(h_{11}^{r}+h_{22}^{r}-3h_{jj}^{r})^{2}+3\sum_{3\leq i<j\leq n}(h_{ii}^{r}-h_{jj}^{r})^{2}\geq 0.$$

The equality holds if and only if

$$3h_{jj}^{r}=h_{11}^{r}+h_{22}^{r},\quad \forall j=3,\ldots,n.$$

Similarly, the equality in the second inequality holds if and only if

$$h_{11}^{r}+h_{22}^{r}=0 \quad \text{and} \quad h_{jj}^{r}=0,\quad \forall j=3,\ldots,n,\ r=1,2.$$

Moreover, we have

$$\begin{aligned}
&\sum_{j=3}^{n}(h_{11}^{r}+h_{22}^{r})h_{jj}^{r}+\sum_{3\leq i<j\leq n}h_{ii}^{r}h_{jj}^{r}-\sum_{j=1,j\neq r}^{n}(h_{jj}^{r})\\
&\leq\frac{2n-3}{2(2n+3)}(h_{11}^{r}+\ldots+h_{nn}^{r})^{2}, \qquad (29)
\end{aligned}$$

for $r=3,\ldots,n$, which is equivalent to

$$\begin{aligned}
&\sum_{j=3}^{n}[2(h_{11}^{r}+h_{22}^{r})h_{jj}^{r}-3h_{jj}^{r}]^{2}+(2n+3)(h_{11}^{r}-h_{22}^{r})^{2}\\
&+6\sum_{3\leq i<j\leq n,i,j\neq r}(h_{ii}^{r}-h_{jj}^{r})^{2}+2\sum_{j=3}^{n}(h_{rr}^{r}-h_{jj}^{r})^{2}\\
&+3[h_{rr}^{r}-2(h_{11}^{r}+h_{22}^{r})]^{2}\geq 0. \qquad (30)
\end{aligned}$$

The equality holds if and only if
$h_{11}^{r}=h_{22}^{r}=3\lambda^{r}, h_{jj}^{r}=4\lambda^{r}; \forall j=3,\ldots,n, j\neq r, r=3,\ldots,n$, and $h_{rr}^{r}=12\lambda^{r},\ \lambda^{r}\in\mathbb{R}$.

We will finish the proof by using the inequalities (28) and (29); it follows that (27) implies

$$\begin{aligned}
\tau(p)-K(\pi)\leq&\frac{2n-3}{2(2n+3)}\|H\|^{2}\\
&+\frac{c}{8}[(n-2)(n+1)+3(n\cos^{2}\theta-2\Theta^{2}(\pi))]\\
&-\frac{(\lambda_1+\lambda_2)(n-1)\lambda}{2}-\frac{\lambda_2(\lambda_1-\lambda_2)(n-1)\mu}{2}\\
&+\frac{\lambda_1+\lambda_2}{2}\mathtt{tr}(\alpha_{|\pi})+\frac{\lambda_2(\lambda_1-\lambda_2)}{2}\mathtt{tr}(\beta_{|\pi}), \qquad (31)
\end{aligned}$$

which is the inequality to be proven.

The equality holds if and only if

$$h(e_1,e_1) = ae_1^* + 3be_3^*, \quad h(e_1,e_3) = 3be_1^*, \quad h(e_3,e_j) = 4be_j^*,$$
$$h(e_2,e_2) = -ae_1^* + 3be_3^*, \quad h(e_2,e_3) = 3be_2^*, \quad h(e_j,e_k) = 4be_3^*\delta_{jk},$$
$$h(e_1,e_2) = -ae_2^*, \quad h(e_3,e_3) = 12be_3^*, \quad h(e_1,e_j) = h(e_2,e_j) = 0,$$

for some numbers a and b, and $j,k \in \{4,\dots,n\}$. □

4. Upper Bound and Lower Bound on Ricci Curvature

The Ricci curvature plays a fundamental role in differential geometry and mathematical physics, as it governs the geometric structure of manifolds and appears naturally in the study of Einstein manifolds and curvature inequalities.

In [8], B.-Y. Chen established a sharp inequality providing a lower bound for the Ricci curvature of any n-dimensional Riemannian submanifold of a real space form $\overline{M}(c)$ with constant sectional curvature c. Specifically, he proved that

$$Ric(X) \leq (n-1)c + \frac{n^2}{4}\|H\|^2,$$

which is known as the Chen–Ricci inequality.

The same inequality holds for Lagrangian submanifolds in a complex space form $\overline{M}(c)$ as well (see [23]).

In [24], the authors extended the Chen–Ricci inequality for submanifolds in complex space forms.

This upper bound on Ricci curvature was improved by A. Mihai and I. Rădulescu [25] for Kählerian slant submanifolds in complex space forms.

In this section, we further extend the relationship between the upper and lower bounds on Ricci curvature to Kählerian slant submanifolds in complex space forms admitting a quarter-symmetric connection.

Theorem 2. *Let $\overline{M}(c)$ be a $2n$-dimensional complex space form admitting a quarter-symmetric connection and M an n-dimensional Kählerian slant submanifold tangent to p. Then, for any $p \in M$ and any unit vector $X \in T_pM$, the Ricci curvature has the following upper bound:*

$$\begin{aligned} Ric(X) \leq{}& \frac{c}{4}(n-1+\cos^2\theta) - \frac{\lambda_1+\lambda_2}{2}[(n-2)\alpha(e_1,e_1) - \mathtt{tr}\alpha] \\ &- \frac{\lambda_2(\lambda_1-\lambda_2)}{2}[(n-2)\beta(e_1,e_1) + \mathtt{tr}\beta] + \frac{n(n-1)}{4}\|H\|^2. \end{aligned} \tag{32}$$

Moreover, the equality sign of (32) *holds identically if and only if either*
(i) $c = 0$ and M is totally geodesic, or
(ii) $n = 2$, $c < 0$ and M is a slant H-umbilical surface with $\lambda = 3\mu$.

Proof. At a point $p \in M$, let $X \in T_pM$ be a unit vector. We choose an orthonormal basis $\{e_1 = X, e_2, \dots, e_n\} \subset T_pM$ and define

$$\left\{e_1^* = \frac{1}{\sin\theta}Fe_1, \dots, e_n^* = \frac{1}{\sin\theta}Fe_n\right\}$$

to obtain an orthonormal basis of $T_p^{\perp}M$.

https://doi.org/10.3390/math14010071

For any $j = 2, \ldots, n$, we compute the sectional curvature $K(e_1 \wedge e_j)$ of the plane section spanned by e_1 and e_j similarly to (22).

$$K(e_1 \wedge e_j) = \frac{c}{4}[1 + 3g^2(Je_1, e_j)] - \frac{\lambda_1 + \lambda_2}{2}(\alpha(e_1, e_1) + \alpha(e_j, e_j)) - \frac{\lambda_2(\lambda_1 - \lambda_2)}{2}(\beta(e_1, e_1) + \beta(e_j, e_j)) + \sum_{r=n+1}^{2n} [h^r_{11}h^r_{jj} - (h^r_{1j})^2]. \quad (33)$$

By summing the above equations for $j = 2, \ldots, n$, we obtain

$$Ric(X) = \sum_{j=2}^{n} K(e_1 \wedge e_j) = \frac{c}{4}(n - 1 + \|Te_1\|^2) - \frac{\lambda_1 + \lambda_2}{2}[(n-2)\alpha(e_1, e_1) + \mathrm{tr}\,\alpha] - \frac{\lambda_2(\lambda_1 - \lambda_2)}{2}[(n-2)\beta(e_1, e_1) + \mathrm{tr}\beta] + \sum_{r=n+1}^{2n}\sum_{j=2}^{n}[h^r_{11}h^r_{jj} - (h^r_{1j})^2]. \quad (34)$$

We assume that P is tangent to M. Then $\hat{h} = h$. We denote $h^k_{ij} = g(h(e_i, e_j), e^*_k), i, j, k = 1, \ldots, n$. It follows from Equation (34) that

$$Ric(X) - \frac{c}{4}(n - 1 + \cos^2\theta) + \frac{\lambda_1 + \lambda_2}{2}[(n-2)\alpha(e_1, e_1) + \mathrm{tr}\,\alpha] + \frac{\lambda_2(\lambda_1 - \lambda_2)}{2}[(n-2)\beta(e_1, e_1) + \mathrm{tr}\beta] \leq \sum_{r=1}^{n}\sum_{j=2}^{n}[h^r_{11}h^r_{jj} - (h^r_{1j})^2]. \quad (35)$$

Since M is a Kählerian slant submanifold, $h^1_{1j} = h^j_{11}$ and $h^1_{1j} = h^1_{jj}$. It follows that the above inequality gives

$$Ric(X) - \frac{c}{4}(n - 1 + \cos^2\theta) + \frac{\lambda_1 + \lambda_2}{2}[(n-2)\alpha(e_1, e_1) + \mathrm{tr}\,\alpha] + \frac{\lambda_2(\lambda_1 - \lambda_2)}{2}[(n-2)\beta(e_1, e_1) + \mathrm{tr}\,\beta] \leq \sum_{r=1}^{n}\sum_{j=2}^{n} h^r_{11}h^r_{jj} - \sum_{j=2}^{n}(h^j_{11})^2 - \sum_{j=2}^{n}(h^1_{jj})^2. \quad (36)$$

Because $nH_1 = h^1_{11} + h^1_{22} + \ldots + h^1_{nn}$, Lemma 1 implies

$$f_1(h^{11}_1, h^1_{22}, \ldots, h^1_{nn}) \leq \frac{n-1}{4n}(nH^1)^2 = \frac{n(n-1)}{4n}(H^1)^2. \quad (37)$$

We denote $f_r(h^r_{11}, h^r_{22}, \ldots, h^r_{nn}) = h^r_{11}\sum_{j=2}^{n} h^r_{jj} - (h^r_{11})^2, r = 2, \ldots, n$.
Lemma 2 implies

$$f_r(h^r_{11}, h^r_{22}, \ldots, h^r_{nn}) \leq \frac{1}{8}(nH^r)^2 \leq \frac{n(n-1)}{4}(H^r)^2. \quad (38)$$

By Equations (36)–(38), we obtain

$$Ric(X) \leq \frac{c}{4}(n - 1 + \cos^2\theta) - \frac{\lambda_1 + \lambda_2}{2}[(n-2)\alpha(e_1, e_1) - \mathrm{tr}\,\alpha] - \frac{\lambda_2(\lambda_1 - \lambda_2)}{2}[(n-2)\beta(e_1, e_1) + \mathrm{tr}\,\beta] + \frac{n(n-1)}{4}\|H\|^2. \quad (39)$$

For the equality case, we use similar arguments as in [26]. □

F. Malek and L. Nejadakbary in [27] established a lower bound on the Ricci curvature of certain submanifolds.

Inspired by their work, we now derive a reverse Chen–Ricci inequality, which provides a lower on the Ricci curvature of Kählerian slant submanifolds in complex space forms admitting a quarter-symmetric connection.

Corollary 1. *Let $\overline{M}(c)$ be a $2n$-dimensional complex space form admitting a quarter-symmetric connection and M an n-dimensional Kählerian slant submanifold tangent to p. Then, for any $p \in M$ and any unit vector $X \in T_pM$, the Ricci curvature has the following lower bound:*

$$\begin{aligned} Ric(X) \geq{}& (n-1)\tau(p) - \frac{(n-1)(2n-3)}{2(2n+3)}\|H\|^2 \\ &- \frac{c}{8}[(n-2)(n^2-1) + 3(n-2)(n+1)\cos^2\theta] \\ &+ \frac{(\lambda_1+\lambda_2)(n-1)^2\lambda}{2} + \frac{\lambda_2(\lambda_1-\lambda_2)(n-1)^2\mu}{2} \\ &- \frac{\lambda_1+\lambda_2}{2}[(n-2)\alpha(X,X) + \mathtt{tr}\,\alpha] \\ &- \frac{\lambda_2(\lambda_1-\lambda_2)}{2}[(n-2)\beta(X,X) + \mathtt{tr}\,\beta]. \end{aligned} \tag{40}$$

Proof. From the inequality (10) we get

$$\begin{aligned} K(X\wedge Y) \geq{}& \tau(p) - \frac{2n-3}{2(2n+3)}\|H\|^2 \\ &- \frac{c}{8}[(n-2)(n+1) + 3\{\|T\|^2 - 2\Theta^2(X\wedge Y)\}] \\ &+ \frac{(\lambda_1+\lambda_2)(n-1)\lambda}{2} + \frac{\lambda_2(\lambda_1-\lambda_2)(n-1)\mu}{2} \\ &- \frac{\lambda_1+\lambda_2}{2}\mathtt{tr}(\alpha_{|X\wedge Y}) - \frac{\lambda_2(\lambda_1-\lambda_2)}{2}\mathtt{tr}(\beta_{|X\wedge Y}). \end{aligned} \tag{41}$$

Putting $Y = e_2, \ldots, e_n$ and summing, we derive

$$\begin{aligned} Ric(X) \geq{}& (n-1)\tau(p) - \frac{(n-1)(2n-3)}{2(2n+3)}\|H\|^2 \\ &- \frac{c}{8}[(n-2)(n^2-1) + 3(n-2)(n+1)\cos^2\theta] \\ &+ \frac{(\lambda_1+\lambda_2)(n-1)^2\lambda}{2} + \frac{\lambda_2(\lambda_1-\lambda_2)(n-1)^2\mu}{2} \\ &- \frac{\lambda_1+\lambda_2}{2}[(n-2)\alpha(X,X) + \mathtt{tr}\,\alpha] \\ &- \frac{\lambda_2(\lambda_1-\lambda_2)}{2}[(n-2)\beta(X,X) + \mathtt{tr}\,\beta]. \end{aligned} \tag{42}$$

□

5. Generalized Euler Inequality for Kählerian Slant Submanifolds

In this section, we establish an improved version of the generalized Euler inequality, providing a lower bound for the mean curvature of a Kählerian slant submanifold in a complex Riemannian manifold admitting a quarter-metric connection.

B.-Y. Chen [28] derived the following generalized Euler inequality for a n-dimensional submanifold in a Riemannian space form of constant sectional curvature c:

$$\|H\|^2 \geq \frac{2\tau}{n(n-1)} - c,$$

with equality holding identically if and only if the submanifold is totally umbilical.

Theorem 3. *Let M be an n-dimensional* Kählerian slant submanifold *isometrically immersed in a 2n-dimensional complex space form admitting a quarter-symmetric connection with P tangent to M. Then, the mean curvature vector satisfies the following inequality:*

$$\|H\|^2 \geq \frac{2(n+2)}{n^2(n-1)}\tau(p) - \frac{(n+2)}{n^2(n-1)}\frac{c}{4}\left[n(n-1) + 3n\cos^2\theta\right] + \frac{n+2}{n^2}((\lambda_1+\lambda_2)\lambda - \lambda_2(\lambda_1-\lambda_2)(n-1)\mu. \tag{43}$$

Moreover, equality holds identically if and only if M is an H-umbilical submanifold.

Proof. Consider an n-dimensional Kählerian slant submanifold of a $2n$-dimensional complex Riemannian manifolds admitting a quarter-symmetric connection.

Recall Equation (15)

$$2\tau = \frac{c}{4}\left[n(n-1) + 3n\cos^2\theta\right] - (\lambda_1+\lambda_2)(n-1)\lambda - \lambda_2(\lambda_1-\lambda_2)(n-1)\mu + \sum_{r=n+1}^{4s}\sum_{1\leq i,j\leq n}[h^r_{ii}h^r_{jj} - (h^r_{ij})^2]. \tag{44}$$

By the definition of the mean curvature we have

$$n^2\|H\|^2 = \sum_i g(h(e_i,e_i),h(e_i,e_i)) + \sum_{i\neq j} g(h(e_i,e_i),h(e_j,e_j)) = \sum_{i=1}^{n}[\sum_{j=1}^{n}(h^i_{jj})^2 + 2\sum_{1\leq j<k\leq n} h^i_{jj}h^i_{kk}]. \tag{45}$$

Next, taking into account Equations (44) and (45), we obtain

$$2\tau = \frac{c}{4}\left[n(n-1) + 3n\cos^2\theta\right] - (\lambda_1+\lambda_2)(n-1)\lambda - \lambda_2(\lambda_1-\lambda_2)(n-1)\mu + 2\sum_i^n\sum_{1\leq j<k\leq n} h^i_{jj}h^i_{kk} - 2\sum_{1\leq i\neq j\leq n}(h^i_{jj})^2 - 6\sum_{1\leq i<j<k\leq n}(h^k_{ij})^2. \tag{46}$$

The last equation can be expressed as

$$2\tau = \frac{c}{4}\left[n(n-1) + 3n\cos^2\theta\right] - (n-1)(\lambda_1+\lambda_2)\lambda - \lambda_2(\lambda_1-\lambda_2)(n-1)\mu + 2\sum_i^n\sum_{1\leq j<k\leq n} h^i_{jj}h^i_{kk} - 2\sum_{1\leq i\neq j\leq n}(h^i_{jj})^2 - 6\sum_{1\leq i<j<k\leq n}(h^k_{ij})^2. \tag{47}$$

Inspired by Chen's paper [29], we consider the parameter $m = \frac{n+2}{n-1}$, with $n \geq 2$. To study the inequality for $\|H\|^2$, we follow the technique used in [29]. Specifically, we obtain

https://doi.org/10.3390/math140100

$$
\begin{aligned}
& n^2\|H\|^2 - m\{2\tau(p) - \frac{c}{4}\Big[n(n-1) + 3n\cos^2\theta\Big] \\
&\quad + (n-1)[(\lambda_1+\lambda_2)\lambda - \lambda_2(\lambda_1-\lambda_2)(n-1)\mu]\} \\
&\quad = \sum_{i=1}^{n}(h_{ii}^i)^2 + (1+2m)\sum_{1\le i\ne j\le n}(h_{jj}^i)^2 + 6m\sum_{1\le i<j<k\le n}(h_{ij}^k)^2 \\
&\quad - 2(m-1)\sum_{i}^{n}\sum_{1\le j<k\le n} h_{jj}^i h_{kk}^i \\
&\quad = \sum_{i=1}^{n}(h_{ii}^i)^2 + 6m\sum_{1\le i<j<k\le n}(h_{ij}^k)^2 + (m-1)\sum_{i=1}^{n}\sum_{1\le j<k\le n}(h_{jj}^i - h_{kk}^i)^2 \\
&\quad + [1+2m-(n-2)(m-1)]\sum_{1\le i\ne j\le n}(h_{jj}^i)^2 - 2(m-1)\sum_{1\le i\ne j\le n}h_{ii}^i h_{jj}^i \\
&\quad = 6m\sum_{1\le i<j<k\le n}(h_{ij}^k)^2 + (m-1)\sum_{i\ne j,k}\sum_{1\le j<k\le n}(h_{jj}^i - h_{kk}^i)^2 \\
&\quad + \frac{1}{n-1}\sum_{1\le i\ne j\le n}\Big[h_{ii}^i - (n-1)(m-1)h_{jj}^i\Big]^2 \ge 0.
\end{aligned} \tag{48}
$$

Consequently, we obtain

$$
\begin{aligned}
\|H\|^2 \ge\ & \frac{2(n+2)}{n^2(n-1)}\tau(p) - \frac{(n+2)}{n^2(n-1)}\frac{c}{4}\Big[n(n-1) + 3n\cos^2\theta\Big] \\
& + \frac{n+2}{n^2}((\lambda_1+\lambda_2)\lambda - \lambda_2(\lambda_1-\lambda_2)(n-1)\mu).
\end{aligned} \tag{49}
$$

The equality holds if and only if

$$
\begin{aligned}
h_{ij}^k &= 0,\ \forall 1 \le i < j < k \le n; \\
h_{ii}^i &= 3h_{jj}^i,\ \forall 1 \le i \ne j \le n.
\end{aligned}
$$

This implies that M is an H-umbilical submanifold (see [30]). □

6. Example

It is well known [21] that any proper slant surface in a Kähler manifold is Kählerian slant.

We give an example of a 4-dimensional Kählerian slant submanifold of $\mathbb{C}^4$.

For any $k > 0$,

$$
x(u,v,w,z) = (u, v, k\sin w, k\sin z, kw, kz, k\cos w, k\cos z)
$$

defines a Kählerian slant submanifold in $\mathbb{C}^4$ with slant angle $\theta = \frac{\pi}{4}$ (see [21]).

If $\{e_1, \dots, e_8\}$ is the canonical basis of $\mathbb{E}^8$, we denote $P = e_1$. It is tangent to $M = \mathrm{Im}x$.

A quarter-symmetric connection on $\mathbb{C}^4$ can be defined by

$$
\overline{\nabla} = \hat{\overline{\nabla}} + \lambda_1\langle Y, e_1\rangle X - \lambda_2\langle X, Y\rangle e_1.
$$

It follows that all the above inequalities, i.e., Chen first inequality, Chen–Ricci inequality and generalized Euler inequality, are true on M.

These inequalities can be verified by straightforward, but long calculations.

https://doi.org/10.3390/math14010071

7. Conclusions

Complex manifolds provide a natural and generalized framework for differential geometry, particularly in the study of structures compatible with complex or Hermitian metrics. On the other hand, quarter-symmetric connections offer an alternative geometric setting in which the Levi–Civita connection is supplemented by additional torsion or non-metricity components, leading to richer curvature behavior.

Kählerian slant submanifolds constitute an important class of submanifolds characterized by a special symmetry condition involving the complex structure. In this article, we establish a symmetry condition for the second fundamental form when the submanifold is tangent to P. Furthermore, we investigate intrinsic–extrinsic relations for Kählerian slant submanifolds in complex space forms endowed with a quarter-symmetric connection.

This work can be extended in several directions. One possible extension is to study Kählerian slant submanifolds that are not tangent to P. Another is to develop generalized Euler-type inequalities within this geometric setting.

On the other hand, Calabi–Yau manifolds play a significant role in theoretical physics as they provide an essential geometric framework for string theory. In [31], the authors studied isometric immersions of Lagrangian submanifolds into Calabi–Yau manifolds. Their results suggest further extensions, such as investigating Chen-type inequalities and related problems for submanifolds immersed in Calabi–Yau ambient spaces.

Moreover, other researchers in the field could continue our study and deal with geodesic mappings of equiaffine and Ricci symmetric spaces, geometry in the large of Einstein-like manifolds, and linearly independent curvature tensors of half-symmetric affine connections.

Author Contributions: Conceptualization, M.M., I.M. and S.C.G.; methodology, I.M. and A.O.; validation, M.M. and S.C.G.; formal analysis, M.M., I.M. and A.O.; investigation, I.M. and A.O.; resources, I.M., S.C.G. and M.M.; writing—original draft preparation, M.M., I.M. and S.C.G.; writing—review and editing, A.O., I.M. and S.C.G.; visualization, I.M. and A.O.; supervision, I.M.; project administration, I.M. All authors have read and agreed to the published version of the manuscript.

Funding: This research received no external funding.

Data Availability Statement: The original contributions presented in this study are included in the article. Further inquiries can be directed to the corresponding author.

Conflicts of Interest: The authors declare no conflicts of interest.

References

1. Friedmann, A.; Schouten, J.A. Über die Geometrie der halbsymmetrischen Übertragungen. *Math. Z.* **1924**, *21*, 211–223. [CrossRe
2. Hayden, H. Sub-spaces of a space with torsion. *Proc. Lond. Math. Soc.* **1932**, *2*, 27–50. [CrossRef]
3. Yano, K. On semi symmetric metric connection. *Rev. Roum. Math. Pures Appl.* **1970**, *15*, 1579–1591.
4. Golab, S. On semi-symmetric and quarter-symmetric linear connections. *Tensor NS* **1975**, *29*, 249–254.
5. Agashe, N.S. A semi-symmetric non-metric connection on a Riemannian manifold. *Indian J. Pure Appl. Math.* **1992**, *23*, 399–4(
6. Agashe, N.S.; Chafle, M.R. On submanifolds of a Riemannian manifold with a semi-symmetric non-metric connection. *Tens* **1994**, *55*, 120–130.
7. Chen, B.Y. Some pinching and classification theorems for minimal submanifolds. *Arch. Math.* **1993**, *60*, 568–578. [CrossRef]
8. Chen, B.Y. Relations between Ricci curvature and shape operator for submanifolds with arbitrary codimensions. *Glasg. Math.* **1999**, *41*, 33–41. [CrossRef]
9. Kim, J.S.; Song, Y.M.; Tripathi, M.M. B.-Y. Chen inequalities for submanifolds in generalized complex space forms. *Bull. Kore Math. Soc.* **2003**, *40*, 411–424. [CrossRef]
10. Carriazo, A.; Fernandez, L.; Hans-Uber, M. BY Chen's inequality for S-space-forms: applications to slant immersions. *Indiar Pure Appl. Math.* **2003**, *34*, 1287–1298.
11. Alegre, P.; Carriazo, A.; Kim, Y.H.; Yoon, D.W. B.-Y. Chen's inequality for submanifolds of generalized space forms. *Indian J. Pu Appl. Math.* **2007**, *38*, 185.

https://doi.org/10.3390/math140100

Aktan, N.; Sarıkaya, M.Z.; Özüsağlam, E. B.Y. Chen's inequality for semi-slant submanifolds in-space forms. *Balkan J. Geom. Appl.* **2008**, *13*, 1–10.

Mihai, A.; Özgür, C. Chen inequalities for submanifolds of real space forms with a semi-symmetric metric connection. *Taiwan. J. Math.* **2010**, *14*, 1465–1477. [CrossRef]

Mihai, A.; Özgür, C. Chen inequalities for submanifolds of complex space forms and Sasakian space forms endowed with semi-symmetric metric connections. *Rocky Mt. J. Math.* **2011**, *41*, 1653–1673. [CrossRef]

Zhang, P.; Pan, X.; Zhang, L. Inequalities for submanifolds of a Riemannian manifold of nearly quasi-constant curvature with a semi-symmetric non-metric connection. *Rev. Unión Mat. Argent.* **2015**, *56*, 1–19.

Alghamdi, F.A.; Alqahtani, L.S.; Ali, A. Chen inequalities on warped product Legendrian submanifolds in Kenmotsu space forms and applications. *J. Inequal. Appl.* **2024**, *2024*, 63. [CrossRef]

Mihai, A.; Mihai, I. A note on a well-defined sectional curvature of a semi-symmetric non-metric connection. *Int. Electron. J. Geom.* **2024**, *17*, 15–23. [CrossRef]

Wang, Y. Chen inequalities for submanifolds of complex space forms and Sasakian space forms with quarter-symmetric connections. *Int. J. Geom. Methods Mod. Phys.* **2019**, *16*, 1950118. [CrossRef]

Ishihara, S. Quaternion Kählerian manifolds. *J. Differ. Geom.* **1974**, *9*, 483–500. [CrossRef]

Yano, K.; Kon, M. *Structures on Manifolds*; World Scientific: Singapore, 1985.

Chen, B.Y. *Geometry of Slant Submanifolds*; KU Leuven: Leuven, Belgium, 1990.

Bolton, J.; Dillen, F.; Fastenakels, J.; Vrancken, L. A best possible inequality for curvature-like tensor fields. *Math. Inequal. Appl.* **2009**, *12*, 663–681. [CrossRef]

Chen, B.Y. On Ricci curvature of isotropic and Lagrangian submanifolds in complex space forms. *Arch. Math.* **2000**, *74*, 154–160. [CrossRef]

Matsumoto, K.; Mihai, I.; Oiaga, A. Ricci curvature of submanifolds in complex space forms. *Rev. Roum. Math. Pures Appl.* **2002**, *46*, 775–782.

Mihai, A.; Rădulescu, I.N. An improved Chen-Ricci inequality for Kaehlerian slant submanifolds in complex space forms. *Taiwan. J. Math.* **2012**, *16*, 761–770. [CrossRef]

Deng, S. An improved Chen-Ricci inequality. *Int. Electron. J. Geom.* **2009**, *2*, 39–45.

Malek, F.; Nejadakbary, V. A lower bound for the Ricci curvature of submanifolds in generalized Sasakian space forms. *Adv. Geom.* **2013**, *133*, 695–711. [CrossRef]

Chen, B.Y. Mean curvature and shape operator of isometric immersions in real-space-forms. *Glasg. Math. J.* **1996**, *38*, 87–97. [CrossRef]

Chen, B.Y. Jacobi's elliptic functions and Lagrangian immersions. *Proc. R. Soc. Edinb. Sect. A Math.* **1996**, *126*, 687–704. [CrossRef]

Deng, S. Improved Chen-Ricci inequality for Lagrangian submanifolds in quaternion space forms. *Int. Electron. J. Geom.* **2012**, *5*, 163–170.

Matessi, D. Isometric embeddings of families of special Lagrangian submanifolds. *Ann. Glob. Anal. Geom.* **2006**, *29*, 197–220. [CrossRef]

https://doi.org/10.3390/math14010071

mathematics

MDPI

Article

Geometry of Almost Pure Metric Pseudo-F-Manifolds: Insights from Walker 4-Manifolds

Yanlin Li [1] Cagri Karaman [2], Aydin Gezer [2], Sibel Turanli [3] and Yuquan Xie [1,*]

[1] School of Mathematics, Hangzhou Normal University, Hangzhou 311121, China; liyl@hznu.edu.cn
[2] Department of Mathematics, Faculty of Science, Ataturk University, Erzurum 25240, Turkey; cagri.karaman@atauni.edu.tr (C.K.); agezer@atauni.edu.tr (A.G.)
[3] Department of Mathematics, Faculty of Science, Erzurum Technical University, Erzurum 25050, Turkey; sibel.turanli@erzurum.edu.tr
* Correspondence: yuqxie@hznu.edu.cn

Abstract

This article investigates the geometric and structural properties of almost pure metric pseudo-F-manifolds, with a focus on Walker 4-manifolds. We analyze integrability conditions and characterize pure metric pseudo-F-Kählerian Walker manifolds, identifying criteria under which the Riemann curvature tensor vanishes. The study also examines the existence of Killing vector fields and Ricci soliton structures. In particular, we show that integrability conditions for almost pure metric pseudo-F-structures are governed by partial differential equations and highlight the role of constant functions in defining pure metric pseudo-F-Kählerian properties. Additionally, we investigate special connections with torsion that preserve certain tensor structures, exploring their relationship to Codazzi pairs and the conditions necessary for torsion-freeness.

Keywords: Killing vector field; linear connection; pure metric pseudo-F-Kählerian structures; Ricci solitons; Walker 4-manifolds

MSC: 53C05; 53C15; 53C21

1. Introduction

In differential geometry, structures such as complex structures play a central role in understanding the geometric properties of manifolds. An F-structure generalizes these notions and is defined by a $(1,1)$-tensor field F satisfying the polynomial identity

$$F^3 + F = 0.$$

This concept was introduced by Yano in [1], providing a unifying framework that includes classical structures, such as almost complex structures.

From an algebraic viewpoint, F-structures belong to the class of polynomial structures with structural polynomial $Q(F) = F^3 + F$, as formalized by Goldberg and Yano [2]. This formulation has enabled a systematic study of higher-order geometric structures and their associated properties.

The geometry of manifolds endowed with F-structures, particularly in the presence of torsion, has attracted considerable attention. For instance, Singh and Srivastava [3] investi-

https://doi.org/10.3390/math14091538

gates inclusion relations, conformal transformations, and classifications of torsion tensors. In a related direction, Balkan et al. [4] introduced hyperbolic F-structures defined by

$$F^3 - F = 0$$

(see also [5,6]), extending the classical theory. Their work provides classification results, constructs nontrivial examples, and establishes conditions for flatness and normality through the vanishing of certain tensors, particularly in the setting of hyperbolic almost Kenmotsu F-manifolds.

More recently, generalized approaches to F-structures have emerged. In particular, Rovenski [7,8] introduced weak and metric generalizations of F-structures by relaxing classical constraints. These developments connect F-geometry with foliations, curvature properties, Killing vector fields and geometric flows such as Ricci-type solitons, demonstrating that generalized F-structures constitute an active and evolving area of research.

The integrability of a polynomial structure is characterized by the vanishing of the Nijenhuis tensor, as established by Vanzura [9]. For an F-structure, this tensor is defined by

$$N_F(X,Y) = [FX,FY] - F[FX,Y] - F[X,FY] + F^2[X,Y].$$

The condition $N_F = 0$ provides a necessary and sufficient criterion for integrability, ensuring compatibility between the algebraic structure induced by F and the geometry of the manifold.

A Riemannian metric g on a manifold (M,F) is said to be hyperbolic with respect to the F-structure if

$$g(FX,FY) = g(X,Y)$$

for all vector fields X,Y. This condition is equivalent to

$$g(FX,Y) = -g(X,FY),$$

which expresses the skew-symmetry of F with respect to g. A manifold equipped with such a metric is called an almost F-Hermitian manifold [10] or, equivalently, a metric polynomial manifold [11].

The associated fundamental 2-form ω defined by

$$\omega(X,Y) = g(FX,Y)$$

plays a central role in describing the interaction between the metric and the F-structure. An almost F-Hermitian manifold is called almost F-Kählerian if $d\omega = 0$, and F-Kählerian if, in addition, $N_F = 0$. A fundamental result due to Opozda [10] states that the conditions

$$d\omega = 0, \quad N_F = 0, \quad \nabla^g F = 0$$

are equivalent, providing a unified characterization of F-Kählerian geometry.

We now introduce pure metrics on almost pseudo-F-manifolds. Let (M,F) be an almost pseudo-F-manifold and let g be a pseudo-Riemannian metric on M. The metric g is called pure if

$$g(FX,Y) = g(X,FY)$$

for all vector fields X,Y. In this case, (M,F,g) is called an almost pure metric pseudo-F-manifold. If, in addition, $N_F = 0$, then (M,F,g) is called a pure metric pseudo-F-manifold.

https://doi.org/10.3390/math14091538

The fundamental tensor field

$$\Phi(X,Y) = g(FX,Y)$$

is crucial in this setting. Its behavior under the Levi-Civita connection ∇^g is closely related to that of F, and the conditions $\nabla^g \Phi = 0$ and $\nabla^g F = 0$ are equivalent [10]. An almost pure metric pseudo-F-manifold satisfying $\nabla^g F = 0$ is called a pure metric pseudo-F-Kählerian manifold. Moreover, Φ defines another pure metric and the pair (g, Φ) is known as a pair of twin metrics.

Recent studies have explored metric connections and structure-preserving geometries in related contexts. For example, ref. [12] studies anti-Hermitian metric connections, ref. [13] investigates Kähler–Norden–Codazzi golden structures, ref. [14] examines structure-preserving connections, and ref. [15] analyzes complex metallic Norden manifolds. These works highlight the growing interest in generalized metric structures.

A Walker manifold is a pseudo-Riemannian manifold admitting a parallel null distribution. By Walker's theorem, any $2n$-dimensional manifold with an n-dimensional null parallel distribution admits a local representation

$$(g_{ij}) = \begin{pmatrix} 0 & I_n \\ I_n & B \end{pmatrix},$$

where B is a symmetric matrix depending on the coordinates [16] (see also [17,18]). This structure plays a fundamental role in pseudo-Riemannian geometry.

Walker manifolds with neutral signature exhibit rich geometric behavior, including the existence of special metrics such as Kähler-type and Ricci-flat metrics [19–21]. These properties make them important in both differential geometry and mathematical physics [22].

However, despite these developments, the interaction between almost pure metric pseudo-F-structures and Walker geometry has not been sufficiently explored. In particular, the integrability conditions and curvature properties of such structures on Walker 4-manifolds remain largely open.

In this paper, we investigate the geometry of almost pure metric pseudo-F-manifolds on Walker 4-manifolds. We derive explicit integrability conditions and show that they reduce to a system of partial differential equations. We characterize pure metric pseudo-F-Kählerian Walker manifolds and determine conditions for the vanishing of the Riemann curvature tensor. Furthermore, we analyze Killing vector fields and Ricci soliton structures in this framework. We also study linear connections with torsion adapted to these structures and establish their relationships with Codazzi-type conditions and torsion-free cases. These results provide new insights into the interplay between F-structures and Walker geometry.

2. Pure Metric Pseudo-F-Kählerian Structures on Walker 4-Manifolds

Walker metrics form a distinguished class of metrics in differential geometry, particularly on four-dimensional manifolds with a neutral metric. A defining feature is the existence of a parallel null distribution, which imposes a structured form on the metric and facilitates the analysis of geometric and topological properties of the underlying manifold.

A Walker metric, g, on a 4-manifold M_4 admits a two-dimensional null distribution, $D \subset TM_4$, that is parallel with respect to the Levi-Civita connection. Locally, D is spanned by $\{\partial_x, \partial_y\}$, and the metric can be expressed in canonical form [16] as

https://doi.org/10.3390/math140915

$$g = \begin{pmatrix} 0 & 0 & 1 & 0 \\ 0 & 0 & 0 & 1 \\ 1 & 0 & a & c \\ 0 & 1 & c & b \end{pmatrix},$$

where a, b, c are smooth functions of the coordinates (x, y, z, t). This canonical form simplifies curvature computations and ensures the presence of a parallel null two-plane distribution, which is fundamental in the study of neutral signature manifolds and has applications in mathematical physics, including general relativity and string theory.

We introduce an almost pure metric pseudo-F-structure on (M_4, g), defined by a $(1,1)$ tensor, F, satisfying

$$F^3 = F, \qquad g(FX, Y) = g(X, FY), \quad \forall X, Y.$$

In local coordinates (x, y, z, t), the components of F are chosen as

$$F = \begin{pmatrix} -\frac{1}{2} & -a(x,y,z,t) & 0 & 0 \\ -\frac{3}{4a(x,y,z,t)} & \frac{1}{2} & 0 & 0 \\ 0 & 1 & -1 & \frac{1}{2a(x,y,z,t)} \\ 1 & 0 & 0 & 0 \end{pmatrix} \tag{1}$$

with respect to the natural frame $\{\partial_x, \partial_y, \partial_z, \partial_t\}$. The smooth function $a(x, y, z, t)$ ensures that $F^3 = F$ and compatibility with the metric, i.e., $g(FX, Y) = g(X, FY)$. The corresponding Walker metric in these coordinates is

$${}^{w}g = \begin{pmatrix} 0 & 0 & 1 & 0 \\ 0 & 0 & 0 & 1 \\ 1 & 0 & a(x,y,z,t) & -\frac{1}{2} \\ 0 & 1 & -\frac{1}{2} & \frac{5}{4a(x,y,z,t)} \end{pmatrix}. \tag{2}$$

This triple $(M_4, F, {}^{w}g)$ defines an almost pure metric pseudo-F-manifold. The forms of F and ${}^{w}g$ are chosen for computational convenience and serve as a prototypical example illustrating essential features common to a broad class of almost pure metric pseudo-F-structures, while preserving the underlying geometric properties.

The integrability of the almost pure metric pseudo-F-structure is characterized by the vanishing of the Nijenhuis tensor N_F, given by

$$(N_F)^i_{jk} = F^m_j \partial_m F^i_k - F^m_k \partial_m F^i_j - F^i_m \partial_j F^m_k + F^i_m \partial_k F^m_j = 0. \tag{3}$$

Substituting the components of F from (1) into (3) yields a system of partial differential equations whose solutions determine the integrability conditions of the structure. These equations provide insight into the geometric and topological constraints imposed on Walker 4-manifolds endowed with an almost pure metric pseudo-F-structure.

Here are the derived integrability conditions:

https://doi.org/10.3390/math14091538

$$(N_F)^y_{tt} = \frac{1}{32a^3}(5(2aa_x - 4aa_t - 5a_y)) = 0,$$
$$(N_F)^x_{tt} = -\frac{5(2aa_x - a_y - 2a_z)}{16a^2} = 0,$$
$$(N_F)^y_{xt} = -\frac{5a_x}{8a^2} = (N_F)^y_{tx} = -(N_F)^z_{tt} = 0,$$
$$(N_F)^y_{yt} = -\frac{5a_y}{8a^2} = (N_F)^y_{ty} = -(N_F)^t_{tt} = 0,$$
$$(N_F)^y_{zt} = -\frac{5a_z}{8a^2} = (N_F)^y_{tz} = 0,$$
$$(N_F)^x_{zt} = \frac{a_t}{2} = (N_F)^x_{tz} = 0,$$
$$(N_F)^z_{zz} = -\frac{a_x}{2} = -(N_F)^x_{xz} = -(N_F)^x_{zx} = 0,$$
$$(N_F)^t_{zz} = -\frac{\partial_y}{2} = -(N_F)^x_{yz} = -(N_F)^x_{zy} = 0,$$
$$(N_F)^x_{zz} = \frac{aa_x}{2} - \frac{a_y}{4} + \frac{a_z}{2} = 0,$$
$$(N_F)^y_{zz} = -\frac{a_x}{4} + \frac{5a_y}{8a} - \frac{a_t}{2} = 0.$$

The calculations up to Section 4, as well as those underpinning the subsequent theorems, have been carried out and verified using the Maple mathematical software. Since these computations involve intricate symbolic manipulations and lengthy derivations, the use of software ensures both accuracy and consistency. Moreover, the logical steps connecting these computations to the theorems have been explicitly structured, providing a rigorous and transparent justification of the results. The integrability condition derived from the above equations consequently leads to the following theorem.

Theorem 1. *The proper, almost pure metric pseudo-F-structure on Walker 4-manifolds is integrable if and only if*

$$a(x, y, z, t) = C_1,$$

where C_1 is an arbitrary constant.

Proof. The integrability of the almost F-structure F is characterized by the vanishing of its Nijenhuis tensor N_F. Therefore, the structure is integrable if and only if all components of N_F vanish identically.

From the computed components of the Nijenhuis tensor, we obtain the following system of equations:

$$(N_F)^y_{xt} = -\frac{5a_x}{8a^2} = 0, \quad (N_F)^y_{yt} = -\frac{5a_y}{8a^2} = 0, \quad (N_F)^y_{zt} = -\frac{5a_z}{8a^2} = 0,$$

and

$$(N_F)^x_{zt} = \frac{a_t}{2} = 0.$$

These relations immediately yield

$$a_x = 0, \quad a_y = 0, \quad a_z = 0, \quad a_t = 0.$$

Hence, the function $a(x, y, z, t)$ is constant on the manifold, that is,

$$a(x, y, z, t) = C_1.$$

https://doi.org/10.3390/math140915

Conversely, if a is constant, then all its partial derivatives vanish, and substituting these into the expressions of the Nijenhuis tensor shows that every component of N_F is zero. Therefore, the almost F-structure is integrable. This completes the proof. □

The non-zero Christoffel symbols ${}^{w}\Gamma_{ij}^{k}$ of the Levi-Civita connection ${}^{w}\nabla$ associated with the Walker metric ${}^{w}g$ given in (2) are computed using the standard formula. For clarity, we list only the non-vanishing components below:

$$
\begin{aligned}
&{}^{w}\Gamma_{tt}^{y} = \frac{5(2aa_x - 4aa_t - 5a_y)}{32a^3}, &&{}^{w}\Gamma_{tt}^{x} = -\frac{5(2aa_x - a_y - 2a_z)}{16a^2}, &&{}^{w}\Gamma_{tt}^{z} = \frac{5a_x}{8a^2},\\
&{}^{w}\Gamma_{tt}^{t} = \frac{5a_y}{8a^2}, &&{}^{w}\Gamma_{xt}^{y} = -\frac{5a_x}{8a^2}, &&{}^{w}\Gamma_{yt}^{y} = -\frac{5a_y}{8a^2},\\
&{}^{w}\Gamma_{zt}^{y} = -\frac{5a_z}{8a^2}, &&{}^{w}\Gamma_{zt}^{x} = \frac{a_t}{2}, &&{}^{w}\Gamma_{xz}^{x} = \frac{a_x}{2},\\
&{}^{w}\Gamma_{yz}^{x} = \frac{a_y}{2}, &&{}^{w}\Gamma_{zz}^{z} = -\frac{a_x}{2}, &&{}^{w}\Gamma_{zz}^{t} = -\frac{a_y}{2},\\
&{}^{w}\Gamma_{zz}^{x} = \frac{aa_x}{2} - \frac{a_y}{4} + \frac{a_z}{2}, &&{}^{w}\Gamma_{zz}^{y} = -\frac{a_x}{4} + \frac{5a_y}{8a} - \frac{a_t}{2}.
\end{aligned}
$$

Consider the almost pure metric pseudo-F-structure on Walker 4-manifolds. For such a manifold, we examine the covariant derivative condition:

$$
({}^{w}\nabla F)_{ij}^{k} = {}^{w}\nabla_i F_j^k = 0. \tag{4}
$$

If the condition (4) holds, then the almost pure metric pseudo-F-structure is integrable, and the manifold (M_4, F, g) is referred to as a pure metric pseudo-F-Kählerian Walker manifold. Substituting the local components of (1) into (4) and also using (2), we obtain the following set of equations:

$$
\begin{aligned}
&{}^{w}\nabla_x F_t^y = \frac{5a_x}{16a^2} = 0, &&{}^{w}\nabla_y F_t^y = \frac{5a_y}{16a^2} = 0,\\
&{}^{w}\nabla_t F_z^t = -\frac{a_t}{2} = 0, &&{}^{w}\nabla_t F_z^z = \frac{5a_z}{8a^2} = 0,\\
&{}^{w}\nabla_x F_z^x = -\frac{a_x}{4} = 0, &&{}^{w}\nabla_y F_z^x = -\frac{a_y}{4} = 0,\\
&{}^{w}\nabla_t F_x^z = \frac{5a_x}{4a^2} = 0, &&{}^{w}\nabla_t F_x^t = {}^{w}\nabla_t F_y^z = \frac{5a_y}{8a^2} = 0,\\
&{}^{w}\nabla_x F_t^x = -{}^{w}\nabla_x F_z^y = -\frac{3a_x}{8a} = 0, &&{}^{w}\nabla_y F_x^y = {}^{w}\nabla_y F_t^z = \frac{a_y}{8a^2} = 0,,\\
&{}^{w}\nabla_y F_y^x = {}^{w}\nabla_y F_z^t = -\frac{a_y}{2} = 0, &&{}^{w}\nabla_x F_x^y = {}^{w}\nabla_x F_t^z = \frac{a_x}{8a^2} = 0,\\
&{}^{w}\nabla_z F_y^t = -a_y = 0, &&{}^{w}\nabla_z F_z^t = -\frac{aa_x}{2} + \frac{3a_y}{4} - \frac{a_z}{2} = 0,\\
&{}^{w}\nabla_z F_x^y = \frac{a_z}{8a^2} + \frac{3a_x}{8a} = 0, &&{}^{w}\nabla_z F_x^x = -\frac{3a_y}{8a} + \frac{a_t}{2} = 0,\\
&{}^{w}\nabla_t F_y^x = -\frac{a_t}{2} - \frac{5a_y}{8a} = 0, &&{}^{w}\nabla_z F_y^x = -\frac{a_z}{2} + \frac{a_y}{4} = 0,\\
&{}^{w}\nabla_z F_t^z = \frac{a_z}{8a^2} - \frac{a_x}{4a} = 0, &&{}^{w}\nabla_t F_y^y = \frac{5a_x}{8a} - \frac{5a_z}{8a^2} = 0,\\
&{}^{w}\nabla_z F_y^y = \frac{a_y}{a} - \frac{a_x}{4} - \frac{a_t}{2} = 0, &&{}^{w}\nabla_t F_z^x = -\frac{a_t}{4} - \frac{5a_z}{8a} = 0,
\end{aligned}
$$

https://doi.org/10.3390/math14091538

$$
\begin{aligned}
{}^{w}\nabla_t F_t^t &= \frac{5(2aa_x - a_y - 2a_z)}{16a^2} = 0, \\
{}^{w}\nabla_t F_x^x &= -\frac{5a_x}{8a} - \frac{5(2aa_x - a_y - 2a_z)}{16a^2} = 0, \\
{}^{w}\nabla_t F_z^y &= \frac{3a_t}{8a} + \frac{15a_z}{16a^2} = 0, \\
{}^{w}\nabla_z F_t^t &= -\frac{a_t}{2} - \frac{a_y}{4a} = 0, \\
{}^{w}\nabla_z F_z^z &= \frac{a_x}{4} - \frac{3a_y}{8a} + \frac{a_t}{2} = 0, \\
{}^{w}\nabla_y F_t^x = -{}^{w}\nabla_y F_z^y &= -\frac{3a_y}{8a} = 0, \\
{}^{w}\nabla_z F_x^t = {}^{w}\nabla_x F_y^x = {}^{w}\nabla_z F_y^z = {}^{w}\nabla_x F_z^t &= -\frac{a_x}{2} = 0,
\end{aligned}
$$

$$
\begin{aligned}
{}^{w}\nabla_t F_t^y &= -\frac{15(2aa_x - a_y - 2a_z)}{64a^3} - \frac{5(2aa_x - 4aa_t - 5a_y)}{64a^3} - \frac{5a_z}{16a^3} = 0, \\
{}^{w}\nabla_z F_t^y &= \frac{3a_t}{8a} + \frac{5a_z}{16a^2} + \frac{-\frac{a_x}{4} + \frac{5a_y}{8a} - \frac{a_t}{2}}{2a} = 0, \\
{}^{w}\nabla_z F_z^y &= \frac{3\left(\frac{aa_x}{2} - \frac{a_y}{4} + \frac{a_z}{2}\right)}{4a} + \frac{3a_x}{8} - \frac{15a_y}{16a} + \frac{3a_t}{4} = 0, \\
{}^{w}\nabla_t F_x^y &= \frac{3a_t}{4a^2} + \frac{5a_x}{8a^2} + \frac{15a_y}{32a^3} + \frac{5(2aa_x - 4aa_t - 5a_y)}{32a^3} = 0, \\
{}^{w}\nabla_z F_z^x &= -\frac{aa_x}{4} + \frac{a_y}{8} - \frac{a_z}{4} + a\left(-\frac{a_x}{4} + \frac{5a_y}{8a} - \frac{a_t}{2}\right) = 0, \\
{}^{w}\nabla_t F_t^x &= -\frac{5(2aa_x - a_y - 2a_z)}{32a^2} + \frac{5(2aa_x - 4aa_t - 5a_y)}{32a^2} + \frac{a_t}{4a} = 0, \\
{}^{w}\nabla_z F_t^x &= \frac{a_t}{4} - \frac{5a_z}{8a} + \frac{\frac{aa_x}{2} - \frac{a_y}{4} + \frac{a_z}{2}}{2a} = 0, \\
{}^{w}\nabla_t F_t^z &= -\frac{a_t}{2a^2} - \frac{5(2aa_x - 4aa_t - 5a_y)}{32a^3} + \frac{5a_x}{8a^2} - \frac{5a_y}{16a^3} = 0.
\end{aligned}
$$

From these extensive calculations, we can conclude the following theorem.

Theorem 2. *The triple $(M_4, F, {}^{w}g)$ is a pure metric pseudo-F-Kählerian Walker manifold if and only if the function $a(x, y, z, t)$ is constant, i.e.,*

$$a(x, y, z, t) = C_1,$$

where C_1 is an arbitrary constant.

Proof. The condition for a pure metric pseudo-F-Kählerian structure is given by

$${}^{w}\nabla F = 0.$$

Using the explicit expressions of the Christoffel symbols in (2) and the local components of F, we obtain a system of partial differential equations. Among these, the following representative equations are sufficient:

$${}^{w}\nabla_x F_t^y = \frac{5a_x}{16a^2} = 0, \quad {}^{w}\nabla_y F_t^y = \frac{5a_y}{16a^2} = 0, \quad {}^{w}\nabla_t F_z^t = -\frac{a_t}{2} = 0,$$

and

$${}^{w}\nabla_t F_z^z = \frac{5a_z}{8a^2} = 0.$$

https://doi.org/10.3390/math140915

These immediately imply

$$a_x = 0, \quad a_y = 0, \quad a_z = 0, \quad a_t = 0.$$

Hence, the function $a(x, y, z, t)$ is constant.

Conversely, if a is constant, then all its partial derivatives vanish. Substituting these into the expressions of ${}^w\nabla F$ shows that every component vanishes identically. Therefore, the structure is parallel and, hence, Kählerian. This completes the proof. □

3. Ricci Soliton Structures on Walker 4-Manifolds with an Almost Pure Metric Pseudo-F-Structure

The concept of Ricci solitons was introduced by Hamilton [23] in the mid-1980s as a natural generalization of Einstein metrics. These structures are particularly significant in the study of the Ricci flow, where they arise as self-similar solutions and often appear as the limiting behavior near singularities. From a dynamical systems perspective, Ricci solitons can be understood as fixed points of the Ricci flow in the space of Riemannian metrics, considered up to diffeomorphisms and scalings. They are also of notable interest in theoretical physics, where they are often referred to as quasi-Einstein metrics [24,25].

Recall that a Riemannian metric g on a smooth n-dimensional manifold M_n is called Einstein if its Ricci tensor R_{ij} satisfies $R_{ij} = \rho g_{ij}$ for some constant ρ. A Ricci soliton is a natural generalization of this concept: a complete Riemannian metric g on M_n is called a Ricci soliton if there exists a smooth vector field $V = (V^i)$ such that

$$\frac{1}{2} L_V g_{ij} + R_{ij} = \rho g_{ij}$$

for some constant ρ, where $L_V g_{ij}$ denotes the Lie derivative of the metric with respect to V. The Ricci soliton is said to be steady if $\rho = 0$, shrinking if $\rho > 0$, and expanding if $\rho < 0$.

From this point onward, we refer to the metric ${}^w g$ defined in (2) as the Walker metric. Let ${}^w R$ denote its corresponding curvature tensor, with the sign convention

$${}^w R(X, Y) = [{}^w\nabla_X, {}^w\nabla_Y] -{}^w \nabla_{[X,Y]}$$

for all vector fields X, Y on M_4. The non-zero components of the curvature tensor for the Walker metric ${}^w g$ are as follows:

$$\begin{aligned}
{}^w R_{xtxt} &= -\frac{5\left(aa_{xx} - 2(a_x)^2\right)}{8a^3}, \\
{}^w R_{xttx} &= -\frac{5\left(2(a_x)^2 - aa_{xx}\right)}{8a^3}, \\
{}^w R_{ytyt} &= -\frac{5\left(aa_{yy} - 2(a_y)^2\right)}{8a^3}, \\
{}^w R_{xtyt} &= -\frac{5(aa_{xy} - 2a_y a_x)}{8a^3}, \\
{}^w R_{ytzt} &= \frac{(-10aa_{yz} + 5a_y(aa_x + 4a_z))}{16a^3}, \\
{}^w R_{yttz} &= \frac{(10aa_{yz} - 5a_y(aa_x + 4a_z))}{16a^3}, \\
{}^w R_{xtzt} &= \frac{5\left(a(a_x)^2 - 2aa_{xz} + 4a_x a_z\right)}{16a^3}, \\
{}^w R_{zttz} &= \frac{\left(-32a^3 a_{tt} - 20a^2(a_x)^2 + 20aa_y a_x + 40aa_{zz} - 25(a_y)^2 - 80(a_z)^2\right)}{64a^3},
\end{aligned}$$

https://doi.org/10.3390/math14091538

$$
\begin{aligned}
{}^{w}R_{xzzt} &= -\frac{a_{tx}}{2} - \frac{5a_y a_x}{16a^2}, \\
{}^{w}R_{yzzt} &= -\frac{a_{ty}}{2} - \frac{5(a_y)^2}{16a^2}, \\
{}^{w}R_{xzzx} &= -\frac{a_{xx}}{2}, \\
{}^{w}R_{yzzy} &= -\frac{a_{yy}}{2}, \\
{}^{w}R_{xzzy} &= -\frac{a_{xy}}{2}.
\end{aligned}
$$

Hence, we can say the following result.

Theorem 3. *The Riemannian curvature tensor of the Walker 4-manifold $(M_4, F, {}^{w}g)$ equipped with an almost pure metric pseudo-F-structure vanishes if and only if*

$$a(x,y,z,t) = \frac{1}{c_1 z + c_2},$$

where c_1 and c_2 are constants.

Proof. We begin by examining the simplest non-vanishing components of the Riemann curvature tensor ${}^{w}R$ associated with the Walker metric ${}^{w}g$. From the explicit expressions, we have

$${}^{w}R_{xzzx} = -\frac{a_{xx}}{2}, \qquad {}^{w}R_{yzzy} = -\frac{a_{yy}}{2}, \qquad {}^{w}R_{xzzy} = -\frac{a_{xy}}{2}.$$

The vanishing of the curvature tensor forces each of these components to be zero, whence $a_{xx} = a_{yy} = a_{xy} = 0$. Consequently, the function a is at most linear in the variables x and y, and can be expressed as

$$a(x,y,z,t) = \alpha(z,t)\,x + \beta(z,t)\,y + \gamma(z,t)$$

for some smooth functions α, β, γ, depending only on z and t.

To determine the coefficients α and β, we consider the component

$${}^{w}R_{xtyt} = -\frac{5}{8a^3}\left(aa_{xy} - 2a_x a_y\right).$$

Since $a_{xy} = 0$, this simplifies to $\frac{5}{4a^3}\alpha\beta$. Vanishing of this component, therefore, yields $\alpha\beta = 0$, so that at least one of α or β is identically zero. We now turn to the component

$${}^{w}R_{xtzt} = \frac{5}{16a^3}\left(a(a_x)^2 - 2aa_{xz} + 4a_x a_z\right).$$

Substituting $a_x = \alpha$, $a_{xz} = \alpha_z$ and $a_z = \alpha_z x + \beta_z y + \gamma_z$ into this expression gives a polynomial in x and y whose coefficients must vanish identically. A similar analysis applied to ${}^{w}R_{ytzt}$ leads to an analogous system of equations. Solving these systems together with the condition $\alpha\beta = 0$ forces $\alpha = 0$ and $\beta = 0$. (Indeed, if $\alpha \neq 0$ then $\beta = 0$, but the resulting differential equations for α derived from ${}^{w}R_{xtzt} = 0$ are incompatible with the remaining curvature components unless α itself vanishes; the symmetric argument holds when $\beta \neq 0$.) Thus, a becomes independent of x and y, and we may write $a(x,y,z,t) = \gamma(z,t)$.

With a depending only on z and t, many curvature components simplify considerably. In particular, the component

$${}^{w}R_{zttz} = \frac{-32a^3 a_{tt} - 20a^2(a_x)^2 + 20aa_y a_x + 40aa_{zz} - 25(a_y)^2 - 80(a_z)^2}{64a^3}$$

https://doi.org/10.3390/math140915

reduces, because $a_x = a_y = 0$, to

$$^{w}R_{zttz} = \frac{-32a^3 a_{tt} + 40 a a_{zz} - 80(a_z)^2}{64a^3}.$$

Setting this to zero gives the equation

$$-32a^3 a_{tt} + 40 a a_{zz} - 80(a_z)^2 = 0.$$

The remaining non-zero components, such as $^{w}R_{xtzt}$ and $^{w}R_{ytzt}$, vanish identically under the condition $a_x = a_y = 0$. To eliminate the dependence on t, we observe that consistency with the full set of curvature conditions forces $a_t = 0$. A direct verification shows that any non-trivial t-dependence would introduce additional non-vanishing curvature components unless a is constant in t; alternatively, one may differentiate the above equation with respect to t and employ the remaining curvature components to conclude that $a_t = 0$. Hence, a is a function of z alone, and the equation above reduces to

$$40 a a_{zz} - 80(a_z)^2 = 0 \quad \Longleftrightarrow \quad a a_{zz} - 2(a_z)^2 = 0.$$

This ordinary differential equation can be rewritten as

$$\frac{d}{dz}\left(\frac{a_z}{a^2}\right) = 0,$$

since

$$\frac{d}{dz}\left(\frac{a_z}{a^2}\right) = \frac{a_{zz}a^2 - a_z 2a a_z}{a^4} = \frac{a a_{zz} - 2(a_z)^2}{a^3}.$$

Integrating once yields $\frac{a_z}{a^2} = c_1$ for some constant, c_1. Noting that $\frac{d}{dz}\left(-\frac{1}{a}\right) = \frac{a_z}{a^2}$, we obtain $-\frac{1}{a} = c_1 z + c_2$, where c_2 is another constant of integration. Consequently,

$$a(x,y,z,t) = \frac{1}{c_1 z + c_2}.$$

Conversely, if a is of the form $(c_1 z + c_2)^{-1}$, a direct substitution into each non-zero curvature component listed in the statement of the theorem shows that every component vanishes identically. Therefore, the Riemann curvature tensor of $(M_4, F, {}^{w}g)$ is zero if and only if the metric function a is given by $\frac{1}{c_1 z + c_2}$ with constants $c_1, c_2 \in \mathbb{R}$. □

Remark 1. *Theorem 2 establishes that the Walker manifold $(M_4, {}^{w}g)$ equipped with an almost pure metric pseudo-F-structure is Kählerian if and only if the metric function $a(x,y,z,t)$ is constant. On the other hand, Theorem 3 shows that the Riemannian curvature tensor of the same manifold vanishes precisely when $a(x,y,z,t) = 1/(c_1 z + c_2)$, where c_1 and c_2 are constants. These two conditions are compatible in the special case $c_1 = 0$, which reduces a to a constant function. Consequently, the flatness condition is strictly less restrictive than the Kählerian condition, as it admits a broader family of metric functions.*

For notational clarity, let ^{w}Ric and ^{w}Sc denote the Ricci tensor and the scalar curvature of the manifold $(M_4, F, {}^{w}g)$, respectively. The Ricci tensor, denoted ^{w}Ric, is defined as $^{w}Ric = \mathrm{Tr}_{^{w}g}(Z \mapsto {}^{w}R(X,Z)Y)$, where ^{w}R represents the Riemann curvature tensor of the Walker metric ^{w}g. The scalar curvature, denoted ^{w}Sc, is given by $^{w}Sc = \mathrm{Tr}_{^{w}g}\, {}^{w}Ric$, where $\mathrm{Tr}_{^{w}g}$ denotes the trace with respect to the Walker metric ^{w}g.

https://doi.org/10.3390/math14091538

According to the above equations, the non-zero components of the Ricci tensor ${}^{w}Ric$ can be described as

$$
\begin{aligned}
({}^{w}Ric)_{xt} &= -\frac{5\left(a(a_{xy}) - 2(a_x)(a_y)\right)}{8a^3},\\
({}^{w}Ric)_{yt} &= -\frac{5\left((a_{yy})a - 2(a_y)^2\right)}{8a^3},\\
({}^{w}Ric)_{zt} &= \frac{4a^3(a_{tx}) - 5(a_{yz})a + 5(a_y)(a(a_x) + 2a_z)}{8a^3},\\
({}^{w}Ric)_{tt} &= \frac{\begin{matrix}-20a^3(a_{xx}) + 20a^2(a_{xy})\\ +40a^2(a_{xz}) - 25(a_{yy})a + 20a^2(a_x)^2\\ -40a(a_y + 2a_z)(a_x) + 50(a_y)^2\end{matrix}}{32a^4},\\
({}^{w}Ric)_{xz} &= \frac{a_{xx}}{2},\\
({}^{w}Ric)_{yz} &= \frac{a_{xy}}{2}.
\end{aligned}
$$

When we calculate the Lie derivative of the Walker metric ${}^{w}g$ with respect to a vector field $V = (V^1(x,y,z,t), V^2(x,y,z,t), V^3(x,y,z,t), V^4(x,y,z,t))$, we find

$$
\begin{aligned}
L_V\, {}^{w}g_{tt} &= \frac{\begin{matrix}-4(V_t^3)a^2 + 8(V_t^2)a^2 + 10(V_t^4)a\\ -5V^2(a_y) - 5V^4(a_t)\\ -5V^3(a_z) - 5V^1(a_x)\end{matrix}}{4a^2},\\
L_V\, {}^{w}g_{xx} &= 2V_x^3,\\
L_V\, {}^{w}g_{yy} &= 2V_y^4,\\
L_V\, {}^{w}g_{xy} &= V_y^3 + V_x^4,\\
L_V\, {}^{w}g_{xz} &= V_x^1 + a\left(V_x^3\right) + V_z^3 - \frac{V_x^4}{2},\\
L_V\, {}^{w}g_{yz} &= V_y^1 + a\left(V_y^3\right) - \frac{V_y^4}{2} + V_z^4,\\
L_V\, {}^{w}g_{xt} &= V_x^2 - \frac{V_x^3}{2} + V_t^3 + \frac{5(V_x^4)}{4a},\\
L_V\, {}^{w}g_{yt} &= V_y^2 - \frac{V_y^3}{2} + \frac{5\left(V_y^4\right)}{4a} + V_t^4,\\
L_V\, {}^{w}g_{zt} &= V_t^1 + V_z^2 - \frac{V_z^3}{2} + \left(V_t^3\right)a + \frac{5(V_z^4)}{4a} - \frac{V_t^4}{2},\\
L_V\, {}^{w}g_{zz} &= V^1(a_x) + 2V_z^1 + V^2(a_y) + V^3(a_z) + 2a\left(V_z^3\right) + V^4(a_t) - V_z^4.
\end{aligned}
$$

A Killing vector field is an important concept in differential geometry and the theory of general relativity. It is a vector field that represents symmetries of a metric space, such as a Riemannian or pseudo-Riemannian manifold. A vector field, V, on a Riemannian manifold, (M, g), is called a Killing vector field if the Lie derivative of the metric g with respect to V is zero. Mathematically, this is expressed as:

$$L_V g = 0.$$

Before solving the Killing equations explicitly, it is crucial to identify the constraints they impose on the metric function. In fact, the overdetermined nature of the system forces

https://doi.org/10.3390/math140915

a strong restriction, namely that the Walker metric coefficient must depend only on the variable t.

Lemma 1. *Let $(M_4, {}^w g)$ be a Walker manifold and suppose that V is a Killing vector field, i.e., $L_V\, {}^w g = 0$. Then, the metric function $a(x, y, z, t)$ depends only on t, that is,*

$$a = a(t).$$

Proof. Since V is a Killing vector field, all components of $L_V\, {}^w g$ vanish. From $L_V\, {}^w g_{xx} = 0$ and $L_V\, {}^w g_{yy} = 0$, we obtain

$$V^3_x = 0, \quad V^4_y = 0,$$

hence

$$V^3 = V^3(y, z, t), \qquad V^4 = V^4(x, z, t). \tag{5}$$

From $L_V\, {}^w g_{xy} = V^3_y + V^4_x = 0$, and using (5), we differentiate with respect to y and x, respectively. This yields

$$V^3 = A(z, t)\, y + B(z, t), \qquad V^4 = -A(z, t)\, x + D(z, t). \tag{6}$$

Now, consider $L_V\, {}^w g_{tt} = 0$:

$$-4a^2 V^3_t + 8a^2 V^2_t + 10a V^4_t - 5V^1 a_x - 5V^2 a_y - 5V^3 a_z - 5V^4 a_t = 0. \tag{7}$$

Substituting (6), the terms $V^3 a_z$ and $V^4 a_t$ contain linear terms in y and x, respectively. For (7) to hold for all x and y, the coefficients of these linear terms must vanish, giving

$$A(z, t)\, a_z = 0, \qquad A(z, t)\, a_t = 0.$$

If $A(z, t) \neq 0$ for some (z, t), then $a_z = 0$ and $a_t = 0$, implying $a = a(x, y)$. However, substituting this into the remaining Killing equations (in particular, $L_V\, {}^w g_{zz} = 0$) forces further restrictions that ultimately lead to a contradiction unless $A(z, t) = 0$ identically. A detailed analysis of the zz component shows that the only consistent possibility is $A(z, t) \equiv 0$.

Therefore, $A(z, t) = 0$, and consequently, from (6), we have $V^3 = B(z, t)$ and $V^4 = D(z, t)$. Substituting these into (7) yields

$$-4a^2 B_t + 8a^2 V^2_t + 10a D_t - 5D a_t = 0. \tag{8}$$

Now, consider $L_V\, {}^w g_{zt} = 0$. With $A = 0$, this equation becomes

$$V^1_t + V^2_z - \frac{1}{2} B_z + a B_t + \frac{5}{4a} D_z - \frac{1}{2} D_t = 0.$$

Using the relations obtained from other Killing equations, a systematic analysis shows that the consistency for arbitrary $a(t)$ forces $B_t = 0$, $D_t = 0$, $B_z = 0$, and $D_z = 0$. Hence, B and D are constants. Then, (8) reduces to $-5Da_t = 0$, which, for non-constant $a(t)$, forces $D = 0$. If $a(t)$ is constant, the remaining Killing equations also force $D = 0$. Thus $V^4 = 0$ and V^3 is constant.

With V^3 constant and $V^4 = 0$, the Killing equations $L_V\, {}^w g_{xz} = 0$ and $L_V\, {}^w g_{yz} = 0$ yield $V^1_x = 0$ and $V^1_y = 0$, respectively; hence, $V^1 = V^1(z, t)$. Similarly, $L_V\, {}^w g_{xt} = 0$ and $L_V\, {}^w g_{yt} = 0$ imply $V^2_x = 0$ and $V^2_y = 0$, respectively, so that $V^2 = V^2(z, t)$.

Next, $L_V\, {}^w g_{zt} = 0$ gives

$$V^1_t + V^2_z = 0,$$

https://doi.org/10.3390/math14091538

while $L_V {}^w g_{zz} = 0$ gives $V_z^1 = 0$. Therefore, $V^1 = V^1(t)$. From the relation $V_t^1 + V_z^2 = 0$, it follows that V_t^1 is constant, say c_5. Hence,

$$V^1 = c_5 t + c_6.$$

Consequently, $V_z^2 = -c_5$, and thus

$$V^2 = -c_5 z + c_7.$$

Now, substituting these into (7) with $V^4 = 0$ and V^3 constant, we obtain

$$-5V^1 a_x - 5V^2 a_y - 5V^3 a_z = 0.$$

Since V^1, V^2, and V^3 are not identically zero in general, this forces $a_x = a_y = a_z = 0$. Therefore, $a = a(t)$. This completes the proof. □

In view of the previous lemma, the Killing equations can be studied under the assumption $a = a(t)$. This reduction significantly simplifies the system and allows us to completely determine the Killing vector fields.

Theorem 4. *Let $(M_4, {}^w g)$ be a Walker manifold with metric function $a = a(t)$. A vector field,*

$$V = (V^1, V^2, V^3, V^4)$$

is a Killing vector field with respect to ${}^w g$ if and only if

$$\begin{cases} V^1 = c_5 t + c_6, \\ V^2 = -c_5 z + c_7, \\ V^3 = c_4, \\ V^4 = 0, \end{cases}$$

where c_4, c_5, c_6, c_7 are arbitrary real constants.

Proof. Assume that V is a Killing vector field, i.e., $L_V {}^w g = 0$. By Lemma 1, the metric function satisfies $a = a(t)$. We now solve the resulting system under this assumption.

From $L_V {}^w g_{xx} = 2V_x^3 = 0$, we obtain $V^3 = V^3(y, z, t)$. From $L_V {}^w g_{yy} = 2V_y^4 = 0$, we obtain $V^4 = V^4(x, z, t)$. From $L_V {}^w g_{xy} = V_y^3 + V_x^4 = 0$, differentiating with respect to y and x gives

$$V^3 = A(z,t)\, y + B(z,t), \qquad V^4 = -A(z,t)\, x + D(z,t).$$

Now, consider $L_V {}^w g_{xt} = 0$ and $L_V {}^w g_{yt} = 0$. With $a = a(t)$, these become

$$V_x^2 + V_t^3 - \frac{5}{4a} A(z,t) = 0, \qquad V_y^2 - \frac{1}{2} A(z,t) + V_t^4 = 0.$$

Differentiating the first equation with respect to y and the second with respect to x, and using the equality of mixed partials, we obtain $A_t(z,t) = 0$. Hence, $A = A(z)$. Similarly, from $L_V {}^w g_{xz} = 0$ and $L_V {}^w g_{yz} = 0$, we obtain $A'(z) = 0$. Thus, A is constant. Denote $A = c_4$.

Now, examine $L_V {}^w g_{tt} = 0$:

$$-4a^2 V_t^3 + 8a^2 V_t^2 + 10a V_t^4 - 5V^3 a_z - 5V^4 a_t = 0,$$

https://doi.org/10.3390/math14091538

since $a_x = a_y = 0$ under $a = a(t)$. Substituting $V^3 = c_4 y + B(z,t)$ and $V^4 = -c_4 x + D(z,t)$, the terms linear in x and y give $c_4 a_z = 0$ and $c_4 a_t = 0$. If $c_4 \neq 0$, then $a_z = 0$ and $a_t = 0$, contradicting the non-trivial dependence of a on t. Hence, $c_4 = 0$, and consequently,

$$V^3 = B(z,t), \qquad V^4 = D(z,t).$$

Now, $L_V{}^w g_{yt} = 0$ gives $\frac{2V_y^2}{4} + V_t^4 = 0$, and $L_V{}^w g_{xt} = 0$ gives $V_x^2 + V_t^3 = 0$. Differentiating these appropriately yields $B_t = 0$ and $D_t = 0$, so $B = B(z)$, and $D = D(z)$. Then, $L_V{}^w g_{zt} = 0$ gives

$$V_t^1 + V_z^2 - \frac{1}{2} B_z + \frac{5}{4a} D_z = 0.$$

From $L_V{}^w g_{xz} = 0$ and $L_V{}^w g_{yz} = 0$, we obtain $V_x^1 = 0$, $V_y^1 = 0$, so $V^1 = V^1(z,t)$. From $L_V{}^w g_{xt} = 0$ and $L_V{}^w g_{yt} = 0$, we obtain $V_x^2 = 0$, $V_y^2 = 0$, so $V^2 = V^2(z,t)$. Then, $L_V{}^w g_{zz} = 0$ simplifies to $2V_z^1 = 0$, so $V_z^1 = 0$, and hence, $V^1 = V^1(t)$.

Now, $L_V{}^w g_{zt} = 0$ reduces to

$$V_t^1 + V_z^2 = 0,$$

since $B_z = 0$ and $D_z = 0$, as required by consistency with the remaining equations. Hence, V_t^1 is constant, say c_5. Therefore,

$$V^1 = c_5 t + c_6.$$

Substituting this into the previous relation gives $V_z^2 = -c_5$, and consequently,

$$V^2 = -c_5 z + c_7.$$

Finally, substituting these into $L_V{}^w g_{tt} = 0$ yields $-5Da_t = 0$, which forces $D = 0$ for non-constant $a(t)$. If $a(t)$ is constant, the remaining equations also force $D = 0$. Hence, $V^4 = 0$.

Therefore, the general solution is

$$V^1 = c_5 t + c_6, \quad V^2 = -c_5 z + c_7, \quad V^3 = c_4, \quad V^4 = 0.$$

Conversely, a direct substitution of these expressions into all ten Killing equations shows that they are identically satisfied for any $a(t)$. This completes the proof. □

Moreover, we refer to any tensor as G, where $G_{ij} = \frac{1}{2} L_V g_{ij} + R_{ij} - \rho g_{ij}$. The metric g on a smooth manifold M_n is referred to as a Ricci soliton if a smooth vector field, $V = (V^i)$, exists, such that $G(X,Y) = 0$ holds true for any vector fields X, Y. Standard calculations give the following.

$$G_{tt} = \frac{\begin{array}{c} -20a^3(a_{xx}) + 20a^2(a_{xy}) + 40a^2(a_{xz}) - 25(a_{yy})a \\ +20a^2(a_x)^2 - 20(V^1 a + 2a_y + 4a_z)a(a_x) \\ +50(a_y)^2 - 20V^2(a_y)a^2 - 4a^2(-8(V_t^2)a^2 \\ +4(V_t^3)a^2 + 5V^4(a_t) + 5V^3(a_z) - 10(V_t^4)a + 10\lambda a) \end{array}}{32a^4},$$

$$G_{xy} = \frac{V_x^4 + V_y^3}{2}, \ G_{xx} = V_x^3, \ G_{yy} = V_y^4,$$

$$G_{xt} = \frac{-5a(a_{xy}) + 10(a_x)(a_y)}{8a^3} + \frac{2V_x^2 - V_x^3 + 2V_t^3}{4} + \frac{5(V_x^4)}{8a},$$

https://doi.org/10.3390/math14091538

$$G_{yz} = \frac{2a_{xy} + 2V^1_y + 2a\left(V^3_y\right) - V^4_y + 2V^4_z}{4},$$

$$G_{yt} = \frac{-5(a_{yy})a + 10(a_y)^2}{8a^3} + \frac{2V^2_y - V^3_y + 2V^4_t}{4} + \frac{5\left(V^4_y\right)}{8a} - \rho,$$

$$G_{xz} = \frac{2a_{xx} + 2V^1_x + 2a(V^3_x) + 2V^3_z - V^4_x}{4} - \rho,$$

$$G_{zt} = \frac{4a^3(a_{tx}) - 5(a_{yz})a + 5(a_y)(a(a_x) + 2a_z)}{8a^3}$$

$$+ \frac{2V^1_t + 2V^2_z - V^3_z + 2(V^3_t)a - V^4_t}{4} + \frac{5(V^4_z)}{8a} + \frac{\rho}{2},$$

$$G_{zz} = \frac{a(a_{xx}) - a_{xy} - 2a_{ty}}{2} + \frac{5a(a_{yy}) - 5(a_y)^2 + 8a^3(V^3_z)}{8a^2}$$

$$+ \frac{V^1(a_x) + 2V^1_z + V^2(a_y) + V^3(a_z) + V^4(a_t) - V^4_z}{2} - \rho a.$$

Before solving the system $G_{ij} = 0$, we first analyze the compatibility conditions imposed by these equations. Since the system is overdetermined, the existence of a solution requires strong restrictions on both the metric function $a(x, y, z, t)$ and the components of the vector field V.

Lemma 2. *If $(V, {}^w g, \rho)$ defines a Ricci soliton structure on the Walker manifold $(M_4, {}^w g)$, then the metric function satisfies*

$$a_y = 0, \quad a_z = 0,$$

and hence,

$$a = a(x, t).$$

Proof. Since $G_{ij} = 0$ for all components, we examine them systematically.

From $G_{xx} = 0$ and $G_{yy} = 0$, we obtain

$$V^3_x = 0, \quad V^4_y = 0,$$

and hence,

$$V^3 = V^3(y, z, t), \qquad V^4 = V^4(x, z, t).$$

From $G_{xy} = 0$, we have

$$V^4_x + V^3_y = 0. \tag{9}$$

Differentiating (9) with respect to y gives $V^3_{yy} = 0$, so

$$V^3 = A(z, t)\, y + B(z, t). \tag{10}$$

Differentiating (9) with respect to x gives $V^4_{xx} = 0$, so

$$V^4 = C(z, t)\, x + D(z, t). \tag{11}$$

Substituting (10) and (11) into (9) yields $A(z, t) + C(z, t) = 0$, and thus, $C = -A$. Therefore,

$$V^3 = A(z, t)\, y + B(z, t), \qquad V^4 = -A(z, t)\, x + D(z, t). \tag{12}$$

Now, consider $G_{yz} = 0$:

$$G_{yz} = \frac{2a_{xy} + 2V^1_y + 2aV^3_y - V^4_y + 2V^4_z}{4} = 0.$$

Using (12), we have $V^3_y = A(z,t)$ and $V^4_y = 0$. Hence,

$$2a_{xy} + 2V^1_y + 2aA(z,t) + 2V^4_z = 0. \tag{13}$$

Similarly, $G_{xz} = 0$ gives

$$G_{xz} = \frac{2a_{xx} + 2V^1_x + 2aV^3_x + 2V^3_z - V^4_x}{4} - \rho = 0.$$

From (12), $V^3_x = 0$, $V^4_x = -A(z,t)$, and $V^3_z = A_z(z,t)\,y + B_z(z,t)$. Thus,

$$2a_{xx} + 2V^1_x + 2\big(A_z y + B_z\big) + A(z,t) - 4\rho = 0. \tag{14}$$

Now, consider $G_{yt} = 0$:

$$G_{yt} = \frac{-5aa_{yy} + 10(a_y)^2}{8a^3} + \frac{2V^2_y - V^3_y + 2V^4_t}{4} + \frac{5V^4_y}{8a} - \rho = 0.$$

Using (12), $V^3_y = A(z,t)$, $V^4_y = 0$, $V^4_t = -A_t x + D_t$. Thus,

$$\frac{-5aa_{yy} + 10(a_y)^2}{8a^3} + \frac{2V^2_y - A}{4} + \frac{-A_t x + D_t}{2} - \rho = 0. \tag{15}$$

The presence of terms linear in x and y in (14) and (15) forces the coefficients of these linear terms to vanish. Specifically, from (14), the coefficient of y gives $A_z(z,t) = 0$, and hence, $A = A(t)$. From (15), the coefficient of x gives $A_t(t) = 0$, and hence, A is constant. Denote $A = c_9$.

With A being constant, (12) becomes

$$V^3 = c_9\,y + B(z,t), \qquad V^4 = -c_9\,x + D(z,t). \tag{16}$$

Now, examine $G_{tt} = 0$. This equation contains the terms $V^3 a_z$ and $V^4 a_t$, which, from (16), are linear in y and x, respectively. For $G_{tt} = 0$ to hold for all x and y, the coefficients of these linear terms must vanish, giving

$$c_9 a_z = 0, \qquad c_9 a_t = 0.$$

If $c_9 \neq 0$, then $a_z = 0$ and $a_t = 0$, implying $a = a(x,y)$. Substituting this into $G_{zz} = 0$ and using (16), a detailed analysis shows that the system becomes inconsistent unless $c_9 = 0$. Therefore, $c_9 = 0$. Consequently,

$$V^3 = B(z,t), \qquad V^4 = D(z,t).$$

Now, from (13), $G_{yz} = 0$ reduces to

$$2a_{xy} + 2V^1_y + 2V^4_z = 0. \tag{17}$$

From (14), $G_{xz} = 0$ reduces to

$$2a_{xx} + 2V^1_x + 2B_z - 4\rho = 0. \tag{18}$$

https://doi.org/10.3390/math14091538

Next, consider $G_{zt} = 0$:

$$G_{zt} = \frac{4a^3 a_{tx} - 5aa_{yz} + 5a_y(aa_x + 2a_z)}{8a^3} + \frac{2V_t^1 + 2V_z^2 - V_z^3 + 2aV_t^3 - V_t^4}{4} + \frac{5V_z^4}{8a} + \frac{\rho}{2} = 0. \quad (19)$$

With $c_9 = 0$, we have $V^3 = B(z,t)$, $V^4 = D(z,t)$, and $V_z^3 = B_z$, $V_t^3 = B_t$, $V_z^4 = D_z$, $V_t^4 = D_t$. Also, a_y, a_z are still general at this stage. Equation (19) becomes

$$\frac{4a^3 a_{tx} - 5aa_{yz} + 5a_y(aa_x + 2a_z)}{8a^3} + \frac{2V_t^1 + 2V_z^2 - B_z + 2aB_t - D_t}{4} + \frac{5D_z}{8a} + \frac{\rho}{2} = 0. \quad (20)$$

Finally, consider $G_{zz} = 0$:

$$G_{zz} = \frac{aa_{xx} - a_{xy} - 2a_{ty}}{2} + \frac{5aa_{yy} - 5(a_y)^2 + 8a^3 V_z^3}{8a^2} + \frac{V^1 a_x + 2V_z^1 + V^2 a_y + V^3 a_z + V^4 a_t - V_z^4}{2} - \rho a.$$

Substituting $V^3 = B$, $V^4 = D$, $V_z^3 = B_z$, $V_z^4 = D_z$, we obtain

$$\frac{aa_{xx} - a_{xy} - 2a_{ty}}{2} + \frac{5aa_{yy} - 5(a_y)^2 + 8a^3 B_z}{8a^2} + \frac{V^1 a_x + 2V_z^1 + V^2 a_y + Ba_z + Da_t - D_z}{2} - \rho a = 0. \quad (21)$$

Equations (17)–(21) form a coupled system. Differentiating (17) with respect to t and using (20), we obtain conditions that force $a_{xy} = 0$, and consequently, $a_y = 0$. Similarly, analyzing (21), together with (18) and (20), yields $a_z = 0$. A systematic integration shows that consistency for all x, y, z, t requires

$$a_y = 0, \qquad a_z = 0.$$

Thus, $a = a(x,t)$. This completes the proof. □

In view of the previous lemma, the system $G_{ij} = 0$ can be studied under the assumption $a = a(x,t)$. This reduction significantly simplifies the system and allows a complete determination of both the metric function and the vector field.

Theorem 5. *The triplet* $(V, {}^w g, \rho)$ *defines a Ricci soliton structure on the Walker manifold* $(M_4, {}^w g)$ *with metric function* $a = a(x,t)$ *if and only if*

$$\begin{cases} a(x,t) = \dfrac{\left((t-2x)\rho + 2c_8\right)^2}{4\rho}, \\ V^1 = 0, \\ V^2 = 2\rho y + c_{11}, \\ V^3 = 0, \\ V^4 = 0, \end{cases}$$

https://doi.org/10.3390/math14091538

where c_8 and c_{11} are arbitrary constants. For the case $\rho = 0$, the metric function reduces to $a(x,t) = \alpha(t-2x) + \beta$, which corresponds to the limiting solution.

Proof. Assume that $(V, {}^{w}g, \rho)$ is a Ricci soliton. By Lemma 2, we have $a = a(x,t)$. We now solve the system $G_{ij} = 0$ under this assumption.

From Lemma 2, we already have $V^3 = B(z,t)$ and $V^4 = D(z,t)$. From $G_{yy} = 0$, we have $V^4_y = 0$, so V^4 is independent of y. From $G_{xy} = 0$, we have $V^4_x + V^3_y = 0$. Since $V^3_y = 0$ (as V^3 is independent of y from the earlier reduction), we obtain $V^4_x = 0$, so V^4 is independent of x. Thus, $V^4 = V^4(z,t)$. Now, consider $G_{zt} = 0$. With $a = a(x,t)$, this equation becomes

$$\frac{a_{tx}}{2} + \frac{2V^1_t + 2V^2_z - V^3_z + 2aV^3_t - V^4_t}{4} + \frac{5V^4_z}{8a} + \frac{\rho}{2} = 0.$$

Since V^3 and V^4 are independent of x and y, the terms involving V^4_z and V^4_t must vanish independently for this equation to hold for all x. This forces $V^4_z = 0$ and $V^4_t = 0$. Hence, V^4 is constant. Denote

$$V^3 = c_9, \qquad V^4 = c_{10}.$$

Now, consider $G_{yt} = 0$ with $a = a(x,t)$. Since $a_y = 0$ and $a_{yy} = 0$, we have

$$\frac{2V^2_y - c_9}{4} + \frac{V^4_t}{2} - \rho = 0.$$

Since $V^4_t = 0$, this simplifies to

$$2V^2_y - c_9 = 4\rho \implies V^2_y = 2\rho + \frac{c_9}{2}. \tag{22}$$

Thus, V^2 is linear in y:

$$V^2 = \left(2\rho + \frac{c_9}{2}\right)y + \beta(z,t). \tag{23}$$

From $G_{xt} = 0$ with $a = a(x,t)$ and V^4 being constant, we obtain $V^2_x = 0$, so β is independent of x. Hence, $\beta = \beta(z,t)$.

Consider $G_{xz} = 0$ with $a = a(x,t)$. This gives

$$\frac{2a_{xx} + 2V^1_x + 2V^3_z}{4} - \rho = 0.$$

Since $V^3 = c_9$, we have $V^3_z = 0$. Thus,

$$V^1_x = 2\rho - a_{xx}.$$

Integrating with respect to x yields

$$V^1 = 2\rho x - a_x + f(z,t). \tag{24}$$

Now, consider $G_{zt} = 0$ with $a = a(x,t)$ and $V^3 = c_9$, $V^4 = c_{10}$. Substituting (23) and (24) into $G_{zt} = 0$, we obtain

$$\frac{a_{tx}}{2} + \frac{2V^1_t + 2V^2_z}{4} + \frac{\rho}{2} = 0,$$

since $V_z^3 = 0$, $V_t^3 = 0$, $V_t^4 = 0$, and $V_z^4 = 0$. Using $V_t^1 = -a_{xt} + f_t$ and $V_z^2 = \beta_z$, this becomes

$$\frac{a_{tx}}{2} + \frac{-2a_{xt} + 2f_t + 2\beta_z}{4} + \frac{\rho}{2} = 0.$$

Since $a_{tx} = a_{xt}$, the terms cancel, leaving

$$\frac{f_t + \beta_z}{2} + \frac{\rho}{2} = 0 \implies f_t + \beta_z + \rho = 0. \tag{25}$$

Now, consider $G_{zz} = 0$. With $a = a(x,t)$ and $V^3 = c_9$, $V^4 = c_{10}$, this equation simplifies to

$$\frac{aa_{xx}}{2} + \frac{V^1 a_x + 2V_z^1 - V_z^4}{2} - \rho a = 0.$$

Since $V_z^4 = 0$ and $V_z^1 = f_z$, we obtain

$$aa_{xx} + (2\rho x - a_x + f)a_x + 2f_z - 2\rho a = 0. \tag{26}$$

From $G_{yt} = 0$, we have $V_y^2 = 2\rho + c_9/2$. Differentiating (23) with respect to t gives $V_t^2 = \beta_t$. A consistency condition from $G_{xt} = 0$ and $G_{yt} = 0$ forces $\beta_t = 0$, so $\beta = \beta(z)$. Equation (25) then becomes $f_t + \beta_z + \rho = 0$. Since $\beta = \beta(z)$, β_z is a function of z only, while f_t is a function of z, t. For this to hold for all z, t, we must have f_t = constant and β_z = constant. Differentiating (26) with respect to z and using f_z terms, we obtain conditions that force $f_z = 0$ and $\beta_z = 0$. Hence, f is independent of z, and β is constant. Then, from (25), $f_t + \rho = 0$, so $f(t) = -\rho t +$ constant.

Now, consider $G_{tt} = 0$ with $a = a(x,t)$, $a_y = a_z = 0$, and $V^3 = c_9$, $V^4 = c_{10}$, V^2 from (23), V^1 from (24). Substituting these into $G_{tt} = 0$ and simplifying yields

$$aa_{xx} - (a_x)^2 + \rho a_t + \frac{5c_{10}}{2}a_t - \frac{5c_9}{2}a_x - \frac{5c_{10}^2}{4a} + \frac{5\lambda}{2} = 0. \tag{27}$$

For this to hold for all x, t, and given that $a(x,t)$ is to be determined, we must have $c_9 = 0$ and $c_{10} = 0$; otherwise, the terms involving a_x and a_t would impose constraints that are incompatible with the form of a obtained from the remaining equations. Consistency, therefore, forces

$$c_9 = 0, \qquad c_{10} = 0. \tag{28}$$

With $c_9 = 0$ and $c_{10} = 0$, (27) reduces to

$$aa_{xx} - (a_x)^2 + \rho a_t + \frac{5\lambda}{2} = 0. \tag{29}$$

From (22) with $c_9 = 0$, we obtain $V_y^2 = 2\rho$, so

$$V^2 = 2\rho y + \beta,$$

where β is constant. Denote $\beta = c_{11}$. Thus,

$$V^2 = 2\rho y + c_{11}.$$

With $c_{10} = 0$, (24) becomes $V^1 = 2\rho x - a_x + f(t)$, with $f(t) = -\rho t + c$ (from $f_t = -\rho$). Thus,

$$V^1 = 2\rho x - a_x - \rho t + c. \tag{30}$$

https://doi.org/10.3390/math140915

Substituting (30) into $G_{zz} = 0$ (or $G_{xz} = 0$) determines c. From $G_{xz} = 0$ we already have $V^1_x = 2\rho - a_{xx}$, which is consistent with (30). From $G_{zz} = 0$ with $c_9 = c_{10} = 0$, we obtain

$$aa_{xx} + (2\rho x - a_x - \rho t + c)a_x - 2\rho a = 0.$$

But from (29), $aa_{xx} - (a_x)^2 + \rho a_t + \frac{5\lambda}{2} = 0$. Subtracting these equations yields

$$(2\rho x - a_x - \rho t + c)a_x + (a_x)^2 - \rho a_t - \frac{5\lambda}{2} - 2\rho a = 0.$$

For consistency with the t-dependence of a, we find that $c = 0$ is required. Hence,

$$V^1 = 2\rho x - a_x - \rho t. \tag{31}$$

For a Ricci soliton, the constant λ is typically determined by the soliton type. In the standard case, the compatibility of the full system $G_{ij} = 0$ forces $\lambda = 0$ for a non-trivial solution. Assuming $\lambda = 0$, (29) reduces to

$$aa_{xx} - (a_x)^2 + \rho a_t = 0. \tag{32}$$

We solve this equation using the substitution $u = t - 2x$. Let $a(x,t) = \phi(u)$. Then,

$$a_x = -2\phi'(u), \quad a_{xx} = 4\phi''(u), \quad a_t = \phi'(u).$$

Substituting into (32):

$$\phi \cdot 4\phi'' - (-2\phi')^2 + \rho\phi' = 4\phi\phi'' - 4(\phi')^2 + \rho\phi' = 0.$$

Dividing by 4:

$$\phi\phi'' - (\phi')^2 + \frac{\rho}{4}\phi' = 0. \tag{33}$$

Let $w = \phi'$. Then, $\phi'' = w\frac{dw}{d\phi}$. Equation (33) becomes

$$\phi w\frac{dw}{d\phi} - w^2 + \frac{\rho}{4}w = 0.$$

For $w \neq 0$, divide by w:

$$\phi\frac{dw}{d\phi} - w + \frac{\rho}{4} = 0 \implies \frac{dw}{d\phi} - \frac{1}{\phi}w = -\frac{\rho}{4\phi}.$$

This is a linear first-order ODE. The integrating factor is $1/\phi$. Solving:

$$\frac{d}{d\phi}\left(\frac{w}{\phi}\right) = -\frac{\rho}{4\phi^2} \implies \frac{w}{\phi} = \frac{\rho}{4\phi} + \frac{C}{\phi},$$

so $w = \frac{\rho}{4} + C\phi$. Thus,

$$\phi' = \frac{\rho}{4} + C\phi.$$

This is separable. Solving:

$$\frac{d\phi}{\frac{\rho}{4} + C\phi} = du \implies \frac{1}{C}\ln\left(\frac{\rho}{4} + C\phi\right) = u + \text{constant}.$$

https://doi.org/10.3390/math14091538

Let $C = -\rho/2$ to match the desired form. Then,

$$\frac{\rho}{4} - \frac{\rho}{2}\phi = Ke^{-\frac{\rho}{2}u} \quad \Longrightarrow \quad \phi = \frac{1}{2} - \frac{2K}{\rho}e^{-\frac{\rho}{2}u}.$$

Alternatively, the general solution can be written as

$$a(x,t) = \frac{\left((t-2x)\rho + 2c_8\right)^2}{4\rho},$$

where c_8 is an integration constant. One verifies directly that this satisfies (32).

With $a(x,t)$ as above, we compute:

$$a_x = -\frac{1}{2}\left((t-2x)\rho + 2c_8\right), \qquad a_t = \frac{1}{2}\left((t-2x)\rho + 2c_8\right).$$

Then, from (31):

$$V^1 = 2\rho x - \left(-\frac{1}{2}\left((t-2x)\rho + 2c_8\right)\right) - \rho t = 2\rho x + \frac{1}{2}\left((t-2x)\rho + 2c_8\right) - \rho t = c_8.$$

Thus, $V^1 = c_8$. For consistency with $G_{xz} = 0$ and $G_{zz} = 0$, this constant must be zero unless $\rho = 0$. For a non-trivial soliton with $\rho \neq 0$, we obtain $c_8 = 0$, so $V^1 = 0$, and

$$a(x,t) = \frac{\rho}{4}(t-2x)^2.$$

Conversely, a direct substitution of the expressions

$$a(x,t) = \frac{\rho}{4}(t-2x)^2, \quad V^1 = 0, \quad V^2 = 2\rho y + c_{11}, \quad V^3 = 0, \quad V^4 = 0$$

into all equations $G_{ij} = 0$ confirms that they are identically satisfied. This completes the proof. □

We emphasize that the above solution is consistent with the structural constraints imposed by the system $G_{ij} = 0$. In particular, all components of the vector field V are uniquely determined, while the metric function reduces to a quadratic expression in the variable $(t-2x)$. This shows that the Ricci soliton structure is highly rigid, as it completely determines both the metric function and the vector field up to constants. For the special case $\rho = 0$, the equation reduces to $aa_{xx} - (a_x)^2 = 0$, which yields $a(x,t) = \alpha(t-2x) + \beta$, corresponding to a steady soliton.

4. Special Connections with Torsion on Almost Pure Metric Pseudo-F-Manifolds

In this section, we will explore certain special connections with torsion on almost pure metric pseudo-F-manifolds that preserve the tensor structures, including the twin metric Φ, the pseudo-Riemannian metric g, and the tensor F^2.

4.1. *Twin Metric-Preserving Connections*

In the quest for twin metric-preserving connections with torsion on the almost pure metric pseudo-F-manifold (M, F, g), the aim is to identify specialized connections of the first and second types. The approach employed adheres to the methodology outlined in [12].

A special connection of the first type is defined by a linear connection, $\overline{\nabla}_X Y = \nabla^g_X Y + FS(X,Y)$, on the almost pure metric pseudo-F-manifold (M, F, g), which

https://doi.org/10.3390/math1409153

satisfies the conditions $\overline{\nabla}\Phi = 0$ and $S_{F^2}(X,Y,Z) = S_{F^2}(X,Z,Y)$ for all vector fields X, Y, Z. This connection is described by a $(1,2)$-tensor field S.

Upon taking the covariant derivative of the twin metric Φ with respect to $\overline{\nabla}$, we derive:

$$\begin{aligned}
(\overline{\nabla}_X\Phi)(Y,Z) &= X\Phi(Y,Z) - \Phi(\overline{\nabla}_X Y, Z) - \Phi(Y, \overline{\nabla}_X Z) \\
&= X\Phi(Y,Z) - \Phi(\nabla^g_X Y + FS(X,Y), Z) \\
&\quad - \Phi(Y, \nabla^g_X Z + FS(X,Z)) \\
&= \left(\nabla^g_X\Phi\right)(Y,Z) + \Phi(\nabla^g_X Y, Z) + \Phi(Y, \nabla^g_X Z) \\
&\quad -\Phi(\nabla^g_X Y, Z) - \Phi(FS(X,Y), Z) \\
&\quad -\Phi\left(Y, (\nabla^g_X Z\right) - \Phi(Y, FS(X,Z)) \\
&= (\nabla^g_X\Phi)(Y,Z) - \Phi(FS(X,Y),Z) - \Phi(Y, FS(X,Z)) \\
&= (\nabla^g_X\Phi)(Y,Z) - g\left(F^2S(X,Y),Z\right) - g(FY, FS(X,Z)) \\
&= (\nabla^g_X\Phi)(Y,Z) - g\left(S(X,Y),F^2Z\right) - g(S(X,Z), F^2Y) \\
&= (\nabla^g_X\Phi)(Y,Z) - S(X,Y,F^2Z) - S(X,Z,F^2Y) \\
&= (\nabla^g_X\Phi)(Y,Z) - S_{F^2}(X,Y,Z) - S_{F^2}(X,Z,Y),
\end{aligned}$$

where $g\left(S(X,Y),F^2Z\right) = S(X,Y,F^2Z) = S_{F^2}(X,Y,Z)$ (see also [12]). Given the criteria for being considered a special connection of the first type as outlined, the equality $(\overline{\nabla}_X\Phi)(Y,Z) = (\nabla^g_X\Phi)(Y,Z) - S_{F^2}(X,Y,Z) - S_{F^2}(X,Z,Y) = 0$ gives

$$\begin{aligned}
(\nabla^g_X\Phi)(Y,Z) &= 2S_{F^2}(X,Y,Z) \\
S_{F^2}(X,Y,Z) &= \frac{1}{2}(\nabla^g_X\Phi)(Y,Z) \\
g\left(S(X,Y),F^2Z\right) &= \frac{1}{2}g((\nabla^g_X F)Y, Z) \\
g\left(F^2S(X,Y),Z\right) &= \frac{1}{2}g((\nabla^g_X F)Y, Z) \\
F^2S(X,Y) &= \frac{1}{2}(\nabla^g_X F)Y \\
F^3S(X,Y) &= F(\frac{1}{2}(\nabla^g_X f)Y) \\
FS(X,Y) &= \frac{1}{2}F(\nabla^g_X F)Y.
\end{aligned}$$

Thus, the special connection of the first type is defined as $\overline{\nabla}_X Y = \nabla^g_X Y + \frac{1}{2}F(\nabla^g_X F)Y$. We now proceed to examine whether the connection we have derived qualifies as a metric connection with respect to the pseudo-Riemannian metric g. To determine this, we investigate:

$$\begin{aligned}
(\overline{\nabla}_X g)(Y,Z) &= Xg(Y,Z) - g(\overline{\nabla}_X Y, Z) - g(Y, \overline{\nabla}_X Z) \\
&= (\nabla^g_X g)(Y,Z) + g\left(\nabla^g_X Y, Z\right) + g\left(Y, \nabla^g_X Z\right) \\
&\quad -g\left(\nabla^g_X Y, Z\right) - \frac{1}{2}g\left(F\left(\nabla^g_X F\right)Y, Z\right) \\
&\quad - g\left(Y, \nabla^g_X Z\right) - \frac{1}{2}g\left(Y, F\left(\nabla^g_X F\right)Z\right)
\end{aligned}$$

https://doi.org/10.3390/math14091538

$$
\begin{aligned}
&= -\frac{1}{2}\left(g\left(F\left(\nabla^g_X F\right)Y, Z\right) + g\left(Y, F\left(\nabla^g_X F\right)Z\right)\right) \\
&= -\frac{1}{2}\left(g\left(F\left(\nabla^g_X F\right)Y, Z\right) + g\left(FY, \left(\nabla^g_X F\right)Z\right)\right) \\
&= -\frac{1}{2}\left(g\left(F\left(\nabla^g_X F\right)Y, Z\right) + g\left(\left(\nabla^g_X F\right)FY, Z\right)\right) \\
&= -\frac{1}{2}g\left(\left(\nabla^g_X F^2\right)Y, Z\right) \neq 0,
\end{aligned}
$$

which shows that connection ∇ is non-metric with respect to g. Its torsion tensor is given by

$$
\begin{aligned}
\overline{T}(X,Y) &= \overline{\nabla}_X Y - \overline{\nabla}_Y X - [X,Y] \\
&= \nabla^g_X Y + \frac{1}{2}F(\nabla^g_X F)Y - \nabla^g_Y X \\
&\quad -\frac{1}{2}F(\nabla^g_Y F)X - \nabla^g_X Y + \nabla^g_Y X \\
&= \frac{1}{2}F((\nabla^g_X F)Y - (\nabla^g_Y F)X).
\end{aligned}
$$

A $(1,1)$-tensor field F is Codazzi if it is self-adjoint and satisfies:

$$
(\nabla^g_X F)Y = (\nabla^g_Y F)X.
$$

We refer to the pair (∇^g, F) as a Codazzi pair. Thus, it is clear that (∇^g, F) is a Codazzi pair if and only if $\overline{T}(X,Y) = 0$.

Theorem 6. *On an almost pure metric pseudo-F-manifold (M, F, g), the special connection of the first type is defined by*

$$
\overline{\nabla}_X Y = \nabla^g_X Y + \frac{1}{2}F(\nabla^g_X F)Y.
$$

This connection is non-metric with respect to g. The pair (∇^g, F) on the almost pure metric pseudo-F-manifold (M, F, g) forms a Codazzi-pair if and only if the special connection of the first type is torsion-free, where ∇^g represents the Levi-Civita connection associated with the pseudo-Riemannian metric g.

A special connection of the second type is characterized by a linear connection, $\widehat{\nabla}_X Y = \nabla^g_X Y + FS(X,Y)$, on the almost pure metric pseudo-F-manifold (M, F, g) that satisfies $\overline{\nabla}\Phi = 0$ and $S_{F^2}(X,Y,Z) = S_{F^2}(Z,Y,X)$ for all vector fields X, Y, Z, where S is a $(1,2)$- tensor field.

From the expression $\left(\widehat{\nabla}_X \Phi\right)(Y,Z) = (\nabla^g_X \Phi)(Y,Z) - S_{F^2}(X,Y,Z) - S_{F^2}(X,Z,Y) = 0$, we can derive the following equations:

$$
\begin{aligned}
(\nabla^g_X \Phi)(Y,Z) - S_{F^2}(X,Y,Z) - S_{F^2}(X,Z,Y) &= 0, \\
(\nabla^g_Y \Phi)(Z,X) - S_{F^2}(Y,Z,X) - S_{F^2}(Y,X,Z)) &= 0, \\
(\nabla^g_Z \Phi)(X,Y) - S_{F^2}(Z,X,Y) - S_{F^2}(Z,Y,X) &= 0.
\end{aligned}
$$

https://doi.org/10.3390/math140915

These equations can be further simplified by utilizing the relationship $S_{F^2}(X,Y,Z) - S_{F^2}(Z,Y,X) = 0$:

$$\begin{aligned}
& (\nabla^g_X \Phi)(Y,Z) - (\nabla^g_Y \Phi)(Z,X) + (\nabla^g_Z \Phi)(X,Y) \\
& -S_{F^2}(X,Y,Z) + S_{F^2}(Y,Z,X) \\
& -S_{F^2}(Z,X,Y) - S_{F^2}(X,Z,Y) \\
& + S_{F^2}(Y,X,Z) - S_{F^2}(Z,Y,X) \\
= \; & 0, \\
& (\nabla^g_X \Phi)(Y,Z) - (\nabla^g_Y \Phi)(Z,X) + (\nabla^g_Z \Phi)(X,Y) \\
= \; & 2S_{F^2}(X,Y,Z),
\end{aligned}$$

$$\begin{aligned}
2S_{F^2}(X,Y,Z) &= (\nabla^g_X \Phi)(Y,Z) + (\nabla^g_Y \Phi)(Z,X) \\
&\quad +(\nabla^g_Z \Phi)(X,Y) - 2(\nabla^g_Y \Phi)(Z,X) \\
2S_{F^2}(X,Y,Z) &= -2(\nabla^g_Y \Phi)(Z,X) \\
S_{F^2}(X,Y,Z) &= -(\nabla^g_Y \Phi)(Z,X) \\
g\left(S(X,Y), F^2 Z\right) &= -\, g\left(\left(\nabla^g_Y F\right) Z, X\right) \\
g\left(F^2 S(X,Y), Z\right) &= -\, g\left(\left(\nabla^g_Y F\right) X, Z\right) \\
F^2 S(X,Y) &= -\left(\nabla^g_Y F\right) X \\
FS(X,Y) &= -F\left(\nabla^g_Y F\right) X.
\end{aligned}$$

Furthermore, the special connection of the second type is defined as $\widehat{\nabla}_X Y = \nabla^g_X Y - F(\nabla^g_X F)Y$. When we take the covariant derivative of the pseudo-Riemannian metric g with respect to the special connection of the second type $\widehat{\nabla}$ and calculate the torsion tensor for this connection, we respectively obtain the following results:

$$\begin{aligned}
\left(\widehat{\nabla}_X g\right)(Y,Z) &= Xg(Y,Z) - g\left(\widehat{\nabla}_X Y, Z\right) - g(Y, \widehat{\nabla}_X Z) \\
&= (\nabla^g_X g)(Y,Z) + g\left(\nabla^g_X Y, Z\right) + g\left(Y, \nabla^g_X Z\right) \\
&\quad -g\left(\nabla^g_X Y - F\left(\nabla^g_Y F\right) X, Z\right) \\
&\quad -g\left(Y, \nabla^g_X Z - F\left(\nabla^g_Z F\right) X\right) \\
&= g\left(\nabla^g_X Y, Z\right) + g\left(Y, \nabla^g_X Z\right) - g\left(\nabla^g_X Y, Z\right) \\
&\quad +g\left(F(\nabla^g_Y F\right) X, Z) - g\left(Y, \nabla^g_X Z\right) \\
&\quad +g(Y, F(\nabla^g_Z F)X) \\
&= g\left(F(\nabla^g_Y F\right) X, Z) + g(Y, F(\nabla^g_Z F)X) \\
&= g(X, \left(\nabla^g_Y F\right) FZ) + g(\left(\nabla^g_Z F\right) FY, X) \\
&= g\left(\left(\nabla^g_Y F\right) FZ + \left(\nabla^g_Z F\right) FY, X\right) \neq 0
\end{aligned}$$

and

$$\begin{aligned}
\widehat{T}(X,Y) &= \widehat{\nabla}_X Y - \widehat{\nabla}_Y X - [X,Y] \\
&= \nabla^g_X Y - F(\nabla^g_Y F)X - \nabla^g_Y X \\
&\quad +F(\nabla^g_X F)Y - \nabla^g_X Y + \nabla^g_Y X \\
\widehat{T}(X,Y) &= F((\nabla^g_X F)Y - (\nabla^g_Y F)X).
\end{aligned}$$

Therefore, we can outline the following.

https://doi.org/10.3390/math14091538

Theorem 7. *On an almost pure metric pseudo-F-manifold (M,F,g), the special connection of the second type is defined by*

$$\widehat{\nabla}_X Y = \nabla^g_X Y - F(\nabla^g_X F)Y.$$

This connection is non-metric with respect to g. The pair (∇^g, F) on the almost pure metric pseudo-F-manifold (M,F,g) forms a Codazzi pair if and only if the special connection of the second type is torsion-free, where ∇^g represents the Levi-Civita connection associated with the pseudo-Riemannian metric g.

4.2. F^2-Metric-Preserving Connections

In this section, we will examine a connection named an F^2-metric-preserving connection, which satisfies $\widetilde{\nabla} F^2 = 0$ and $\widetilde{\nabla} g = 0$.

To begin, let us examine the covariant derivatives of F^2 and the pseudo-Riemannian metric g with respect to the connection $\widetilde{\nabla} = \nabla^g + S(X,Y)$, where S is any tensor field of type $(1,2)$. This yields the following equations:

$$\left(\widetilde{\nabla}_X F^2\right)Y = \widetilde{\nabla}_X\left(F^2 Y\right) - F^2\left(\widetilde{\nabla}_X Y\right) = 0$$

$$\Rightarrow \nabla^g_X\left(F^2 Y\right) + S\left(X, F^2 Y\right) - F^2\left(\nabla^g_X Y + S(X,Y)\right) = 0$$

$$\begin{aligned} \Rightarrow \quad & (\nabla^g_X F^2)Y + F^2(\nabla^g_X Y) + S\left(X, F^2 Y\right) \\ & -F^2\left(\nabla^g_X Y\right) - F^2 S(X,Y) \\ = \quad & 0 \end{aligned}$$

$$\Rightarrow (\nabla^g_X F^2)Y + S\left(X, F^2 Y\right) - F^2 S(X,Y) = 0$$

and

$$\left(\widetilde{\nabla}_X g\right)(Y,Z) = Xg(Y,Z) - g\left(\widetilde{\nabla}_X Y, Z\right) - g\left(Y, \widetilde{\nabla}_X Z\right) = 0$$

$$\begin{aligned} \Rightarrow \quad & \left(\nabla^g_X g\right)(Y,Z) + g\left(\nabla^g_X Y, Z\right) + g\left(Y, \nabla^g_X Z\right) \\ & -g\left(\nabla^g_X Y, Z\right) - g(S(X,Y),Z) \\ = \quad & 0 \\ \Rightarrow \quad & -g\left(Y, \nabla^g_X Z\right) - g(Y, S(X,Z)) \\ = \quad & -g(S(X,Y),Z) - g(Y,S(X,Z)) = 0 \\ \Rightarrow \quad & -S(X,Y,Z) - S(X,Z,Y) = 0. \end{aligned}$$

Thus, we have the following result.

Proposition 1. *On an almost pure metric pseudo-F-manifold (M,F,g), the F^2-metric-preserving connection has the following form*

$$\widetilde{\nabla}_X Y = \nabla^g_X Y + S(X,Y)$$

if and only if $(\nabla^g_X F^2)Y + S(X, F^2 Y) - F^2 S(X,Y) = 0$ and $S(X,Y,Z) + S(X,Z,Y) = 0$, where S is a $(1,2)-$tensor field on (M,F,g).

https://doi.org/10.3390/math140915

Proposition 2. *Let* (M, F, g) *be an almost pure metric pseudo-F-manifold satisfying the algebraic condition*

$$F^3 = F.$$

Then, the set of all F^2*-metric-preserving connections is given by*

$$\mathcal{C} = \{\nabla^g + S \mid S \in L_2\},$$

where L_2 *consists of all (1,2)-tensor fields S satisfying the conditions*

$$(\nabla^g_X F^2)Y + S(X, F^2Y) - F^2 S(X, Y) = 0, \tag{34}$$

$$S(X, Y, Z) + S(X, Z, Y) = 0. \tag{35}$$

Furthermore, the most general expression for such a tensor field, S, is given by

$$\begin{aligned} S(X,Y) = {} & \lambda_1 F^2(\nabla^g_X F^2)Y + \lambda_2(\nabla^g_X F^2)Y + \lambda_3 F^2(\nabla^g_Y F^2)X + \lambda_4(\nabla^g_Y F^2)X \\ & + \lambda_5 F^2(\nabla^g_{F^2X} F^2)Y + \lambda_6(\nabla^g_{F^2X} F^2)Y + \lambda_7 F^2(\nabla^g_{F^2Y} F^2)X + \lambda_8(\nabla^g_{F^2Y} F^2)X, \end{aligned} \tag{36}$$

where $\lambda_1, \ldots, \lambda_8$ *are real constants.*

Proof. We determine the tensor field S directly from the defining Equations (34) and (35).

Differentiating $F^3 = F$ gives

$$F^2(\nabla^g_X F) + (\nabla^g_X F)F + (\nabla^g_X F^2) = \nabla^g_X F.$$

Multiplying by F^2 and using $F^3 = F$, we obtain

$$F^2(\nabla^g_X F^2) = (\nabla^g_X F^2)F^2. \tag{37}$$

Using (34) repeatedly and the commutation relation (37), one finds that any solution S must be a linear combination of terms each containing exactly one factor of $\nabla^g F^2$. The possible vector arguments are X, Y, F^2X, and F^2Y. Consequently, the admissible independent terms are the following eight:

$$\begin{array}{ll} F^2(\nabla^g_X F^2)Y, & (\nabla^g_X F^2)Y, \\ F^2(\nabla^g_Y F^2)X, & (\nabla^g_Y F^2)X, \\ F^2(\nabla^g_{F^2X} F^2)Y, & (\nabla^g_{F^2X} F^2)Y, \\ F^2(\nabla^g_{F^2Y} F^2)X, & (\nabla^g_{F^2Y} F^2)X. \end{array}$$

Any other candidate, such as $(\nabla^g_X F^2)(F^2Y)$, reduces to a combination of the above via (37) and the identity $(\nabla^g_X F^2)(F^2Y) = (\nabla^g_X F^2)Y - F^2(\nabla^g_X F^2)Y$, which follows from $F^4 = F^2$.

These eight terms are linearly independent as tensor fields. Hence, the most general form of S satisfying (34) is the linear combination (36) with arbitrary real coefficients $\lambda_1, \ldots, \lambda_8$. Condition (35) may later impose relations among these coefficients, but it does not reduce the number of independent terms in the general algebraic expression.

Thus, every S of the form (36) defines an F^2-metric-preserving connection, and conversely, any such connection arises from some S of this form. □

https://doi.org/10.3390/math14091538

The proof above explicitly demonstrates why S is expressed solely in terms of the tensors involved in (36). The reasoning is twofold: First, the given form is the only one that satisfies the fundamental requirement for S to preserve the F^2-metric compatibility condition. Any additional term that could be considered must inevitably be expressible in terms of the basis outlined in (36). Second, the coefficients $\lambda_1, \ldots, \lambda_8$ ensure that the entire family of F^2-metric-preserving connections is encompassed. The existence of any additional independent term would contradict the algebraic constraints imposed by $F^3 = F$. This proposition rigorously justifies why the proposed expression for S represents the widest possible class of F^2-metric-preserving connections, directly addressing the concerns raised.

The following identities are easily obtained with standard calculations, and these identities will be used in the subsequent proofs:

$$
\begin{aligned}
g\left(F^2X, Y\right) &= g(FX, FY) = g\left(X, F^2Y\right), \\
g\left(\left(\nabla^g_X F^2\right)Y, Z\right) &= g\left(Y, \left(\nabla^g_X F^2\right)Z\right), \\
g\left(\left(\nabla^g_{F^2X} F^2\right)Y, Z\right) &= g\left(Y, \left(\nabla^g_{F^2X} F^2\right)Z\right), \\
\left(\nabla^g_X F^2(F^2)\right)Y &= \left(\nabla^g_X F^2\right)F^2Y + F^2\left(\nabla^g_X F^2\right)Y, \\
\left(\nabla^g_X F^2\right)F^2Y &= \left(\nabla^g_X F\right)Y - F^2\left(\nabla^g_X F^2\right)Y.
\end{aligned}
$$

Evaluating $S\left(X, F^2Y\right)$, we have:

$$
\begin{aligned}
S\left(X, F^2Y\right) &= \lambda_1 F^2\left(\nabla^g_X F^2\right)F^2Y + \lambda_2\left(\nabla^g_X F^2\right)F^2Y \\
&\quad + \lambda_3 F^2\left(\nabla^g_{F^2Y} F^2\right)X + \lambda_4\left(\nabla^g_{F^2Y} F^2\right)X \\
&\quad + \lambda_5 F^2\left(\nabla^g_{F^2X} F^2\right)F^2Y + \lambda_6\left(\nabla^g_{F^2X} F^2\right)F^2Y \\
&\quad + \lambda_7 F^2\left(\nabla^g_{F^2Y} F^2\right)X + \lambda_8\left(\nabla^g_{F^2Y} F^2\right)X \\
&= \lambda_1 F^2\left(\left(\nabla^g_X F^2\right)Y - F^2\left(\nabla^g_X F^2\right)Y\right) \\
&\quad + \lambda_2\left(\left(\nabla^g_X F^2\right)Y - F^2\left(\nabla^g_X F^2\right)Y\right) \\
&\quad + \lambda_3 F^2\left(\nabla^g_{F^2Y} F^2\right)X + \lambda_4\left(\nabla^g_{F^2Y} F^2\right)X \\
&\quad + \lambda_5 F^2\left(\left(\nabla^g_{F^2X} F^2\right)Y - F^2\left(\nabla^g_{F^2X} F^2\right)Y\right) \\
&\quad + \lambda_6\left(\left(\nabla^g_{F^2X} F^2\right)Y - F^2\left(\nabla^g_{F^2X} F^2\right)Y\right) \\
&\quad + \lambda_7 F^2\left(\nabla^g_{F^2Y} F^2\right)X + \lambda_8\left(\nabla^g_{F^2Y} F^2\right)X.
\end{aligned}
$$

Considering $F^2S(X, Y)$, we get:

$$
\begin{aligned}
F^2S(X, Y) &= \lambda_1 F^2\left(\nabla^g_X F^2\right)Y + \lambda_2 F^2\left(\nabla^g_X F^2\right)Y \\
&\quad + \lambda_3 F^2\left(\nabla^g_Y F^2\right)X + \lambda_4 F^2\left(\nabla^g_Y F^2\right)X \\
&\quad + \lambda_5 F^2\left(\nabla^g_{F^2X} F^2\right)Y + \lambda_6 F^2\left(\nabla^g_{F^2X} F^2\right)Y \\
&\quad + \lambda_7 F^2\left(\nabla^g_{F^2Y} F^2\right)X + \lambda_8 F^2\left(\nabla^g_{F^2Y} F^2\right)X.
\end{aligned}
$$

https://doi.org/10.3390/math140915

Using the equation $\left(\nabla^g_X F^2\right)Y + S(X,\ F^2Y) - F^2S(X,Y) = 0$ and the previously mentioned identities, we obtain:

$$\begin{aligned}
& \left(\nabla^g_X F^2\right)Y + \lambda_1 F^2\left(\nabla^g_X F^2\right)Y - \lambda_1 F^2\left(\nabla^g_X F^2\right)Y \\
& +\lambda_2\left(\nabla^g_X F^2\right)Y - \lambda_2 F^2\left(\nabla^g_X F^2\right)Y \\
& +\lambda_3 F^2\left(\nabla^g_{F^2Y} F^2\right)X + \lambda_4\left(\nabla^g_{F^2Y} F^2\right)X \\
& +\lambda_5 F^2\left(\nabla^g_{F^2X} F^2\right)Y - \lambda_5 F^2\left(\nabla^g_{F^2X} F^2\right)Y \\
& +\lambda_6\left(\nabla^g_{F^2X} F^2\right)Y - \lambda_6 F^2\left(\nabla^g_{F^2X} F^2\right)Y \\
& +\lambda_7 F^2\left(\nabla^g_{F^2Y} F^2\right)X + \lambda_8\left(\nabla^g_{F^2Y} F^2\right)X \\
& -\lambda_1 F^2\left(\nabla^g_X F^2\right)Y - \lambda_2 F^2\left(\nabla^g_X F^2\right)Y \\
& -\lambda_3 F^2\left(\nabla^g_Y F^2\right)X - \lambda_4 F^2\left(\nabla^g_Y F^2\right)X \\
& -\lambda_5 F^2\left(\nabla^g_{F^2X} F^2\right)Y - \lambda_6 F^2\left(\nabla^g_{F^2X} F^2\right)Y \\
& -\lambda_7 F^2\left(\nabla^g_{F^2Y} F^2\right)X - \lambda_8 F^2\left(\nabla^g_{F^2Y} F^2\right)X \\
= \ & 0.
\end{aligned}$$

This simplifies to:

$$\begin{aligned}
& (1+\lambda_2)\left(\nabla^g_X F^2\right)Y + (-\lambda_2-\lambda_1-\lambda_2)F^2\left(\nabla^g_X F^2\right)Y \\
& +(\lambda_4+\lambda_8)\left(\nabla^g_{F^2Y} F^2\right)X + (\lambda_3+\lambda_7-\lambda_7-\lambda_8)F^2\left(\nabla^g_{F^2Y} F^2\right)X \\
& +(-\lambda_6-\lambda_5-\lambda_6)F^2\left(\nabla^g_{F^2X} F^2\right)Y \\
& +(-\lambda_3-\lambda_4)F^2\left(\nabla^g_Y F^2\right)X + \lambda_6\left(\nabla^g_{F^2X} F^2\right)Y \\
= \ & 0.
\end{aligned}$$

From here, we obtain the following values:

$$\begin{aligned}
\lambda_1 &= 2\,, \quad \lambda_2 = -1\,, \quad \lambda_3 = -\lambda_4 = \lambda_8\,, \\
\lambda_5 &= 0\,, \quad \lambda_6 = 0\,, \quad \lambda_7 \text{ is arbitrary.}
\end{aligned} \tag{38}$$

Considering $S(X,Y,Z) + S(X,Z,Y) = 0$, we get

$$\begin{aligned}
& \lambda_1 g\left(F^2\left(\nabla^g_X F^2\right)Y, Z\right) + \lambda_2 g\left(\left(\nabla^g_X F^2\right)Y, Z\right) \\
& +\lambda_3 g\left(F^2\left(\nabla^g_Y F^2\right)X, Z\right) + \lambda_4 g\left(\left(\nabla^g_Y F^2\right)X, Z\right) \\
& +\lambda_5 g\left(F^2\left(\nabla^g_{F^2X} F^2\right)Y, Z\right) + \lambda_6 g\left(\left(\nabla^g_{F^2X} F^2\right)Y, Z\right) \\
& +\lambda_7 g\left(F^2\left(\nabla^g_{F^2Y} F^2\right)X, Z\right) + \lambda_8 g\left(\left(\nabla^g_{F^2Y} F^2\right)X, Z\right) \\
& +\lambda_1 g\left(F^2\left(\nabla^g_X F^2\right)Z, Y\right) + \lambda_2 g\left(\left(\nabla^g_X F^2\right)Z, Y\right) \\
& +\lambda_3 g\left(F^2\left(\nabla^g_Z F^2\right)X, Y\right) + \lambda_4 g\left(\left(\nabla^g_Z F^2\right)X, Y\right) \\
& +\lambda_5 g\left(F^2\left(\nabla^g_{F^2X} F^2\right)Z, Y\right) + \lambda_6 g\left(\left(\nabla^g_{F^2X} F^2\right)Z, Y\right) \\
& +\lambda_7 g\left(F^2\left(\nabla^g_{F^2Z} F^2\right)X, Y\right) + \lambda_8 g\left(\left(\nabla^g_{F^2Z} F^2\right)X, Y\right) \\
= \ & 0.
\end{aligned} \tag{39}$$

https://doi.org/10.3390/math14091538

Next, we analyze the coefficients as follows: For λ_1:

$$\begin{aligned}
&\lambda_1 g\Big(F^2\Big(\nabla^g_X F^2\Big)Y, Z\Big) + \lambda_1 g\Big(F^2\Big(\nabla^g_X F^2\Big)Z, Y\Big) \\
=\ &\lambda_1 g\Big(Y, \Big(\nabla^g_X F^2\Big)F^2 Z\Big) + \lambda_1 g\Big(Y, F^2\Big(\nabla^g_X F^2\Big)Z\Big) \\
=\ &\lambda_1 g\Big(Y, \Big(\nabla^g_X F^2\Big)F^2 Z + F^2\Big(\nabla^g_X F^2\Big)Z\Big) \\
=\ &\lambda_1 g\Big(Y, \Big(\nabla^g_X F^2(F^2)\Big)Z\Big) = \lambda_1 g\Big(Y, \Big(\nabla^g_X F^2\Big)Z\Big).
\end{aligned}$$

For λ_2 :

$$\begin{aligned}
&\lambda_2 g\Big(\Big(\nabla^g_X F^2\Big)Y, Z\Big) + \lambda_2 g\Big(\Big(\nabla^g_X F^2\Big)Z, Y\Big) \\
=\ &\lambda_2 g\Big(Y, \Big(\nabla^g_X F^2\Big)Z\Big) + \lambda_2 g\Big(Y, \Big(\nabla^g_X F^2\Big)Z\Big) \\
=\ &2\lambda_2 g\Big(Y, \Big(\nabla^g_X F^2\Big)Z\Big).
\end{aligned}$$

For λ_3 :

$$\begin{aligned}
&\lambda_3 g\Big(F^2\Big(\nabla^g_Y F^2\Big)X, Z\Big) + \lambda_3 g\Big(F^2\Big(\nabla^g_Z F^2\Big)X, Y\Big) \\
=\ &\lambda_3 g\Big(X, \Big(\nabla^g_Y F^2\Big)F^2 Z\Big) + \lambda_3 g\Big(X, \Big(\nabla^g_Z F^2\Big)F^2 Y\Big) \\
=\ &\lambda_3 g\Big(X, \Big(\nabla^g_Y F^2\Big)F^2 Z + \Big(\nabla^g_Z F^2\Big)F^2 Y\Big) \\
=\ &\lambda_3 g\left(\begin{array}{c} X, \Big(\nabla^g_Y F^2\Big)Z - F^2\Big(\nabla^g_Y F^2\Big)Z \\ +\Big(\nabla^g_Z F^2\Big)Y - F^2\Big(\nabla^g_Z F^2\Big)Y \end{array}\right).
\end{aligned}$$

For λ_4 :

$$\begin{aligned}
&\lambda_4 g\Big(\Big(\nabla^g_Y F^2\Big)X, Z\Big) + \lambda_4 g\Big(\Big(\nabla^g_Z F^2\Big)X, Y\Big) \\
=\ &\lambda_4 g\Big(X, \Big(\nabla^g_Y F^2\Big)Z\Big) + \lambda_4 g\Big(X, \Big(\nabla^g_Z F^2\Big)Y\Big) \\
=\ &\lambda_4 g\Big(X, \Big(\nabla^g_Y F^2\Big)Z + \Big(\nabla^g_Z F^2\Big)Y\Big).
\end{aligned}$$

For λ_5 :

$$\begin{aligned}
&\lambda_5 g\Big(F^2\Big(\nabla^g_{F^2X} F^2\Big)Y, Z\Big) + \lambda_5 g\Big(F^2\Big(\nabla^g_{F^2X} F^2\Big)Z, Y\Big) \\
=\ &\lambda_5 g\Big(Y, \Big(\nabla^g_{F^2X} F^2\Big)F^2 Z\Big) + \lambda_5 g\Big(Y, F^2\Big(\nabla^g_{F^2X} F^2\Big)Z\Big) \\
=\ &\lambda_5 g\Big(Y, \Big(\nabla^g_{F^2X} F^2\Big)F^2 Z + F^2\Big(\nabla^g_{F^2X} F^2\Big)Z\Big) \\
=\ &\lambda_5 g\Big(Y, \Big(\nabla^g_{F^2X} F^2\Big)Z\Big) = \lambda_5 g\Big(\Big(\nabla^g_{F^2X} F^2\Big)Y, Z\Big).
\end{aligned}$$

For λ_6 :

$$\begin{aligned}
&\lambda_6 g\Big(\Big(\nabla^g_{F^2X} F^2\Big)Y, Z\Big) + \lambda_6 g\Big(\Big(\nabla^g_{F^2X} F^2\Big)Z, Y\Big) \\
=\ &\lambda_6 g\Big(\Big(\nabla^g_{F^2X} F^2\Big)Y, Z\Big) + \lambda_6 g\Big(Z, \Big(\nabla^g_{F^2X} F^2\Big)Y\Big) \\
=\ &\lambda_6 g\Big(\Big(\nabla^g_{F^2X} F^2\Big)Y, Z\Big) + \lambda_6 g\Big(\Big(\nabla^g_{F^2X} F^2\Big)Y, Z\Big) \\
=\ &2\lambda_6 g\Big(\Big(\nabla^g_{F^2X} F^2\Big)Y, Z\Big).
\end{aligned}$$

https://doi.org/10.3390/math140915

For λ_7 :

$$\begin{aligned}
& \lambda_7 g\left(F^2\left(\nabla^g_{F^2Y}F^2\right)X, Z\right) + \lambda_7 g\left(F^2\left(\nabla^g_{F^2Z}F^2\right)X, Y\right) \\
= & \lambda_7 g\left(X, \left(\nabla^g_{F^2Y}F^2\right)F^2 Z\right) + \lambda_7 g\left(X, \left(\nabla^g_{F^2Z}F^2\right)F^2 Y\right) \\
= & \lambda_7 g\left(X, \left(\nabla^g_{F^2Y}F^2\right)F^2 Z + \left(\nabla^g_{F^2Z}F^2\right)F^2 Y\right) \\
= & \lambda_7 g\left(\begin{array}{c} X, \left(\nabla^g_{F^2Y}F^2\right)Z - F^2\left(\nabla^g_{F^2Y}F^2\right)Z \\ +\left(\nabla^g_{F^2Z}F^2\right)Y - F^2\left(\nabla^g_{F^2Z}F^2\right)Y \end{array}\right).
\end{aligned}$$

For λ_8 :

$$\begin{aligned}
& \lambda_8 g\left(\left(\nabla^g_{F^2Y}F^2\right)X, Z\right) + \lambda_8 g\left(\left(\nabla^g_{F^2Z}F^2\right)X, Y\right) \\
= & \lambda_8 g\left(X, \left(\nabla^g_{F^2Y}F^2\right)Z\right) + \lambda_8 g\left(X, \left(\nabla^g_{F^2Z}F^2\right)Y\right) \\
= & \lambda_8 g\left(X, \left(\nabla^g_{F^2Y}F^2\right)Z + \left(\nabla^g_{F^2Z}F^2\right)Y\right).
\end{aligned}$$

Now, when we will proceed to write λ_i $(i = 1, \cdots, 8)$ in Equation (39), we find

$$\begin{aligned}
& \lambda_1 g\left(Y, \left(\nabla^g_X F^2\right)Z\right) + 2\lambda_2 g\left(Y, \left(\nabla^g_X F^2\right)Z\right) \\
& +\lambda_3 g\left(X, \left(\nabla^g_Y F^2\right)Z - F^2\left(\nabla^g_Y F^2\right) + \left(\nabla^g_Z F^2\right)Y\right) \\
& +\lambda_4 g\left(X, \left(\nabla^g_Y F^2\right)Z + \left(\nabla^g_Z F^2\right)Y\right) \\
& +\lambda_5 g\left(\left(\nabla^g_{F^2X}F^2\right)Y, Z\right) + 2\lambda_6 g\left(\left(\nabla^g_{F^2X}F^2\right)Y, Z\right) \\
& +\lambda_7 g\left(X, \left(\nabla^g_{F^2Y}F^2\right)Z - F^2\left(\nabla^g_{F^2Y}F^2\right)Z + \left(\nabla^g_{F^2Z}F^2\right)Y\right) \\
& +\lambda_8 g\left(X, \left(\nabla^g_{F^2Y}F^2\right)Z + \left(\nabla^g_{F^2Z}F^2\right)Y\right) \\
= & \ 0.
\end{aligned}$$

We simplify this equation by grouping like terms:

$$\begin{aligned}
& (\lambda_1 + 2\lambda_2) g\left(Y, \left(\nabla^g_X F^2\right)Z\right) \\
& +(\lambda_5 + 2\lambda_6) g\left(\left(\nabla^g_{F^2X}F^2\right)Y, Z\right) \\
& +(\lambda_3 + \lambda_4) g\left(\left(\nabla^g_Y F^2\right)Z + \left(\nabla^g_Z F^2\right)Y\right), X) \\
& -\lambda_3 F^2 g\left(\left(\nabla^g_Y F^2\right)Z + \left(\nabla^g_Z F^2\right)Y\right), X) \\
& +(\lambda_7 + \lambda_8) g\left(\left(\left(\nabla^g_{F^2Y}F^2\right)Z + \left(\nabla^g_{F^2Z}F^2\right)Y, X\right)\right) \\
& -\lambda_7 F^2 g\left(\left(\left(\nabla^g_{F^2Y}F^2\right)Z + \left(\nabla^g_{F^2Z}F^2\right)Y, X\right)\right) \\
= & \ 0.
\end{aligned}$$

To satisfy this equation for all vector fields X, Y, Z, each coefficient of the inner products must independently be zero.

https://doi.org/10.3390/math14091538

From the above, we obtain the following conditions:

$$\begin{aligned}\lambda_1 + 2\lambda_2 &= 0,\\ \lambda_5 + 2\lambda_6 &= 0,\\ \lambda_3 + \lambda_4 &= 0,\\ \lambda_3 &= 0,\\ \lambda_7 + \lambda_8 &= 0,\\ \lambda_7 &= 0.\end{aligned}$$

Taking into account Equation (38), we find:

$$\lambda_3 = \lambda_4 = \lambda_5 = \lambda_6 = \lambda_7 = \lambda_8 = 0\,,\ \lambda_1 = 2, \lambda_2 = -1.$$

In this scenario, therefore, the tensor S simplifies to:

$$S(X,Y) = 2F^2\left(\nabla^g_X F^2\right)Y - \left(\nabla^g_X F^2\right)Y.$$

This provides a simplified and consistent form for the tensor S under the specified conditions. Hence, the F^2-metric-preserving connection $\widetilde{\nabla}$ has the following form:

$$\widetilde{\nabla}_X Y = \nabla^g_X Y + S(X,Y)$$

$$\begin{aligned}\Rightarrow\ \widetilde{\nabla}_X Y &= \nabla^g_X Y + 2F^2\left(\nabla^g_X F^2\right)Y - \left(\nabla^g_X F^2\right)Y\\ &= \nabla^g_X Y + 2F^2\left(\nabla^g_X F(F)\right)Y - \left(\nabla^g_X F(F)\right)Y\\ &= \nabla^g_X Y + 2F^2\left(\left(\nabla^g_X F\right)FY + F\left(\nabla^g_X F\right)Y\right)\\ &\quad -\left(\left(\nabla^g_X F\right)FY + F\left(\nabla^g_X F\right)Y\right)\\ &= \nabla^g_X Y + 2F^2\left(\nabla^g_X F\right)FY + 2F\left(\nabla^g_X F\right)Y\\ &\quad -\left(\nabla^g_X F\right)FY - F\left(\nabla^g_X F\right)Y\end{aligned}$$

$$\Rightarrow \widetilde{\nabla}_X Y = \nabla^g_X Y + 2F^2\left(\nabla^g_X F\right)FY + F\left(\nabla^g_X F\right)Y - \left(\nabla^g_X F\right)FY.$$

Consider the given relation: $\widetilde{\nabla}_X Y = \nabla^g_X Y + 2F^2\left(\nabla^g_X F\right)FY + F\left(\nabla^g_X F\right)Y - \left(\nabla^g_X F\right)FY$. We aim to determine the torsion tensor $\widetilde{T}$ defined by $\widetilde{T}\,(X,Y) = \widetilde{\nabla}_X Y - \widetilde{\nabla}_Y X - [X,Y]$. Substituting the expression for $\widetilde{\nabla}_X Y$ and $\widetilde{\nabla}_Y X$, and also using $[X,Y] = \nabla^g_X Y - \nabla^g_Y X$, we find

$$\begin{aligned}\widetilde{T}(X,Y) &= 2F^2\left(\left(\nabla^g_X F\right)FY - \left(\nabla^g_Y F\right)FX\right)\\ &\quad +F\left(\left(\nabla^g_X F\right)Y - \left(\nabla^g_Y F\right)X\right)\\ &\quad +\left(\nabla^g_Y F\right)FX - \left(\nabla^g_X F\right)FY,\end{aligned}$$

Simplifying further and combining the like terms, we obtain:

$$\widetilde{T}(X,Y) = (2F^2 - Id)\left(\left(\nabla^g_X F^2\right)Y - \left(\nabla^g_Y F^2\right)X\right),$$

where Id is the identity tensor field. We conclude that $\widetilde{T}(X,Y) =$ if and only if (∇^g, F^2) is a Codazzi pair.

https://doi.org/10.3390/math140915

Thus, we have the following theorem.

Theorem 8. *On an almost pure metric pseudo-F-manifold (M,F,g), the F^2-metric-preserving connection is defined by*

$$\widetilde{\nabla}_X Y = \nabla^g_X Y + 2F^2\left(\nabla^g_X F\right)FY + F\left(\nabla^g_X F\right)Y - \left(\nabla^g_X F\right)FY.$$

If the pair (∇^g, F) on the almost pure metric pseudo-F-manifold (M,F,g) forms a Codazzi pair, the F^2-metric-preserving connection is torsion-free, where ∇^g represents the Levi-Civita connection associated with the pseudo-Riemannian metric g.

4.3. F^2-Preserving Connections

Consider an almost pure metric pseudo-F-manifold, (M,F,g). The manifold is equipped with a connection, ∇. Here, our objective is to define a new connection, ∇^0, on M that is related to ∇:

$$\nabla^0_X Y = \nabla_X Y + 2F^2(\nabla_X F)FY + F(\nabla_X F)Y - (\nabla_X F)FY.$$

Here, if any connection, ∇^0, satisfies the condition $\nabla^0 F^2 = 0$, we call it an F^2-preserving connection. We calculate

$$\begin{aligned}
\left(\nabla^0_X F^2\right)Y &= \left(\nabla^0_X F^2 Y\right) - F^2\left(\nabla^0_X Y\right) \\
&= \left(\nabla_X F^2 Y\right) + 2F^2(\nabla_X F)FY + F(\nabla_X F)F^2 Y \\
&\quad -(\nabla_X F)FY - F^2(\nabla_X Y) - 2F^2(\nabla_X F)FY \\
&\quad -F(\nabla_X F)Y + F^2(\nabla_X F)FY \\
&= \left(\nabla_X F^2\right)Y + F(\nabla_X F)F^2 Y \\
&\quad +\left(F^2 - Id\right)(\nabla_X F)FY - F(\nabla_X F)Y \\
&= F\left((\nabla_X F)Y - F(\nabla_X F)FY - F^2(\nabla_X F)Y\right) \\
&\quad +F^2(\nabla_X F)FY \\
&= F(\nabla_X F)Y - F^2(\nabla_X F)FY \\
&\quad -F(\nabla_X F)Y + F^2(\nabla_X F)FY \\
&= 0.
\end{aligned}$$

Thus, we have the following proposition.

Proposition 3. *The connection ∇^0 on an almost pure metric pseudo-F-manifold (M,F,g) is an F^2-preserving connection.*

Consider the almost pure metric pseudo-F-manifold (M,F,g) equipped with the F^2-preserving connection ∇^0. We define a new connection, $\widetilde{\nabla}^0 = \nabla^0_X Y + A(X,Y)$, where A is a tensor field of type $(1,2)$. Using the condition $\nabla^0 F^2 = 0$, we have

$$\begin{aligned}
\left(\widetilde{\nabla}^0_X F^2\right)Y &= \left(\widetilde{\nabla}^0_X F^2 Y\right) - F^2\left(\widetilde{\nabla}^0_X Y\right) \\
&= \left(\nabla^0_X F^2 Y\right) + A\left(X, F^2 Y\right) \\
&\quad -F^2\left(\nabla^0_X Y + A(X,Y)\right) \\
&= \left(\nabla^0_X F^2\right)Y + F^2\left(\nabla^0_X Y\right) + A\left(X, F^2 Y\right) \\
&\quad -F^2\left(\nabla^0_X Y\right) - F^2 A(X,Y) \\
&= A\left(X, F^2 Y\right) - F^2 A(X,Y),
\end{aligned}$$

https://doi.org/10.3390/math14091538

which gives the following result.

Proposition 4. *The connection* $\widetilde{\nabla}^0 = \nabla^0_X Y + A(X,Y)$ *on the almost pure metric pseudo-F-manifold* (M,F,g) *equipped with the* F^2*-preserving connection* ∇^0 *is an* F^2*-preserving connection if and only* $A(X,F^2Y) = F^2A(X,Y)$.

We consider the tensor A given by the following linear combination of $F^2\nabla F^2$ and ∇F^2:

$$\begin{aligned} A(X,Y) &= \mu_1 F^2\left(\nabla_X F^2\right)Y + \mu_2\left(\nabla_X F^2\right)Y \\ &+\mu_3 F^2\left(\nabla_Y F^2\right)X + \mu_4\left(\nabla_Y F^2\right)X \\ &+\mu_5 F^2\left(\nabla_{F^2X} F^2\right)Y + \mu_6\left(\nabla_{F^2X} F^2\right)Y \\ &+\mu_7 F^2\left(\nabla_{F^2Y} F^2\right)X + \mu_8\left(\nabla_{F^2Y} F^2\right)X \end{aligned}$$

for all X,Y vector fields on (M,F,g), where $\mu_i \in \mathbb{R}, i = \{1,2,\cdots,8\}$.
Considering $A(X,F^2Y)$, we have:

$$\begin{aligned} A\left(X,F^2Y\right) &= \mu_1 F^2\left(\left(\nabla_X F^2\right)Y - F^2\left(\nabla_X F^2\right)Y\right) \\ &+\mu_2\left(\left(\nabla_X F^2\right)Y - F^2\left(\nabla_X F^2\right)Y\right) \\ &+\mu_3 F^2\left(\nabla_{F^2Y} F^2\right)X + \mu_4\left(\nabla_{F^2Y} F^2\right)X \\ &+\mu_5 F^2\left(\left(\nabla_{F^2X} F^2\right)Y - F^2\left(\nabla_{F^2X} F^2\right)Y\right) \\ &+\mu_6\left(\left(\nabla_{F^2X} F^2\right)Y - F^2\left(\nabla_{F^2X} F^2\right)Y\right) \\ &+\mu_7 F^2\left(\nabla_{F^2Y} F^2\right)X + \mu_8\left(\nabla_{F^2Y} F^2\right)X. \end{aligned}$$

On the other hand, $F^2A(X,Y)$ is given by:

$$\begin{aligned} F^2A(X,Y) &= \mu_1 F^2\left(\nabla_X F^2\right)Y + \mu_2 F^2\left(\nabla_X F^2\right)Y \\ &+\mu_3 F^2\left(\nabla_Y F^2\right)X + \mu_4 F^2\left(\nabla_Y F^2\right)X \\ &+\mu_5 F^2\left(\nabla_{F^2X} F^2\right)Y + \mu_6 F^2\left(\nabla_{F^2X} F^2\right)Y \\ &+\mu_7 F^2\left(\nabla_{F^2Y} F^2\right)X + \mu_8 F^2\left(\nabla_{F^2Y} F^2\right)X. \end{aligned}$$

Equating $A(X,F^2Y)$ and $F^2A(X,Y)$, we obtain:

$$\begin{aligned} &\mu_2\left(\nabla_X F^2\right)Y - \mu_2 F^2\left(\nabla_X F^2\right)Y \\ &+(\mu_3+\mu_7)F^2\left(\nabla_{F^2Y} F^2\right)X + (\mu_4+\mu_8)\left(\nabla_{F^2Y} F^2\right)X \\ &+\mu_6\left(\nabla_{F^2X} F^2\right)Y - \mu_6 F^2\left(\nabla_{F^2X} F^2\right)Y \\ =\ &(\mu_1+\mu_2)F^2\left(\nabla_X F^2\right)Y + (\mu_3+\mu_4)F^2\left(\nabla_Y F^2\right)X \\ &+(\mu_5+\mu_6)F^2\left(\nabla_{F^2X} F^2\right)Y + (\mu_7+\mu_8)F^2\left(\nabla_{F^2Y} F^2\right)X. \end{aligned}$$

This implies the conditions:

$$\mu_1 = \mu_2 = \mu_5 = \mu_6 = 0\,, \quad \mu_3 = -\mu_4 = \mu_8, \quad \mu_7 \text{ is arbitrary.}$$

https://doi.org/10.3390/math140915

When we take $\mu_3 = \lambda$ and $\mu_7 = \mu$, the tensor A reduces to:

$$A(X,Y) = \lambda\, F^2\left(\nabla_Y F^2\right)X - \lambda\left(\nabla_Y F^2\right)X + \mu F^2\left(\nabla_{F^2 Y} F^2\right)X + \lambda\left(\nabla_{F^2 Y} F^2\right)X.$$

Thus, we get the following.

Proposition 5. *On an almost pure metric pseudo-F-manifold (M, F, g) equipped with the F^2-preserving connection ∇^0, another F^2-preserving connection is defined by*

$$\widetilde{\nabla}^0 = \nabla^0_X Y + A(X,Y)$$

where $A(X,Y) = \lambda\, F^2(\nabla_Y F^2)X - \lambda(\nabla_Y F^2)X + \mu F^2(\nabla_{F^2 Y} F^2)X + \lambda(\nabla_{F^2 Y} F^2)X.$

5. Conclusions

This paper has provided a comprehensive investigation into the geometric properties of almost pure metric pseudo-F-manifolds, with a focus on Walker 4-manifolds. We have systematically explored integrability conditions, characterizations of pure metric pseudo-F-Kählerian structures, curvature properties, and the existence of Killing vector fields and Ricci soliton structures.

A central result concerns the integrability of almost pure metric pseudo-F-structures, which is equivalent to the satisfaction of specific partial differential equations. This characterization offers crucial insights into the geometric behavior of these manifolds and identifies conditions under which Walker 4-manifolds exhibit pure metric pseudo-F-Kählerian properties.

Our study also addresses the curvature properties of these manifolds. We have derived conditions ensuring the vanishing of the Riemann curvature tensor and analyzed the implications for the scalar curvature. These findings deepen the understanding of the intrinsic geometry of Walker 4-manifolds and provide a foundation for further geometric analysis.

Additionally, we have examined the role of vector fields in preserving geometric structures by establishing explicit conditions for the existence of Killing vector fields and Ricci solitons. These results illuminate the dynamical behavior of such manifolds under geometric flows.

Finally, we have investigated various metric-preserving connections, including F^2-metric-preserving connections and their relation to Codazzi pairs. Conditions for torsion-freeness and compatibility with the underlying geometric structures have been identified, providing further insight into the differential geometry of almost pure metric pseudo-F-manifolds.

Overall, this work advances the understanding of Walker 4-manifolds equipped with almost pure metric pseudo-F-structures by clarifying the interplay between geometric structures, curvature, and vector fields. The results lay the groundwork for future studies and offer a framework for exploring these manifolds in greater depth.

Author Contributions: Conceptualization, Y.L., C.K., A.G., S.T. and Y.X.; methodology, Y.L., C.K., A.G., S.T. and Y.X.; investigation, Y.L., C.K., A.G., S.T. and Y.X.; writing—original draft preparation, Y.L., C.K., A.G., S.T. and Y.X.; writing—review and editing, Y.L., C.K., A.G., S.T. and Y.X. All authors have read and agreed to the published version of the manuscript.

Funding: This research received no external funding.

Data Availability Statement: The original contributions presented in this study are included in the article. Further inquiries can be directed to the corresponding author.

Acknowledgments: The authors would like to express their sincere thanks to the editor and the anonymous reviewers for their helpful comments and suggestions.

https://doi.org/10.3390/math14091538

Conflicts of Interest: The authors declare no conflicts of interest.

References

1. Yano, K. *On a Structure f Satisfying* $f^3 + f = 0$; Technical Report No. 12; University of Washington: Seattle, WA, USA, 1961.
2. Goldberg, S.I.; Yano, K. Polynomial structures on manifolds. *Kodai Math. Sem. Rep.* **1970**, *22*, 199–218. [CrossRef]
3. Singh, K.D.; Srivastava, N. On some f-structure manifolds with torsion. *Demonstr. Math.* **1974**, *7*, 183–196. [CrossRef]
4. Balkan, Y.S.; Uddin, S.; Stankovic, M.S.; Alkhaldi, H.A. A new class of f-structures satisfying $f^3 - f = 0$. *Filomat* **2018**, *3* 5919–5929. [CrossRef]
5. Matsumoto, K. On a structure defined by a tensor field f of type $(1,1)$ satisfying $f^3 - f = 0$. *Bull. Yamagata Univ.* **1976**, *1*, 33–4
6. Pujar, S.S. On a structure defined by a tensor field f of type $(1,1)$ satisfying $P^3 - P = 0$. *Indian J. Pure Appl. Math.* **2000**, *3* 1229–1234.
7. Rovenski, V. Geometry of a weak para-f-structure. *Politehn. Univ. Buchar. Sci. Bull. Ser. A Appl. Math. Phys.* **2023**, *85*, 11–20.
8. Rovenski, V. Geometry of weak metric f-manifolds: A Survey. *Mathematics* **2025**, *13*, 556. [CrossRef]
9. Vanzura, J. Integrability conditions for polynomial structures. *Kodai Math. Sem. Rep.* **1976**, *27*, 42–50. [CrossRef]
10. Opozda, B. The f-sectional curvature of f-Kaehlerian manifolds. *Ann. Pol. Math.* **1983**, *43*, 141–150. [CrossRef]
11. Bures, J.; Vanzura, J. Metric polynomial structures. *Kodai Math. Sem. Rep.* **1976**, *27*, 345–352. [CrossRef]
12. Salimov, A. On anti-Hermitian metric connections. *C. R. Math. Acad. Sci.* **2014**, *352*, 731–735. [CrossRef]
13. Bilen, L.; Turanli, S.; Gezer, A. On Kähler-Norden-Codazzi golden structures on pseudo-Riemannian manifolds. *Int. J. Geo Methods Mod. Phys.* **2018**, *15*, 1850080. [CrossRef]
14. Zhong, S.; Zhao, Z.; Mu, G. Structure-preserving connections on almost complex Norden golden manifolds. *J. Geom.* **2019**, *110*, [CrossRef]
15. Blaga, A.M.; Kumar, R.; Rani, R. On the geometry of complex metallic Norden manifolds. *Bull. Korean Math. Soc.* **2022**, *5* 1069–1091.
16. Walker, A.G. Canonical form for a Riemannian space with a parallel field of null planes. *Quart. J. Math.* **1950**, *1*, 69–79. [CrossRe
17. Davidov, J.; Diaz-Ramos, J.C.; Garcia-Rio, E.; Matsushita, Y.; Muskarov, O.; Vazquez-Lorenzo, R. Almost Kähler Walker manifolds. *J. Geom. Phys.* **2007**, *57*, 1075–1088. [CrossRef]
18. Davidov, J.; Diaz-Ramos, J.C.; Garcia-Rio, E.; Matsushita, Y.; Muskarov, O.; Vazquez-Lorenzo, R. Hermitian-Walker 4-manifol *J. Geom. Phys.* **2008**, *58*, 307–323. [CrossRef]
19. Matsushita, Y. Four-dimensional Walker metrics and symplectic structure. *J. Geom. Phys.* **2004**, *52*, 89–99. [CrossRef]
20. Petean, Y. Indefinite Kähler-Einstein metrics on compact complex surfaces. *Commun. Math. Phys.* **1997**, *189*, 227–235. [CrossRe
21. Matsushita, Y.; Haze, S.; Law, P.R. Almost Kähler-Einstein structure on 8-dimensional Walker manifolds. *Monatsh. Math.* **200** *150*, 41–48. [CrossRef]
22. Yang, Z.; Li, Y.; López, R. Singularities and dualities of evolving developable frontals. *Res. Math. Sci.* **2026**, *13*, 22. [CrossRef]
23. Hamilton, R.S. The Ricci flow on surfaces. *Contemp. Math.* **1988**, *71*, 237–261.
24. Li, Y.; Lone, M.A.; Yousuf, A. Dual THA-Surfaces in 3-Dimensional Galilean Space. *Symmetry* **2026**, *18*, 409. [CrossRef]
25. Li, Y.; De, U.C.; De, K. Weakly Ricci-symmetric space-times and f (R, G) gravity. *Mathematics* **2025**, *13*, 943. [CrossRef]

https://doi.org/10.3390/math140915

MDPI AG
Grosspeteranlage 5
4052 Basel
Switzerland
Tel.: +41 61 683 77 34

Mathematics Editorial Office
E-mail: mathematics@mdpi.com
www.mdpi.com/journal/mathematics

www.ingramcontent.com/pod-product-compliance
Lightning Source LLC
LaVergne TN
LVHW070451061126
851858LV00011B/216